"十三五"国家重点出版物出版规划项目

持久性有机污染物
POPs研究系列专著

汞的分子转化与长距离传输

史建波 阴永光 江桂斌／著

科学出版社
北京

内 容 简 介

汞是重金属中唯一能以气态长距离传输的全球污染物，具有特殊的物理化学性质。分子转化和长距离传输是汞的两个重要特征，各形态汞之间的转化直接影响其环境行为及毒性，而长距离传输导致汞污染在全球范围广泛分布，并对环境和人体健康造成潜在危害。本书系统总结了有关汞的氧化/还原、甲基化/去甲基化、大气长距离传输、高山及极地地区生物地球化学循环等方面研究进展，同时介绍了同位素技术在汞分子转化及长距离传输研究中的应用。

本书可供环境化学、地球科学、污染控制及环境管理等相关领域的科研和技术人员参考，也可作为相关学科本科生及研究生的学习参考书。

审图号：GS(2019)2415 号
图书在版编目（CIP）数据

汞的分子转化与长距离传输/史建波，阴永光，江桂斌著. —北京：
科学出版社，2019.7
（持久性有机污染物(POPs)研究系列专著）
"十三五"国家重点出版物出版规划项目　国家出版基金项目
ISBN 978-7-03-061905-1

Ⅰ．①汞⋯　Ⅱ．①史⋯　②阴⋯　③江⋯　Ⅲ．①汞污染–污染控制
Ⅳ．①X5

中国版本图书馆 CIP 数据核字（2019）第 147039 号

责任编辑：朱　丽　杨新改 / 责任校对：杜子昂
责任印制：肖　兴 / 封面设计：黄华斌

科 学 出 版 社 出版
北京东黄城根北街 16 号
邮政编码：100717
http://www.sciencep.com

北京画中画印刷有限公司印刷
科学出版社发行　各地新华书店经销
*

2019 年 7 月第 一 版　　开本：720×1000　1/16
2019 年 7 月第一次印刷　　印张：20　插页：4
字数：403 000
定价：138.00 元
（如有印装质量问题，我社负责调换）

《持久性有机污染物（POPs）研究系列专著》丛书编委会

主　编　江桂斌

编　委（按姓氏汉语拼音排序）

蔡亚岐　陈景文　李英明　刘维屏

刘咸德　麦碧娴　全　燮　阮　挺

王亚韡　吴永宁　尹大强　余　刚

张爱茜　张　干　张庆华　郑明辉

周炳升　周群芳　朱利中

丛 书 序

持久性有机污染物（persistent organic pollutants，POPs）是指在环境中难降解（滞留时间长）、高脂溶性（水溶性很低），可以在食物链中累积放大，能够通过蒸发–冷凝、大气和水等的输送而影响到区域和全球环境的一类半挥发性且毒性极大的污染物。POPs 所引起的污染问题是影响全球与人类健康的重大环境问题，其科学研究的难度与深度，以及污染的严重性、复杂性和长期性远远超过常规污染物。POPs 的分析方法、环境行为、生态风险、毒理与健康效应、控制与削减技术的研究是最近 20 年来环境科学领域持续关注的一个最重要的热点问题。

近代工业污染催生了环境科学的发展。1962 年，*Silent Spring* 的出版，引起学术界对滴滴涕（DDT）等造成的野生生物发育损伤的高度关注，POPs 研究随之成为全球关注的热点领域。1996 年，*Our Stolen Future* 的出版，再次引发国际学术界对 POPs 类环境内分泌干扰物的环境健康影响的关注，开启了环境保护研究的新历程。事实上，国际上环境保护经历了从常规大气污染物（如 SO_2、粉尘等）、水体常规污染物［如化学需氧量（COD）、生化需氧量（BOD）等］治理和重金属污染控制发展到痕量持久性有机污染物削减的循序渐进过程。针对全球范围内 POPs 污染日趋严重的现实，世界许多国家和国际环境保护组织启动了若干重大研究计划，涉及 POPs 的分析方法、生态毒理、健康危害、环境风险理论和先进控制技术。研究重点包括：①POPs 污染源解析、长距离迁移传输机制及模型研究；②POPs 的毒性机制及健康效应评价；③POPs 的迁移、转化机理以及多介质复合污染机制研究；④POPs 的污染削减技术以及高风险区域修复技术；⑤新型污染物的检测方法、环境行为及毒性机制研究。

20 世纪国际上发生过一系列由于 POPs 污染而引发的环境灾难事件（如意大利 Seveso 化学污染事件、美国拉布卡纳尔镇污染事件、日本和中国台湾米糠油事件等），这些事件给我们敲响了 POPs 影响环境安全与健康的警钟。1999 年，比利时鸡饲料二噁英类污染波及全球，造成 14 亿欧元的直接损失，导致该国政局不稳。

国际范围内针对 POPs 的研究，主要包括经典 POPs（如二噁英、多氯联苯、含氯杀虫剂等）的分析方法、环境行为及风险评估等研究。如美国 1991～2001 年的二噁英类化合物风险再评估项目，欧盟、美国环境保护署（EPA）和日本环境厅先后启动了环境内分泌干扰物筛选计划。20 世纪 90 年代提出的蒸馏理论和蚂蚱跳效应较好地解释了工业发达地区 POPs 通过水、土壤和大气之间的界面交换而长距离迁移到南北极等极地地区的现象，而之后提出的山区冷捕集效应则更加系统地解释了高山地区随着海拔的增加其环境介质中 POPs 浓度不断增加的迁移机理，从而为 POPs 的全球传输提供了重要的依据和科学支持。

2001 年 5 月，全球 100 多个国家和地区的政府组织共同签署了《关于持久性有机污染物的斯德哥尔摩公约》（简称《斯德哥尔摩公约》）。目前已有包括我国在内的 179 个国家和地区加入了该公约。从缔约方的数量上不仅能看出公约的国际影响力，也能看出世界各国对 POPs 污染问题的重视程度，同时也标志着在世界范围内对 POPs 污染控制的行动从被动应对到主动防御的转变。

进入 21 世纪之后，随着《斯德哥尔摩公约》进一步致力于关注和讨论其他同样具 POPs 性质和环境生物行为的有机污染物的管理和控制工作，除了经典 POPs，对于一些新型 POPs 的分析方法、环境行为及界面迁移、生物富集及放大，生态风险及环境健康也越来越成为环境科学研究的热点。这些新型 POPs 的共有特点包括：目前为正在大量生产使用的化合物、环境存量较高、生态风险和健康风险的数据积累尚不能满足风险管理等。其中两类典型的化合物是以多溴二苯醚为代表的溴系阻燃剂和以全氟辛基磺酸盐（PFOS）为代表的全氟化合物，对于它们的研究论文在过去 15 年呈现指数增长趋势。如有关 PFOS 的研究在 Web of Science 上搜索结果为从 2000 年的 8 篇增加到 2013 年的 323 篇。随着这些新增 POPs 的生产和使用逐步被禁止或限制使用，其替代品的风险评估、管理和控制也越来越受到环境科学研究的关注。而对于传统的生态风险标准的进一步扩展，使得大量的商业有机化学品的安全评估体系需要重新调整。如传统的以鱼类为生物指示物的研究认为污染物在生物体中的富集能力主要受控于化合物的脂–水分配，而最近的研究证明某些低正辛醇–水分配系数、高正辛醇–空气分配系数的污染物（如 HCHs）在一些食物链特别是在陆生生物链中也表现出很高的生物放大效应，这就向如何修订污染物的生态风险标准提出了新的挑战。

作为一个开放式的公约，任何一个缔约方都可以向公约秘书处提交意在将某一化合物纳入公约受控的草案。相应的是，2013 年 5 月在瑞士日内瓦举行的缔约方大会第六次会议之后，已在原先的包括二噁英等在内的 12 类经典 POPs 基础上，新增 13 种包括多溴二苯醚、全氟辛基磺酸盐等新型 POPs 成为公约受控名单。目前正在进行公约审查的候选物质包括短链氯化石蜡（SCCPs）、多氯萘（PCNs）、六氯丁二烯（HCBD）及五氯苯酚（PCP）等化合物，而这些新型有机污染物在我国均有一定规模的生产和使用。

中国作为经济快速增长的发展中国家，目前正面临比工业发达国家更加复杂的环境问题。在前两类污染物尚未完全得到有效控制的同时，POPs 污染控制已成为我国迫切需要解决的重大环境问题。作为化工产品大国，我国新型 POPs 所引起的环境污染和健康风险问题比其他国家更为严重，也可能存在国外不受关注但在我国环境介质中广泛存在的新型污染物。对于这部分化合物所开展的研究工作不但能够为相应的化学品管理提供科学依据，同时也可为我国履行《斯德哥尔摩公约》提供重要的数据支持。另外，随着经济快速发展所产生的污染所致健康问题在我国的集中显现，新型 POPs 污染的毒性与健康危害机制已成为近年来相关研究的热点问题。

随着 2004 年 5 月《斯德哥尔摩公约》正式生效，我国在国家层面上启动了对 POPs 污染源的研究，加强了 POPs 研究的监测能力建设，建立了几十个高水平专业实验室。科研机构、环境监测部门和卫生部门都先后开展了环境和食品中 POPs 的监测和控制措施研究。特别是最近几年，在新型 POPs 的分析方法学、环境行为、生态毒理与环境风险，以及新污染物发现等方面进行了卓有成效的研究，并获得了显著的研究成果。如在电子垃圾拆解地，积累了大量有关多溴二苯醚（PBDEs）、二噁英、溴代二噁英等 POPs 的环境转化、生物富集/放大、生态风险、人体赋存、母婴传递乃至人体健康影响等重要的数据，为相应的管理部门提供了重要的科学支撑。我国科学家开辟了发现新 POPs 的研究方向，并连续在环境中发现了系列新型有机污染物。这些新 POPs 的发现标志着我国 POPs 研究已由全面跟踪国外提出的目标物，向发现并主动引领新 POPs 研究方向发展。在机理研究方面，率先在珠穆朗玛峰、南极和北极地区"三极"建立了长期采样观测系统，开展了 POPs 长距离迁移机制的深入研究。通过大量实验数据证明了 POPs 的冷捕集效应，在新的源汇关系方面也有所发现，为优化 POPs 远距离迁移模型及认识 POPs 的环境归宿做出了贡献。在污染物控制方面，系统地摸清了二噁英类污染物的排放源，获得了我国二噁英类排放因子，相关成果被联合国环境规划署《全球二噁英类污染源识别与定量技术导则》引用，以六种语言形式全球发布，为全球范围内评估二噁英类污染来源提供了重要技术参数。以上有关 POPs 的相关研究是解决我国国家环境安全问题的重大需求、履行国际公约的重要基础和我国在国际贸易中取得有利地位的重要保证。

我国 POPs 研究凝聚了一代代科学家的努力。1982 年，中国科学院生态环境研究中心发表了我国二噁英研究的第一篇中文论文。1995 年，中国科学院武汉水生生物研究所建成了我国第一个装备高分辨色谱/质谱仪的标准二噁英分析实验室。进入 21 世纪，我国 POPs 研究得到快速发展。在能力建设方面，目前已经建成数十个符合国际标准的高水平二噁英实验室。中国科学院生态环境研究中心的二噁英实验室被联合国环境规划署命名为"Pilot Laboratory"。

2001 年，我国环境内分泌干扰物研究的第一个"863"项目"环境内分泌干扰物的筛选与监控技术"正式立项启动。随后经过 10 年 4 期"863"项目的连续资助，形成了活体与离体筛选技术相结合，体外和体内测试结果相互印证的分析内分泌干扰物研究方法体系，建立了有中国特色的环境内分泌污染物的筛选与研究规范。

2003 年，我国 POPs 领域第一个"973"项目"持久性有机污染物的环境安全、演变趋势与控制原理"启动实施。该项目集中了我国 POPs 领域研究的优势队伍，围绕POPs在多介质环境的界面过程动力学、复合生态毒理效应和焚烧等处理过程中POPs的形成与削减原理三个关键科学问题，从复杂介质中超痕量POPs的检测和表征方法学；我国典型区域POPs污染特征、演变历史及趋势；典型POPs的排放模式和运移规律；典型POPs的界面过程、多介质环境行为；POPs污染物的复合生态毒理效应；POPs的削减与控制原理以及POPs生态风险评价模式和预警方法体系七个方面开展了富有成效的研究。该项目以我国POPs污染的演变趋势为主，基本摸清了我国POPs特别是二噁英排放的行业分布与污染现状，为我国履行《斯德哥尔摩公约》做出了突出贡献。2009年，POPs项目得到延续资助，研究内容发展到以POPs的界面过程和毒性健康效应的微观机理为主要目标。2014年，项目再次得到延续，研究内容立足前沿，与时俱进，发展到了新型持久性有机污染物。这3期"973"项目的立项和圆满完成，大大推动了我国POPs研究为国家目标服务的能力，培养了大批优秀人才，提高了学科的凝聚力，扩大了我国POPs研究的国际影响力。

2008年开始的"十一五"国家科技支撑计划重点项目"持久性有机污染物控制与削减的关键技术与对策"，针对我国持久性有机物污染物控制关键技术的科学问题，以识别我国POPs环境污染现状的背景水平及制订优先控制POPs国家名录，我国人群POPs暴露水平及环境与健康效应评价技术，POPs污染控制新技术与新材料开发，焚烧、冶金、造纸过程二噁英类减排技术，POPs污染场地修复，废弃POPs的无害化处理，适合中国国情的POPs控制战略研究为主要内容，在废弃物焚烧和冶金过程烟气减排二噁英类、微生物或植物修复POPs污染场地、废弃POPs降解的科研与实践方面，立足自主创新和集成创新。项目从整体上提升了我国POPs控制的技术水平。

目前我国POPs研究在国际SCI收录期刊发表论文的数量、质量和引用率均进入国际第一方阵前列，部分工作在开辟新的研究方向、引领国际研究方面发挥了重要作用。2002年以来，我国POPs相关领域的研究多次获得国家自然科学奖励。2013年，中国科学院生态环境研究中心POPs研究团队荣获"中国科学院杰出科技成就奖"。

我国POPs研究开展了积极的全方位的国际合作，一批中青年科学家开始在国际学术界崭露头角。2009年8月，第29届国际二噁英大会首次在中国举行，来自世界上44个国家和地区的近1100名代表参加了大会。国际二噁英大会自1980年召开以来，至今已连续举办了38届，是国际上有关持久性有机污染物（POPs）研究领域影响最大的学术会议，会议所交流的论文反映了当时国际POPs相关领域的最新进展，也体现了国际社会在控制POPs方面的技术与政策走向。第29届国际二噁英大会在我国的成功召开，对提高我国持久性有机污染物研究水平、加速国际化进程、推进国际合作和培养优秀人才等方面起到了积极作用。近年来，我国科学家多次应邀在国际二噁英大会上作大会报告和大会总结报告，一些高水平研究工作产生了重要的学术影响。与此同时，我国科学家自己发起的 POPs 研究的国内外学术会议也产生了重要影响。2004 年开始的 "International Symposium on Persistent Toxic Substances" 系列国际会议至今已连续举行 14 届，近几届分别在美国、加拿大、中国香港、德国、日本等国家和地区召开，产生了重要学术影响。每年 5 月 17～18 日定期举行的"持久性有机污染物论坛"已经连续 12 届，在促进我国 POPs 领域学术交流、促进官产学研结合方面做出了重要贡献。

本丛书《持久性有机污染物（POPs）研究系列专著》的编撰，集聚了我国 POPs 研究优秀科学家群体的智慧，系统总结了 20 多年来我国 POPs 研究的历史进程，从理论到实践全面记载了我国 POPs 研究的发展足迹。根据研究方向的不同，本丛书将系统地对 POPs 的分析方法、演变趋势、转化规律、生物累积/放大、毒性效应、健康风险、控制技术以及典型区域 POPs 研究等工作加以总结和理论概括，可供广大科技人员、大专院校的研究生和环境管理人员学习参考，也期待它能在 POPs 环保宣教、科学普及、推动相关学科发展方面发挥积极作用。

我国的 POPs 研究方兴未艾，人才辈出，影响国际，自树其帜。然而，"行百里者半九十"，未来事业任重道远，对于科学问题的认识总是在研究的不断深入和不断学习中提高。学术的发展是永无止境的，人们对 POPs 造成的环境问题科学规律的认识也是不断发展和提高的。受作者学术和认知水平限制，本丛书可能存在不同形式的缺憾、疏漏甚至学术观点的偏颇，敬请读者批评指正。本丛书若能对读者了解并把握 POPs 研究的热点和前沿领域起到抛砖引玉作用，激发广大读者的研究兴趣，或讨论或争论其学术精髓，都是作者深感欣慰和至为期盼之处。

2017 年 1 月于北京

前　　言

汞是重金属中唯一能以气态长距离传输的全球污染物，其环境行为与其他重金属显著不同。同时，汞化合物具有持久性有机污染物（POPs）的四大特征：持久性、长距离传输、高毒性和生物累积，可以看作是一类"特殊"的 POPs。鉴于汞化合物特殊的物理化学性质及环境行为，经过 5 次政府间谈判委员会会议讨论，最终在 2013 年 1 月达成一项专门针对汞的国际公约——《关于汞的水俣公约》（简称《水俣公约》），标志着汞污染已成为当前重要的全球环境问题。

我国是汞使用量和排放量最大的国家，每年通过人为活动向大气排放的汞量为 500～700 t，约占全球人为排放量的 30%左右。如此大量排放的汞在环境中的分布、迁移、转化以及归宿尚不完全清楚，对我国乃至全球造成的环境和健康影响仍不可预知。中国科学院生态环境研究中心是国内较早开展汞污染研究的单位之一，早在 20 世纪 70 年代就对蓟运河、第二松花江等区域的汞污染开展了深入研究。作者所在课题组在 80 年代末建立了气相色谱-原子吸收在线联用系统，应用于环境和生物样品中汞的形态分析研究，并于大气样品中发现了二甲基汞。近年来，课题组在汞的形态分析方法、联用仪器、区域污染、分子转化过程及同位素技术等方面开展持续研究，积累了丰富的经验。

分子转化和长距离传输是汞的两个重要特征。汞的毒性并不完全取决于其总量，而是与其存在的形态密切相关。环境中的汞可以无机和有机多种形态存在，通常有机汞的毒性和生物可利用性远大于无机汞，甲基汞是已知毒性最大的汞形态，也是环境和生物体中分布最广、最主要的有机汞形态。大气、土壤和水体中各种汞形态之间可通过生物作用或者化学过程发生相互转化（氧化/还原、甲基化/去甲基化等），由于各形态汞化合物的环境行为和毒性存在显著差异，其形态转化必然影响汞的生物地球化学循环及毒性。

早期发生的汞污染事件大部分是由化工过程使用和排放汞造成的，属于区域性污染。但随着经济的发展，燃煤、金属冶炼和水泥生产等无意排放逐渐成为环境中汞的主要人为源。由于汞的长距离传输，即使在没有明显汞污染源的高山（如青藏高原）和极地地区，也检测到汞的污染，全球大气、水和土壤中汞含量显著升高，对环境和人体健康造成潜在危害。因此，汞污染已不再是区域性的问题，需要更多地从全球污染物角度去认识汞污染并采取措施。

因此，本书围绕汞的这两个重要特征，系统总结了近年来在汞的氧化/还原、甲基化/去甲基化、大气长距离传输、高山及极地地区生物地球化学循环等方面的研究进展，同时介绍了同位素技术在汞分子转化及长距离传输研究中的应用。

全书共 7 章。第 1 章由中国科学院生态环境研究中心史建波和江桂斌撰写，第 2 章由中国科学院生态环境研究中心阴永光、史建波和胡立刚撰写，第 3 章由南开大学张彤、华中农业大学刘玉荣和南京大学钟寰撰写，第 4 章由天津大学孙若愚撰写，第 5 章由中国科学院地球化学研究所付学吾撰写，第 6 章由中国科学院青藏高原研究所张强弓撰写，第 7 章由中国海洋大学李雁宾撰写。江桂斌院士全程参与了书稿的策划和统稿工作。科学出版社朱丽和杨新改对本书进行了认真细致的编校工作。在此向所有参与本书撰写和出版的人员表示衷心的感谢！

由于汞污染的复杂性以及作者的认知局限，书中内容难免有疏漏和不妥之处，敬请广大读者批评指正。

作 者

2019 年 5 月于北京

目 录

丛书序
前言

第1章 绪论 ··· 1
1.1 汞的物理化学性质 ·· 1
1.2 汞的毒性 ·· 2
    1.2.1 甲基汞的毒性效应 ·· 2
    1.2.2 无机汞的毒性效应 ·· 3
1.3 汞的形态与分子转化 ··· 4
    1.3.1 汞的氧化与还原 ·· 5
    1.3.2 汞的甲基化与去甲基化 ··· 6
1.4 汞的长距离传输 ·· 6
1.5 汞污染与《关于汞的水俣公约》 ·· 8
    1.5.1 汞的使用与排放 ·· 8
    1.5.2 汞污染 ·· 10
    1.5.3 《关于汞的水俣公约》 ·· 11
1.6 我国汞研究进展与挑战 ··· 12
参考文献 ··· 15

第2章 汞的氧化与还原 ··· 19
2.1 单质汞、氧化态汞与汞的循环 ·· 19
2.2 汞的化学氧化与生物氧化 ·· 20
    2.2.1 大气气相汞化学氧化 ··· 20
    2.2.2 大气水相汞氧化 ··· 25
    2.2.3 汞氧化与大气汞"亏损"事件 ·· 27
    2.2.4 表层水汞化学氧化 ·· 28
    2.2.5 土壤中汞的氧化 ··· 31
    2.2.6 冰雪中汞的氧化 ··· 31
    2.2.7 冰冻引发汞的氧化 ·· 32
    2.2.8 汞氧化过程的生物作用 ··· 33

2.3　汞的化学与生物还原···35
      2.3.1　大气气相汞化学还原··35
      2.3.2　大气水相汞化学还原··36
      2.3.3　表层水均相汞还原···39
      2.3.4　表层水非均相汞还原···45
      2.3.5　土壤表面汞的还原···47
      2.3.6　底泥中汞的还原···52
      2.3.7　冰雪中汞的还原···52
      2.3.8　汞还原过程的生物作用···55
   2.4　展望···59
   参考文献···59
第3章　汞的甲基化与去甲基化··82
   3.1　甲基汞的物理化学性质··82
   3.2　甲基汞的人体暴露与健康风险··83
   3.3　甲基汞的环境行为与归趋··85
      3.3.1　甲基汞的环境分布···85
      3.3.2　甲基汞的生物积累与放大效应·································94
   3.4　汞的甲基化研究···99
      3.4.1　汞的微生物甲基化···99
      3.4.2　汞的非生物甲基化过程··108
   3.5　甲基汞的去甲基化···110
      3.5.1　甲基汞的微生物去甲基化··110
      3.5.2　甲基汞在其他生物体内的去甲基化过程················111
      3.5.3　甲基汞的非生物去甲基化过程·································112
   3.6　展望···117
   参考文献···118
第4章　同位素技术在汞分子转化及区域传输研究中的应用·····································139
   4.1　汞同位素分馏的基本概念和理论··139
      4.1.1　质量分馏··141
      4.1.2　非质量分馏··142
   4.2　汞分子转化过程中同位素分馏规律及应用··144
      4.2.1　汞在无机转化过程中的同位素分馏规律················148
      4.2.2　汞在无机-有机转化过程中的同位素分馏规律······150
      4.2.3　汞分子转化过程中同位素分馏效应的应用············152

  4.3 汞同位素在大气汞转化和跨区域传输方面的应用 ·········· 155
    4.3.1 大气形态汞同位素样品的采集与预处理技术 ·········· 155
    4.3.2 实地观测的大气形态汞同位素组成 ·········· 157
    4.3.3 人为/自然源大气汞排放的同位素数据库 ·········· 163
    4.3.4 大气形态汞转化过程中的同位素分馏效应 ·········· 169
    4.3.5 大气形态汞同位素组成对汞来源的解析 ·········· 174
  4.4 基于汞同位素的全球汞循环传输模型 ·········· 176
  参考文献 ·········· 177

第5章 大气汞的长距离传输 ·········· 185
  5.1 大气汞形态和理化性质 ·········· 185
    5.1.1 大气汞形态分类 ·········· 185
    5.1.2 大气汞物理化学性质和形态转化 ·········· 186
    5.1.3 不同形态大气汞测定方法 ·········· 191
  5.2 大气汞的来源和分布特征 ·········· 194
    5.2.1 大气汞的来源 ·········· 194
    5.2.2 全球大气汞区域分布特征 ·········· 195
    5.2.3 我国大气汞区域分布特征 ·········· 198
  5.3 大气汞长距离传输研究方法 ·········· 201
    5.3.1 基于监测数据的源汇关系模型 ·········· 201
    5.3.2 区域大气汞循环模型 ·········· 205
    5.3.3 全球大气汞循环模型 ·········· 206
  5.4 国内外大气汞长距离传输研究进展 ·········· 207
    5.4.1 我国大气汞传输过程和特征 ·········· 207
    5.4.2 亚洲其他地区大气汞传输过程和特征 ·········· 214
    5.4.3 北美大气汞传输过程和特征 ·········· 216
    5.4.4 欧洲大气汞传输过程和特征 ·········· 219
    5.4.5 其他地区大气汞长距离传输过程和特征 ·········· 221
  5.5 展望 ·········· 222
  参考文献 ·········· 223

第6章 高山地区汞的生物地球化学循环 ·········· 239
  6.1 高山地区生态环境与汞的生物地球化学循环 ·········· 239
    6.1.1 高山地区生态环境 ·········· 239
    6.1.2 高山地区汞的生物地球化学循环 ·········· 240
  6.2 高山地区汞的重要环境行为 ·········· 240

　　　　6.2.1　大气汞形态和特征 ···240
　　　　6.2.2　大气汞沉降 ···243
　　　　6.2.3　海拔梯度分布及效应 ···246
　　6.3　高山地区汞的历史记录 ··248
　　　　6.3.1　湖泊沉积物 ···248
　　　　6.3.2　冰芯 ··250
　　　　6.3.3　其他环境介质 ···253
　　　　6.3.4　高山多环境介质汞历史记录的集成研究 ···254
　　6.4　高山地区汞的关键生物地球化学过程 ···255
　　　　6.4.1　山地冰川消融和汞的迁移释放 ···255
　　　　6.4.2　冻土的汞库效应 ···258
　　　　6.4.3　高山森林汞的生物地球化学过程 ···260
　　6.5　高山汞循环的复杂性及其变化趋势 ··262
　　　　6.5.1　高山地区汞循环的复杂性和重要性 ··262
　　　　6.5.2　高山地区汞循环的变化趋势 ···263
　　　　6.5.3　高山地区汞循环研究的未来趋势 ··264
　　参考文献 ···264
第7章　极地地区汞的生物地球化学循环 ··272
　　7.1　极地地区汞的主要来源 ··272
　　　　7.1.1　极地汞的主要来源与收支 ··272
　　　　7.1.2　极地大气汞"亏损"事件对汞输入贡献 ···273
　　7.2　极地地区环境介质中汞的分布与影响因素 ··275
　　　　7.2.1　极地地区大气汞及其影响因素 ··275
　　　　7.2.2　极地地区陆地和水环境汞及其影响因素 ···282
　　　　7.2.3　极地地区生物及人体中汞及其影响因素 ···284
　　7.3　极地地区汞的关键生物地球化学过程 ···285
　　　　7.3.1　汞大气化学过程 ···285
　　　　7.3.2　水环境汞关键生物地球化学过程 ··286
　　　　7.3.3　雪、海冰中汞生物地球化学过程 ··289
　　7.4　展望 ···290
参考文献 ··290
附录　缩略语（英汉对照） ···301
索引 ···303
彩图

第1章 绪　　论

> **本章导读**
> - 汞是常温常压下唯一以液态存在的金属，具有特殊的物理化学性质。
> - 汞的毒性与其形态密切相关，不同形态在环境中可以发生相互转化。
> - 汞可以随大气进行长距离传输，为一种全球污染物。重点介绍汞的使用、排放及全球污染。
> - 《关于汞的水俣公约》的签署使我国汞污染研究与控制面临国内环境保护和国际履约的双重压力。

1.1　汞的物理化学性质

汞，俗称水银，元素符号为 Hg，在元素周期表中位于第 6 周期，第ⅡB 族，原子序数为 80。汞是常温常压下唯一以液态存在的金属，沸点为 356.6℃，熔点为 –38.87℃，密度为 13.59 g/cm^3。这些特性使得元素汞易于进入气相，并随大气环流进行长距离传输和全球分布。

汞是一种自然产生的地壳元素，在各种环境和生物介质（如岩石、土壤、大气、水和生物）中都有广泛的存在。作为一种剧毒非必需元素，汞在地壳中的平均含量很小，被认为是稀有金属。它有 7 种稳定同位素，平均丰度分别为：^{196}Hg（0.15%）、^{198}Hg（9.97%）、^{199}Hg（16.87%）、^{200}Hg（23.10%）、^{201}Hg（13.18%）、^{202}Hg（29.86%）、^{204}Hg（6.87%）（Cohen et al.，2008）。

单质汞的化学性质较为稳定，其金属活跃性低于锌和镉，不能与大多数的酸（例如稀硫酸）和碱等发生反应；但是氧化性较强的酸，例如浓硫酸、浓硝酸和王水可以溶解汞从而形成+1 价或+2 价的硫酸盐、硝酸盐和氯化物。除了氧化性酸，汞还可以与一些强氧化剂如氯气等发生反应生成相应的含汞盐类。汞具有强烈的亲硫性和亲铜性，在常温下即可与硫或铜单质发生化合反应生成稳定的化合物，汞与单质硫反应会生成硫化汞（俗称朱砂或辰砂），在实验室中通常利用这一反应去处理泄漏的汞。此外，汞易与大部分普通金属（如金、银、钠、铝、锰、铜和锌等）形成合金，这些汞合金统称为汞齐。例如，当汞和铝的纯金属接触时，它们易于形成铝汞齐，由于铝汞齐可以破坏金属铝的氧化层，所以即使很少量的汞也能严重腐蚀金属铝；而钠汞齐是有机合成中常用的还原剂，也被用于高压钠灯中。

1.2 汞的毒性

汞是一种具有持久性的高毒性物质，它具有极强的遗传毒性和神经毒性，严重时可导致人体行动迟缓、记忆力衰减和语言功能障碍等，并可以在母婴之间传递从而对新生儿产生严重的毒性作用。汞化合物对生物体的毒性不仅与其存在形态密切相关，也与暴露的浓度、途径以及生物体自身的特征差异有关。汞及其化合物很容易与一些含硫的基团如巯基、二硫基等结合，因此各种形态的汞在进入生物体后可使蛋白质的三级、四级结构发生改变从而在生物体的各个水平（亚细胞、细胞、组织、器官、个体）产生毒性作用。环境中的汞可以分为无机汞（如汞单质、含汞盐类等）和有机汞（如甲基汞、乙基汞和苯基汞等）。整体看来，有机汞的毒性显著高于无机汞，其中甲基汞被认为是毒性最大的汞形态。

1.2.1 甲基汞的毒性效应

甲基汞具有较强的急性毒性［啮齿动物半数致死量（median lethal dose，LD_{50}）约为 10~40 mg/kg 体重（JECFA，2004）］、肾毒性、肝毒性、神经毒性、生殖毒性、发育毒性和免疫毒性等，并可能具有致癌性。甲基汞对人体最主要的致毒机制是作用于神经系统而引起神经毒性和神经发育毒性（JECFA，2000；WHO，1990），以及生殖毒性、免疫毒性和心脏/心血管毒性（Choi et al.，2009；Grandjean et al.，2004；JECFA，2004；Stern，2005）。而且，甲基汞对成人和儿童的毒性效应有所不同（JECFA，2000，2004；WHO，1990），胎儿以及儿童处于发育中的中枢神经系统比成年人的神经系统对甲基汞毒性损伤更为敏感（WHO，1990）。出生前的甲基汞暴露，会影响胎儿神经元的正常发育，进而改变大脑结构、减小脑体积；甲基汞也可能在中枢神经系统形成的关键期通过抑制微管系统的功能而影响细胞分裂，从而抑制大脑的正常发育（WHO，1990）。此外，甲基汞的毒性作用可能存在潜伏期，使得甲基汞对儿童发育影响的表现更为复杂（Davidson et al.，2006），成年后的某些神经认知功能障碍甚至可能与出生前甲基汞暴露有关（Debes et al.，2016；Oulhote et al.，2017）。

甲基汞可以经消化道、呼吸道和皮肤黏膜等方式进入生物体，并随血液传输到达各个组织器官。此外，甲基汞有很强的生物富集和生物放大能力，可以沿食物链传递从而对高营养级的动物和人体产生严重的毒性作用。作为一种神经毒素，甲基汞会对生物体造成以神经系统为主的全身性损伤，表现为精神和行为障碍，并可能破坏肾脏、肝脏、胃肠道等器官的正常功能，出现诸如语言和记忆力功能障碍、运动共济失调、昏迷、震颤、脑瘫、胃肠炎和肾功能衰竭等中毒症状。更为严重的是，甲基汞可以穿透胎盘屏障进入胎儿体内，导致新生儿发生先天性疾病，出现智力低下、生长缓慢等神经系统疾病（Spyker et al.，1972）。研究人员通过计算分析发现，孕妇长期处于低剂量甲基汞暴露会对婴儿的智力发育产生严重影响，孕妇头发中总汞的含量每升高 1 mg/kg，新生儿的平均智商水平会有约 0.18 的下降（Trasande et al.，2005）。

进入生物体内的甲基汞可以到达身体各个组织器官，通过与含巯基的蛋白质相互作用，一方面会直接破坏蛋白质的正常功能，另一方面也使甲基汞很难发生代谢降解和排泄。甲基汞暴露会对生物体的一些细胞器造成不同程度的损伤，可以破坏细胞膜的完整性，并会对血红蛋白中亚铁血红素的合成产生负面影响。甲基汞可以干扰生物体 DNA 的转录和蛋白质的合成（包括脑发育过程中蛋白质的合成）等重要生命活动。在神经毒性方面，甲基汞会对中枢神经系统和相关组织器官的众多亚细胞结构产生破坏作用，如破坏神经递质的正常功能，干扰自由基的产生等，从而对脑组织和周围神经系统产生毒性作用。

在自然环境中，甲基汞之所以是对人体和其他生物体毒性危害最大的汞形态，除了因为其本身的高毒性，更与其在食物链中的富集和放大有关。经过生物富集与放大，在作为捕食者的鱼类体内，其甲基汞浓度可以是周围水环境中汞浓度万倍以上。对于普通居民，尤其是沿海地区的居民，他们对软体动物、鱼类等水产品的摄入是暴露甲基汞的主要途径，当摄入较多水产品尤其是甲基汞含量较高的鱼类时，甲基汞可能会对这部分居民产生更高的毒性作用（Weihe and Grandjean，1998）。

1.2.2 无机汞的毒性效应

对于无机汞而言，离子态含汞盐类和元素态的单质汞都可以对生物体产生一定的毒性作用。含汞盐类主要以+1 价和+2 价的形态存在，大部分含汞盐类溶解性较低，被生物吸收的程度较小，最终可以经肾脏、肠道等器官排出体外，而且无机汞离子难以穿透血脑屏障，因此对生物体的毒性相对较小。

急性中毒时，二价汞盐（如氯化汞）主要会对胃肠道和肾脏等器官造成损伤，可能会引发严重的肠上皮细胞蛋白沉淀并伴随着肠道黏膜的坏死，表现出呕吐、腹痛、腹泻等症状，更严重者，可能会引起败血症、腹膜炎和低血容量性休克进而导致生物体的死亡，幸存的患者通常会发展为肾小管坏死并伴有无尿等症状。二价汞盐的慢性中毒基本上都发生在直接暴露汞的职业人群中，主要表现为对肾脏的毒性作用，容易引起单发或并发的自身免疫性肾小球肾炎和肾小管坏死等疾病。此外，二价汞盐还可能引起免疫系统的功能障碍，包括对汞暴露的过敏性反应（如皮炎、哮喘和各种类型自身免疫反应），对生物体自然杀伤细胞的抑制作用和对各种其他淋巴细胞亚群的破坏作用等。二价汞盐也可以对生物甲状腺的正常功能造成破坏，这可能与其对 5′-脱碘酶的抑制作用有关，并可能会抑制雄性动物体内精子的产生造成生殖发育毒性。在对脑组织的功能性损伤方面，由于难以穿透血脑屏障，因此二价汞盐不如其他形态汞（如有机态汞）的毒性效应明显。相比于二价汞盐，一价无机汞化合物具有更低的溶解性，因此在生物体内的吸收更加困难，但是一价汞盐类能够通过各种途径被氧化生成二价汞盐而被生物体吸收，进而引起后续的如二价汞化合物的毒性效应。

单质汞由于具有易挥发性、高扩散性和脂溶性，并可以穿透血脑屏障，因此具有比含汞盐类更高的毒性作用。单质汞挥发形成的汞蒸气，其主要的目标器官是大脑，也可能引起一些皮炎疾病。另外，汞蒸气同样会对末梢神经、免疫系统、肾、内分泌系统和肌肉功能造成一定程度的损伤。短时间内大量吸入汞蒸气会导致汞中毒性震颤及一系列过敏反应，引起支气管炎和细支气管炎等呼吸系统疾病，并伴随中枢神经系统的损伤，表现为胸闷气短、呼吸困难、行为和性格异常、肌肉频繁异常抖动、记忆力缺失、抑郁失眠、躁动不安等症状，严重时可导致精神错乱甚至引发幻觉。而慢性较低剂量的汞蒸气暴露也可以导致生物体神经功能的障碍，表现出诸如身体虚弱、头痛发晕、倦怠麻木、胃肠道功能紊乱、恶心厌食和体重下降等症状。

需要指出的是，由于环境中的汞存在着复杂的迁移转化过程，因此环境中汞及其化合物的毒性效应是非常复杂的，在评价汞及其化合物的毒性时应当综合考虑各种形态汞之间可能的相互作用。事实上，以各种方式（如皮肤接触扩散、呼吸道吸入或消化道摄入）进入生物体中的元素态单质汞可以被氧化生成二价汞盐从而对生物产生毒性效应，无机态的汞也可以通过微生物和化学的方式被转化为毒性更高的甲基汞，进一步随食物链进行富集放大从而对高营养级生物及人体的健康造成更加严重的毒性作用（Poulain and Barkay，2013；Yin et al.，2014）。

1.3 汞的形态与分子转化

大气、土壤和水体中各种汞形态之间可通过生物或者化学过程发生相互转化

(如图 1-1 所示)。由于各形态汞化合物的物理化学性质存在显著差异，其形态转化必然影响汞的环境过程及毒性。

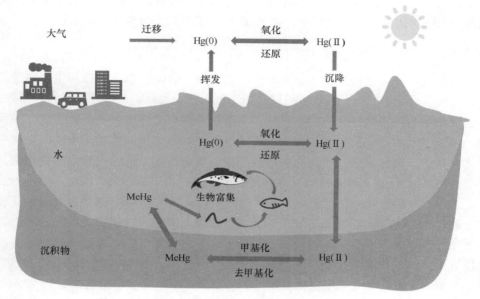

图 1-1　汞的形态转化

1.3.1　汞的氧化与还原

零价汞具有较高的挥发性，其较低的挥发热与较高的饱和蒸气压使得零价汞易于挥发进入气相，并随大气环流进行长距离传输和全球分布。二价汞在水中溶解度较高，因此大气中氧化态二价汞易于通过湿沉降等过程从大气中清除并进入地表。汞的氧化与还原转化导致了汞的大气长距离传输，使汞成为一种全球性的污染物。此外，相较于零价汞，微生物对二价汞的甲基化效率更高。不同氧化还原态的汞在环境迁移、转化、生物摄入、毒性等方面均存在显著差异，因此汞的氧化与还原在汞的生物地球化学循环中起着至关重要的作用。

汞的氧化与还原在多种环境和生物介质中均可发生，涉及一系列化学或生物反应过程。汞的氧化过程在多种环境介质如大气、水体、土壤中普遍存在，可能涉及多种化学或生物微观机制。除了均相反应体系，非均相体系中颗粒物的界面反应可能对汞的氧化有催化作用。大气中零价汞的氧化降低了大气汞驻留时间，增加了汞的大气沉降与地表输入；而水体中溶解性气态汞的氧化则抑制了汞的水-

气交换，增加了水体中汞的浓度，相应促进了水体中汞的甲基化与生物累积。汞的还原过程在多种环境介质如大气、水、土壤、冰雪中均可发生。大气气相/水相中汞的还原增加了汞在大气中的驻留时间与长距离迁移。表层水与土壤中汞的还原增加了零价汞向大气中的释放，降低了水/土壤中的汞浓度。

1.3.2 汞的甲基化与去甲基化

甲基汞是环境中毒性最强的汞形态，在土壤、底泥、水、大气和生物体等多种介质均有广泛的分布。人体可通过食用甲基汞污染的水产品或稻米摄入甲基汞。由于甲基汞的神经毒性与生物累积性，环境中汞的甲基化过程备受关注。一般认为，环境中汞的甲基化主要来自于厌氧环境（如底泥、厌氧水层）汞的微生物甲基化。其中，硫还原菌、铁还原菌以及产甲烷菌是目前发现的主要汞甲基化菌。汞甲基化基因 $hgcA$ 和 $hgcB$ 的发现，揭开了汞微生物甲基化遗传机制，也进一步拓宽了对汞甲基化微生物种类与所处环境的认识。目前，实验室研究证实约50种微生物具有汞甲基化能力，其分布于淡水、河流、湖泊、沉积物、湿地、冰雪甚至肠道等多种生态系统中。目前，人们对微生物汞甲基化的下游机制认知较少，尚待进一步深入研究。关于汞的非生物甲基化目前研究较少。在表层水、大气介质中可能存在汞的非生物甲基化过程，但其重要性有待进一步证实。

环境中甲基汞的去甲基化导致甲基汞的浓度与生物累积的降低。汞的甲基化与去甲基化过程的平衡最终决定环境中甲基汞的净含量及其健康风险。因此，可以说，汞的去甲基化与甲基化过程同等重要。在自然环境中，甲基汞的去甲基化主要通过生物降解和非生物降解两种途径。绝大部分汞的甲基化微生物同时具有去甲基化的能力。微生物去甲基化过程可分为还原去甲基化与氧化去甲基化。在还原去甲基化过程中，甲基汞在微生物还原作用下转化为零价汞和甲烷；在氧化去甲基化过程中，微生物将甲基汞主要降解为二价汞与二氧化碳。目前，对去甲基化的微生物类群研究较少，氧化去甲基化的关键酶和功能基因尚不明确。甲基汞的非生物降解主要分为光化学降解与非光化学降解。近期研究表明，在避光条件下，甲基汞可能存在非生物介导的降解过程，但其机制尚不明确。甲基汞的光化学降解是地表水中甲基汞去除的主要途径，可占底泥扩散至表层水的甲基汞总量的 30%～80%。环境水体中甲基汞的光降解依赖于其水化学参数，可能多途径并存，溶解有机质在甲基汞光降解中起着关键的作用。

1.4 汞的长距离传输

汞被认为是唯一能随大气进行长距离传输的重金属污染物。汞的长距离传输导致汞污染在全球范围内广泛地分布。即使在没有明显汞污染源的偏远地区（如南北极、青藏高原），也可检测到汞的污染。如在青藏高原 4 条河流中采集的鱼类肌肉中的总汞和甲基汞平均分别可高达 819 ng/g dw（dry weight，干重）和 756 ng/g dw（Shao et al., 2016）。偏远地区冰芯、泥炭、湖泊沉积物等介质受人类污染源的直接扰动较少，可反映全球大气成分变化（丛志远等，2010）。冰芯的汞历史记录显示，在近百年里，人为源汞贡献率已经达到 70%（Schuster et al., 2002）。湖泊沉积物中汞记录显示，现代大气汞沉降速率是工业革命前背景值的 3～5 倍（Biester et al., 2007）。从另一方面看，环境介质汞的历史记录也证实了汞的大气长距离传输及其对偏远地区的影响。

　　汞长距离传输的根源在于汞的氧化/还原以及零价汞在大气中较长的半衰期。在生物与化学因素作用下，土壤、表层水中的二价汞可被还原为零价汞。零价汞具有较低的挥发热与较高的饱和蒸气压，导致其具有较高的挥发性，因此土壤、表层水中的零价汞易于释放进入大气。人为过程如燃煤等的汞排放包括零价汞、气态二价汞以及颗粒态汞。二价汞易于通过湿法烟气脱硫等装置有效去除，因此，排放到大气中的汞以零价汞为主（Streets et al., 2005b）。自然源或人为源排放至大气中的零价汞相对惰性，其大气停留时间长达数月，甚至 1～2 年（Lindqvist and Rodhe, 1985）。大气中 90%以上的汞以气态单质汞形式存在，并随大气循环进行长距离的传输。大气中的零价汞可被各种活性卤物种、臭氧、过氧化氢、羟基自由基以及硝酸根自由基等氧化，生成二价汞，并通过干湿沉降进入地表环境。例如，极地地区春季臭氧与溴作用生成反应活性极强的溴自由基，溴自由基对大气零价汞的氧化造成了极地春季大气汞"亏损"事件（Schroeder et al., 1998）。除了大气零价汞的氧化，近期研究发现，植物吸收也是大气零价汞消除的重要途径。陆地植被吸收零价汞，导致大气中零价汞的降低与季节性变化（Jiskra et al., 2018）。北极苔原植被吸收大气零价汞，其凋落物引发的汞输入造成了极地地区汞的污染（Obrist et al., 2017）。以上氧化态汞的干湿沉降以及零价汞的植被吸收降低了大气中汞的浓度与传输，增加了偏远地区汞的输入。与之相反的是，大气气相与水相中二价汞的光还原可增加大气中汞的停留时间与传输距离（Saiz-Lopez et al., 2018）。

　　沉降至偏远地区的汞可被微生物甲基化从而进入食物链。极地汞"亏损"及其他干湿沉降过程输入至地表的汞可被微生物所利用（Scott, 2001）。极地海水、海冰、湿地等环境存在汞的甲基化微生物与汞的甲基化过程（Gionfriddo et al., 2016；Kirk et al., 2008；Lehnherr et al., 2012；Lehnherr et al., 2011；Loseto et al., 2004）。微生物作用生成的甲基汞可进一步随淡水、海水与陆生食物链累积与富集（Douglas et al., 2012）。全球变暖引起的冰冻区融化可引发之前累积汞的再释放并加剧汞的甲基化，这可能对极地生态系统产生重要影响（孙学军等，2017）。在陆地森林生态系统中，叶片吸收汞以凋落物形式输入地表。土壤中凋落物分解伴随着汞的甲基化过程，生成的甲基汞可进入森林生态系统并随食物网累积（Kronberg et al.，2016；Tavshunsky et al.，2017；Zhou et al.，2018）。

1.5 汞污染与《关于汞的水俣公约》

1.5.1 汞的使用与排放

在我国，汞化合物的使用历史悠久。我国出土的公元前 3～公元前 4 世纪青铜文物上可以看到有镀金，据推断当时人们可能已经掌握了汞齐镀金（镏金）法。汉武帝（公元前 140～公元前 87 年）时期，司马迁所著《史记·封禅书》中，就记有"丹砂可化为黄金"之说，说明当时人们已经能够利用汞了。古人为追求"长生"而发展的各种"炼丹术"也是对汞化合物（朱砂，HgS）的使用，其主要是利用汞和硫黄的反应，即"丹砂烧之成水银，积变，又还成丹砂"。由于朱砂具有多种药效，因此古代的许多中药中都含有大量的朱砂成分，现在部分广泛使用的中成药中也可以检出很高含量的汞（Mino and Yamada，2005；汤毅珊等，1999）。需要说明的是，中成药中添加的主要是硫化汞，其生物可利用性和毒性均较小。

如今，随着科学技术的发展，世界各国对汞及其化合物的使用更加广泛，应用于化学工业、医药卫生、交通运输、电器仪器、冶金行业及军工生产等各种领域。目前，我国仍在生产和使用的添汞产品有含汞荧光灯、含汞体温计、含汞血压计、含汞电池和牙科汞合金等。图 1-2 是美国地质勘探局（USGS）统计的 1990～2016 年全球和中国的汞生产总量，其中 2000 年全球汞生产总量为 1360 t，我国约 200 t；2010 年全球汞生产总量为 2180 t，我国约 1600 t；2016 年全球为 2480 t，我国约 2000 t。总体来看，全球汞生产总量基本在 1500～3000 t 之间，我国汞产量在 2003 年之前基本小于 1000 t（占全球汞产量的 20%～35%），但 2003 年之后在 1000～2000 t 之间（占全球汞产量的 60%～80%），并呈现增长的趋势。据统计，2007 年我国汞的需求量在 1527～1563 t 之间，主要用汞工艺是乙炔法聚氯乙烯生产，其汞需求量约占总需求量的 60%；电池、电光源、体温计和血压计需汞量之和约占总需求量的 38%（菅小东等，2009）。

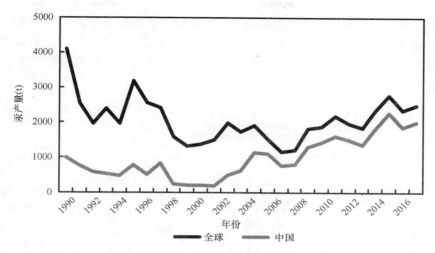

图 1-2　1990～2016 年全球和中国汞产量
（数据引自 USGS 1994～2016 年统计数据，https://minerals.usgs.gov/minerals/pubs/commodity/mercury/）

汞可以通过自然过程和人为排放两种方式进入环境（图 1-3），污染大气、土壤和水体，进而导致食品和生物体中汞含量的升高。汞的自然源包括火山与地热活动、森林火灾，以及从土壤和水体表面蒸发等；人为源主要包括原生汞矿开采、燃煤电厂、燃煤工业锅炉、有色金属冶炼、水泥生产，以及含汞产品的生产和使用。据估算，全球各种人为源每年向大气排放的汞量在 2000 t 左右，如 1983 年为 3560 t（Nriagu and Pacyna，1988），1990 年为 2140 t（Pacyna and Pacyna，2002），1995 年为 1910 t（Pacyna and Pacyna，2002），2000 年为 2190 t（Pacyna et al.，2006），2005 年为 1930 t（UNEP，2008），2010 年为 1960 t（UNEP，2013a），2015 年为 2220 t（UNEP，2019）。

对于我国每年人为大气汞的排放量，不同文献的估算结果有较大的差异，但基本在 500～700 t 范围。如 Streets 等（2005a）估算 1999 年我国人为排放汞总量为 536 t；Wu 等（2006）估算 2003 年为 696 t；Pacyna 等（2010）估算 2005 年高达 825 t；Pirrone 等（2010）估算 2007 年为 609 t；根据联合国环境规划署（UNEP）*Global Mercury Assessment 2013* 中的数据，2010 年我国人为汞排放总量为 575 t，约占全球排放总量（1960 t）的 30%（UNEP，2013a）。近期，Zhang 等（2015）采用 CAME（中国大气汞排放）模型对 2000～2010 年间我国汞排放清单进行了统一评估，最后估算出此期间我国人为排放汞总量以平均年增长率 4.2%的速度持续增加，从 2000 年的 356 t 增加到 2010 年的 538 t。需要注意的是，在早期的研究中由于缺乏我国各行业的排放因子，大多采用的是国外排放因子；而在近期的研究中已经逐渐加入了我国实际测定的排放因子，因此估算的排放量更为准确。同时，汞排放的行业分布也显示出一些新的变化，如燃煤电厂和燃煤工业锅炉虽然仍是我国主要的人为汞排放行业，但水泥生产和有色金属冶炼所占的比例更加突显（Zhang et al., 2015）。

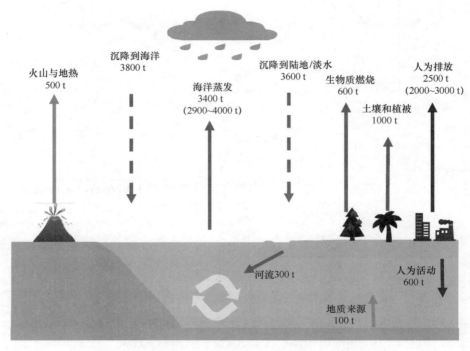

图 1-3　环境中汞的来源［数据引自（UNEP，2019）］

1.5.2　汞污染

人们对汞及其化合物的大量生产、使用和排放造成了全球性的汞污染，在许多国家和地区都发生过汞污染事件。最引人注意的是 20 世纪 50~60 年代发生在日本熊本县水俣镇的汞中毒事件，"水俣病"也因此而得名。这次事件的主要起因是日本 Chisso 公司于 1932 年在水俣湾建造了化工厂，该化工厂从 1949 年开始生产氯乙烯，且将没有经过任何处理的废水排放到水俣湾。由于在氯乙烯生产过程中需要使用含汞的催化剂，导致大量的汞化合物随废水进入到水俣湾，在水体中被甲基化为生物可利用性和毒性更大的甲基汞，进而被生物体吸收并随食物链累积和放大，最终导致当地居民的严重汞中毒。自 1956 年正式报告首例中毒患者，至 2001 年被确诊的患者共有 2265 人（Ministry of the Environment Government of Japan，2002），还有上万人出现水俣病的症状而未得到确认，造成了严重的危害，也引起了世界各国对汞污染的广泛关注。

另一有代表性的事件是 1971～1972 年发生在伊拉克的全国性汞中毒。这次事件主要是由于伊拉克连年干旱，粮食歉收，人们误食了用含烷基汞农药浸泡过的小麦种子，致使 6000 多人住院，500 余人死亡。在患者血液、头发和死亡者的脏器中，都检测出了较高的汞含量。

早期发生的汞污染事件大部分是由于化工过程使用和排放汞造成的，属于区域性污染。但随着经济的发展，燃煤、金属冶炼和水泥生产等无意排放汞逐渐成为环境中汞的主要人为源，而且由于汞具有长距离传输的特性，导致北美和北欧许多偏远地区湖泊中鱼体汞含量超过世界卫生组织建议的标准，并且在青藏高原、远海及南北极等人为活动较少地区生物样品中也检测到较高的汞含量。联合国环境规划署（UNEP）在 2002 年的 *Global Mercury Assessment* 报告中指出，"自工业革命以来，汞在全球大气、水和土壤中的含量已增加了 3 倍左右，汞污染的不断加剧对人类健康和环境造成极大危害，在全球产生了重大的不利影响"。根据美国环境保护署（USEPA）的报告，2000 年美国将近 500 万名妇女体内的汞含量高于安全标准，每年可能有高达 30 万名新生儿因为汞的侵害导致其智力和神经系统受到影响，而联合国环境规划署推测在全世界这一数据可能高达千万。因此，汞污染已不再是区域性的问题，需要更多地从全球角度去认识汞污染并采取措施。

1990 年，在瑞典科学家 Oliver Lindqvist 教授的倡议下，第一届"汞全球污染物国际会议（International Conference on Mercury as a Global Pollutant，ICMGP）"在瑞典召开，正式提出汞是一种全球污染物的概念。之后分别在美国、加拿大、德国、巴西等地连续召开了 13 届会议，其中第九届会议于 2009 年 6 月 7～12 日在我国贵阳召开，第十四届会议将于 2019 年 9 月在波兰召开。参会人数不断增多，会议规模不断扩大。

1.5.3 《关于汞的水俣公约》

鉴于汞的全球污染和危害，联合国环境规划署在 2010～2013 年召开了 5 次政府间谈判委员会会议，并于 2013 年 1 月达成一项具有法律约束力的国际公约——*Minamata Convention on Mercury*（《关于汞的水俣公约》，以下简称《水俣公约》），旨在控制和削减全球人为汞排放和含汞产品的使用。目前共有 128 个国家和地区签署，我国政府于 2013 年 10 月 10 日作为首批签约国签署了《水俣公约》。2016 年 4 月 28 日，第十二届全国人民代表大会常务委员会第二十次会议通过了批准《水俣公约》的决定，并于 8 月 31 日向联合国交存批准文书，成为第三十个批约国。

围绕《水俣公约》的达成和批准，我国政府已经在政策制定、监督管理、技术研发和能力建设等方面开展了大量的工作。

《水俣公约》共包括35条正文和5个附件（A～E）。其中附件A规定了可申请豁免的添汞产品种类和具体淘汰时限，并列举了受限制的添汞产品种类及管控要求；附件B主要规定了可申请豁免的生产工艺和具体淘汰时限，并列举了受限制的生产工艺和管控要求；附件C明确了缔约方关于手工和小规模炼金国家行动计划的编制要求；附件D明确了汞及其化合物的大气排放点源名录；附件E主要是关于公约的仲裁和调解程序。

《水俣公约》具有一定的法律约束力，对汞的供应和贸易、汞的排放、添汞产品和涉汞行业都提出了强制性的条款并明确了淘汰时限。例如，公约规定缔约方应在公约生效之日禁止开采新的原生汞矿，并在公约生效15年内关闭现有原生汞矿；各缔约方应自2021年起停止规定添汞产品的生产、进口或出口，除非该缔约方申请了某项豁免；在公约生效3年内查明向土地和水中释放汞化合物的点源类别；5年内建立汞的排放清单，并使用最佳可得技术和最佳环境实践控制和减少"新来源汞"（"新来源"指《水俣公约》对该国生效1年后开始建设的，或者在生效1年后才开始的重大改造工程的来源）的释放；自公约生效之日起淘汰使用汞或汞化合物作为催化剂的乙醛生产工艺，公约生效5年后不允许在氯乙烯单体的生产中继续使用汞，10年内淘汰使用含汞催化剂进行生产的聚氨酯工艺，自2025年起淘汰使用汞或汞化合物的氯碱生产工艺，除非该缔约方申请某项豁免。

1.6 我国汞研究进展与挑战

丰富的汞矿储存加上快速的经济发展，使我国成为汞生产、使用和排放大国。二十世纪七八十年代在第二松花江和蓟运河都曾出现过严重的汞污染，贵州省由于汞矿开采和燃煤造成部分地区大气、水体和土壤中汞含量严重超标。随着《水俣公约》的签署和实施，我国汞污染研究将面临国内环境保护和国际履约的双重压力。

为了应对汞污染防治及履约需求，环境保护部环境保护对外合作中心于2013年10月将汞工作组正式更名为汞公约履约处，主要负责《水俣公约》和汞公约淘汰国家方案编制及具体实施；汞国际公约履约项目的选择、准备和报批工作，并对项目的实施进行统一协调、管理和监督；协助环保部有关部门拟定汞国际公约国家和行业的政策、法规和管理规章，承担汞国际公约履约的具体事务性工作等。

针对我国汞污染研究基础薄弱的现状和未来履约面临的压力，科技部于2012年批准了首个关于汞的"973"计划项目"我国汞污染特征、环境过程及减排技术原理"(2013~2017年),对我国汞污染的相关基础研究进行了提前部署。该项目主要针对我国大气汞源汇及迁移规律、汞的环境过程与效应、典型行业烟气汞控制与减排技术原理等关键科学问题开展了深入的研究。主要研究内容包括:重点源的汞排放因子、排放清单、形态分布和同位素组成;典型地表与大气间汞交换过程、机制和影响因素,汞在环境中的分子转化;我国大气中不同形态汞的时空变化、沉降通量及迁移转化;典型区域中汞的污染特征、生物富集及演变趋势;燃煤和有色金属冶炼过程汞形态调控机制与减排技术原理。通过项目的研究,初步阐明了我国汞污染来源、归趋、生物累积和环境效应,在汞生物地球化学循环方面取得了原创性的成果,同时也培养了一支汞污染研究的团队,为我国《水俣公约》的谈判和履行提供了重要科技支撑。

在国家需求和相关项目的支持下,我国汞污染研究呈现迅速发展的态势。在 Web of Science 核心合集中检索主题涉及汞的文章,在过去 20 年全球共发表 66270 篇,其中中国学者发表了 11386 篇 (17.2%),仅次于美国的 15163 篇 (22.9%)。在 1999 年全球学者共发表关于汞的文章 2235 篇,其中中国学者发表的仅有 84 篇 (3.8%);而 2018 年全球学者发表的汞文章数为 4282 篇,其中中国学者发表的文章数增加到了 1476 篇,占总数的 34.5%,如图 1-4 所示。

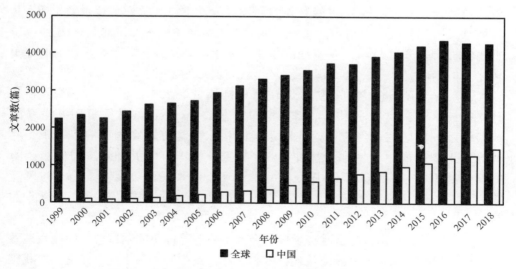

图 1-4 1999~2018 年 Web of Science 中关于汞的文章数

由于汞污染的特殊性和复杂性，目前对我国环境中汞的分布、迁移、转化以及归宿尚不完全清楚，对我国汞污染的环境与健康影响仍不可预知，汞污染研究还存在诸多挑战。

（1）《水俣公约》第 22 条要求，缔约方大会应在本公约生效后 6 年内开始，并按照确定的时间间隔定期对本公约的成效进行评估。目前，对我国大气汞排放源及排放清单已经进行了较为系统的研究，但尚缺乏系统性的环境介质和生物体汞监测数据，相关监测能力有待提高。需要进行全面、系统的调查，建立汞流向数据库，实行动态管理。

（2）长距离传输是汞污染的重要特征。虽然我国大气汞排放量较大，但最近的一些研究已经显示我国部分地区的大气汞含量也受到周边国家和地区的影响。因此，需要广泛开展汞的大气迁移规律研究，建立汞的区域、全国及跨境传输模型，厘清我国排放的汞对境内和境外的影响，以及境外汞排放对我国环境的影响，为我国制定汞减排政策及履行《水俣公约》提供科学依据。

（3）汞形态是研究汞污染的关键环节，汞的迁移、转化、累积和毒性，以及减排工艺和修复技术等都与汞的存在形态密切相关。目前对大气汞物理化学转化过程认识的不足，会导致模型预测结果与实际观测值存在一定偏差。最近的研究在环境中发现了一些"新"的汞形态（如含汞纳米颗粒、一价汞、水体中颗粒结合态零价汞等），可能对汞的环境行为及生物有效性具有重要影响，其环境意义还需要进一步阐明。此外，对于汞在不同环境条件下的氧化/还原、甲基化/去甲基化等反应机制及主控因子还不完全清楚，仍需要深入开展汞的分子转化及机制研究。

（4）由于我国较大的汞生产、使用和排放量，水库、海洋、汞矿区和其他汞敏感区域都可能是甲基汞高累积的生态系统，从而导致人群甲基汞暴露风险。近期的研究已经表明，我国汞矿区稻米中含有较高含量的甲基汞，并成为当地居民的主要甲基汞暴露来源。目前，我国有关人群暴露与健康风险方面的数据非常有限，需要开展典型生态环境中汞的污染现状、趋势与风险研究，系统评估我国汞污染的暴露风险，以及汞减排措施对防控健康风险的效益。

（5）开发新技术，尽量减少燃煤、金属冶炼等过程中的汞排放，削减汞在各个行业中的应用，研发有效的替代技术和产品，是控制汞污染的有效途径。我国这方面的研究还很薄弱，应大力加强减少汞排放的技术研究和应用。

参 考 文 献

丛志远, 康世昌, 郑伟, 张强弓, 2010. 偏远地区铅和汞的现代过程与历史记录研究综述. 地理学报, 65: 351-360.

菅小东, 沈英娃, 姚薇, 王玉晶, 张鑫, 2009. 我国汞供需现状分析及削减对策. 环境科学研究, 2009, 22: 788-792.

孙学军, 康世昌, 张强弓, 丛志远, 2017. 山地冰川消融过程中汞的行为及环境效应综述. 地球科学进展, 32: 589-598.

汤毅珊, 黄志尧, 潘华新, 王培训, 王宁生, 1999. 微波消解-原子荧光法测定中药中的汞、砷. 中药新药与临床药理, 10: 177-179.

Biester H, Bindler R, Martinez-Cortizas A, Engstrom D R, 2007. Modeling the past atmospheric deposition of mercury using natural archives. Environmental Science & Technology, 41: 4851-4860.

Choi A L, Pal W, Budtz-Jorgensen E, Jorgensen P J, Salonen J T, Tuomainen T-P, Murata K, Nielsen H P, Petersen M S, Askham J, Grandjean P, 2009. Methylmercury exposure and adverse cardiovascular effects in Faroese whaling men. Environmental Health Perspectives, 117: 367-372.

Cohen E R, Cvitas T, Frey J G, Holmström B, Kuchitsu K, Marquardt R, Mills I, Pavese F, Quack M, Stohner J, Strauss H L, Takami M, Thor A J, 2008.Quantities, units and symbols in physical chemistry, IUPAC Green Book. 3rd Edition. Cambridge: IUPAC & RSC Publishing.

Davidson P W, Myers G J, Weiss B, Shamlaye C F, Cox C, 2006. Prenatal methyl mercury exposure from fish consumption and child development: A review of evidence and perspectives from the Seychelles Child Development Study. Neurotoxicology, 27: 1106-1109.

Debes F, Weihe P, Grandjean P, 2016. Cognitive deficits at age 22 years associated with prenatal exposure to methylmercury. Cortex, 74: 358-369.

Douglas T A, Loseto L L, Macdonald R W, Outridge P, Dommergue A, Poulain A, Amyot M, Barkay T, Berg T, Chetelat J, Constant P, Evans M, Ferrari C, Gantner N, Johnson M S, Kirk J, Kroer N, Larose C, Lean D, Nielsen T G, Poissant L, Rognerud S, Skov H, Sorensen S, Wang F Y, Wilson S, Zdanowicz C M, 2012. The fate of mercury in Arctic terrestrial and aquatic ecosystems, a review. Environmental Chemistry, 9: 321-355.

Gionfriddo C M, Tate M T, Wick R R, Schultz M B, Zemla A, Thelen M P, Schofield R, Krabbenhoft D P, Holt K E, Moreau J W, 2016. Microbial mercury methylation in Antarctic sea ice. Nature Microbiology, 1: 16127.

Grandjean P, Murata K, Budtz-Jorgensen E, Weihe P, 2004. Cardiac autonomic activity in methylmercury neurotoxicity: 14-year follow-up of a Faroese birth cohort. Journal of Pediatrics, 144: 169-176.

JECFA, 2000. Fifty-third meeting of the Joint FAO/WHO Expert Committee on Food Additives. Methylmercury. *In*: World Health Organization (WHO), ed.WHO food additives series: 44-Safety evaluation of certain food additives and contaminants. Geneva.

JECFA, 2004. Sixty-first meeting of the Joint FAO/WHO Expert Committee on Food Additives. Methylmercury (addendum). *In*: World Health Organization (WHO), ed. WHO food additives series: 52-Safety evaluation of certain food additives and contaminants. Geneva.

Jiskra M, Sonke J E, Obrist D, Bieser J, Ebinghaus R, Myhre C L, Pfaffhuber K A, Wangberg I,

Kyllonen K, Worthy D, Martin L G, Labuschagne C, Mkololo T, Ramonet M, Magand O, Dommergue A, 2018. A vegetation control on seasonal variations in global atmospheric mercury concentrations. Nature Geoscience, 11: 244-250.

Kirk J L, Louis V L S, Hintelmann H, Lehnherr I, Else B, Poissant L, 2008. Methylated mercury species in marine waters of the Canadian high and sub Arctic. Environmental Science & Technology, 42: 8367-8373.

Kronberg R M, Jiskra M, Wiederhold J G, Bjorn E, Skyllberg U, 2016. Methyl mercury formation in hillslope soils of Boreal forests: The role of forest harvest and anaerobic microbes. Environmental Science & Technology, 50: 9177-9186.

Lehnherr I, St Louis V L, Emmerton C A, Barker J D, Kirk J L, 2012. Methylmercury cycling in High Arctic wetland ponds: Sources and sinks. Environmental Science & Technology, 46: 10514-10522.

Lehnherr I, St Louis V L, Hintelmann H, Kirk J L, 2011. Methylation of inorganic mercury in polar marine waters. Nature Geoscience, 4: 298-302.

Lindqvist O, Rodhe H, 1985. Atmospheric mercury: A review. Tellus Series B-Chemical and Physical Meteorology, 37: 136-159.

Loseto L L, Siciliano S D, Lean D R S, 2004. Methylmercury production in High Arctic wetlands. Environmental Toxicology and Chemistry, 23: 17-23.

Ministry of the Environment Government of Japan, 2002. Minamata Disease: The History and Measures. http://www.env.go.jp/en/chemi/hs/minamata2002/.

Mino Y, Yamada Y, 2005. Detection of high levels of arsenic and mercury in some Chinese traditional medicines using X-ray fluorescence spectrometry. Journal of Health Science, 51: 607-613.

Nriagu J O, Pacyna J M, 1988. Quantitative assessment of worldwide contamination of air, water and soils by trace metals. Nature, 333: 134-139.

Obrist D, Agnan Y, Jiskra M, Olson C L, Colegrove D P, Hueber J, Moore C W, Sonke J E, Helmig D, 2017. Tundra uptake of atmospheric elemental mercury drives Arctic mercury pollution. Nature, 547: 201-204.

Oulhote Y, Debes F, Vestergaard S, Weihe P, Grandjean P, 2017. Aerobic fitness and neurocognitive function scores in young Faroese adults and potential modification by prenatal methylmercury exposure. Environmental Health Perspectives, 125: 677-683.

Pacyna E G, Pacyna J M, 2002. Global emission of mercury from anthropogenic sources in 1995. Water, Air, and Soil Pollution, 137: 149-165.

Pacyna E G, Pacyna J M, Steenhuisen F, Wilson S, 2006. Global anthropogenic mercury emission inventory for 2000. Atmospheric Environment, 40: 4048-4063.

Pacyna E G, Pacyna J M, Sundseth K, Munthe J, Kindbom K, Wilson S, Steenhuisen F, Maxson P, 2010. Global emission of mercury to the atmosphere from anthropogenic sources in 2005 and projections to 2020. Atmospheric Environment, 44: 2487-2499.

Pirrone N, Cinnirella S, Feng X, Finkelman R B, Friedli H R, Leaner J, Mason R, Mukherjee A B, Stracher G B, Streets D G, Telmer K, 2010. Global mercury emissions to the atmosphere from anthropogenic and natural sources. Atmospheric Chemistry and Physics, 10: 5951-5964.

Poulain A J, Barkay T, 2013. Cracking the mercury methylation code. Science, 339: 1280-1281.

Saiz-Lopez A, Sitkiewicz S P, Roca-Sanjuan D, Oliva-Enrich J M, Davalos J Z, Notario R, Jiskra M, Xu Y, Wang F, Thackray C P, Sunderland E M, Jacob D J, Travnikov O, Cuevas C A, Acuna A U, Rivero D, Plane J M C, Kinnison D E, Sonke J E, 2018. Photoreduction of gaseous oxidized

mercury changes global atmospheric mercury speciation, transport and deposition. Nature Communications, 9: 4796.
Schroeder W H, Anlauf K G, Barrie L A, Lu J Y, Steffen A, Schneeberger D R, Berg T, 1998. Arctic springtime depletion of mercury. Nature, 394: 331-332.
Schuster P F, Krabbenhoft D P, Naftz D L, Cecil L D, Olson M L, Dewild J F, Susong D D, Green J R, Abbott M L, 2002. Atmospheric mercury deposition during the last 270 years: A glacial ice core record of natural and anthropogenic sources. Environmental Science & Technology, 36: 2303-2310.
Scott K J, 2001. Bioavailable mercury in arctic snow determined by a light-emitting mer-lux bioreporter. Arctic, 54: 92-95.
Shao J, Shi J, Duo B, Liu C, Gao Y, Fu J, Yang R, Jiang G, 2016. Mercury in alpine fish from four rivers in the Tibetan Plateau. Journal of Environmental Sciences (China), 39: 22-28.
Spyker J M, Sparber S B, Goldberg A M, 1972. Subtle consequences of methylmercury exposure: Behavioral deviations in offspring of treated mothers. Science, 177: 621-623.
Stern A H, 2005. A review of the studies of the cardiovascular health effects of methylmercury with consideration of their suitability for risk assessment. Environmental Research, 98: 133-142.
Streets D, Hao J, Wu Y, Jiang J, Chan M, Tian H, Feng X, 2005a. Anthropogenic mercury emissions in China. Atmospheric Environment, 39: 7789-7806.
Streets D G, Hao J M, Wu Y, Jiang J K, Chan M, Tian H Z, Feng X B, 2005b. Anthropogenic mercury emissions in China. Atmospheric Environment, 39: 7789-7806.
Tavshunsky I, Eggert S L, Mitchell C P J, 2017. Accumulation of methylmercury in invertebrates and masked shrews (*Sorex cinereus*) at an upland forest-peatland interface in Northern Minnesota, USA. Bulletin of Environmental Contamination and Toxicology, 99: 673-678.
Trasande L, Landrigan P J, Schechter C, 2005. Public health and economic consequences of methyl mercury toxicity to the developing brain. Environmental health perspectives, 113: 590-596.
UNEP, 2002. Global Mercury Assessment 2002. https://www.unenvironment.org/explore-topics/chemicals-waste/what-we-do/mercury/global-mercury-assessment.
UNEP, 2008. The Global Atmospheric Mercury Assessment: Sources, Emissions and Transport. UNEP Chemicals Branch, Geneva, Switzerland. https://www.unenvironment.org/explore-topics/chemicals-waste/what-we-do/mercury/global-mercury-assessment.
UNEP, 2013a. Global Mercury Assessment 2013: Sources, Emissions, Releases and Environmental Transport. UNEP Chemicals Branch, Geneva, Switzerland. https://www.unenvironment.org/explore-topics/chemicals-waste/what-we-do/mercury/global-mercury-assessment.
UNEP, 2013b. Technical Background Report for the Global Mercury Assessment 2013. Arctic Monitoring and Assessment Programme, Oslo, Norway/UNEP Chemicals Branch Geneva, Switzerland.
UNEP, 2019. Global Mercury Assessment 2018. UNEP Chemicals and Health Branch, Geneva, Switzerland, Geneva, Switzerland. https://www.unenvironment.org/explore-topics/chemicals-waste/what-we-do/mercury/global-mercury-assessment.
Weihe P, Grandjean P, 1998. Methylmercury risks. Science, 279: 635-635.
WHO, 1990. Environmental health criteria 101-Methylmercury. Geneva.
Wu Y, Wang S X, Streets D G, Hao J M, Chan M, Jiang J K, 2006. Trends in anthropogenic mercury emissions in China from 1995 to 2003. Environmental Science & Technology, 40: 5312-5318.
Yin Y G, Li Y B, Tai C, Cai Y, Jiang G B, 2014. Fumigant methyl iodide can methylate inorganic

mercury species in natural waters. Nature Communications, 5: 7.

Zhang L, Wang S, Wang L, Wu Y, Duan L, Wu Q, Wang F, Yang M, Yang H, Hao J, Liu X, 2015. Updated emission inventories for speciated atmospheric mercury from anthropogenic sources in China. Environmental Science & Technology, 49: 3185-3194.

Zhou J, Wang Z W, Zhang X S, 2018. Deposition and fate of mercury in litterfall, litter, and soil in coniferous and broad-leaved forests. Journal of Geophysical Research-Biogeosciences, 123: 2590-2603.

第 2 章 汞的氧化与还原

> **本章导读**
> - 在自然界中无机汞主要以二价汞与零价汞形式存在。二价汞与零价汞形态的不同导致其物理化学特性及迁移转化过程存在极大差异,两者之间的转化(即汞的氧化与还原)是汞生物地球化学循环的重要组成部分。
> - 汞的氧化与还原在多种环境与生物介质中均可发生,涉及一系列化学或生物反应过程。
> - 概述氧化还原对汞的转化、迁移、循环等行为的影响,总结近些年有关汞的氧化/还原过程及其机制的研究进展。

2.1 单质汞、氧化态汞与汞的循环

汞的电子构型为 [Xe] $4f^{14}5d^{10}6s^2$,相对论收缩效应致使零价汞 $6s^2$ 壳层非常稳定,汞原子之间金属键较弱,汞原子之间主要由范德瓦耳斯力相互维系(Gonzalez-Raymat et al., 2017)。因此,零价汞在常温、常压下呈液态。零价汞具有较高的挥发性,其挥发热(73 cal/g)[①]远低于水和乙醇(Zhang, 2006)。25 ℃时,零价汞在水中的溶解度仅为 60 μg/L(Onat, 1974),亨利定律常数为 $4.4×10^7$ Pa(Andersson et al., 2008)。这些特性使得单质汞易于进入气相,并随大气环流进行大尺度的全球分布。

二价汞与零价汞是无机汞在自然界的主要存在形式。大气中的零价汞被氧化为二价汞后易于通过干湿沉降等过程从大气中清除并进入地表。地表中的汞也可通过自然还原过程再次进入大气。据估计,2008 年全球自然源向大气中的释汞量为 5207 t,远高于人为源的 2320 t(Pirrone et al., 2010)。不同氧化还原态的汞在环境迁移、转化、生物摄入、毒性等方面存在显著差异,因此汞的氧化/还原过程在其生物地球化学循环中起着至关重要的作用。相较于零价汞,通常微生物对二

[①] 1 cal=4.184 J。

价汞的甲基化效率更高（Hu et al., 2013a）。在模拟与天然非饱和土壤中，二价汞-有机质络合物的迁移性亦显著高于零价汞（Gai et al., 2016）。因此，二价汞与零价汞形态的差异导致其物理化学特性及迁移转化过程存在极大不同，二价汞与零价汞的转化（即汞的氧化与还原过程）[如式（2-1）至式（2-4）所示]，是汞生物地球化学循环的重要组成部分。

$$2Hg^{2+}+2e \rightleftharpoons Hg_2^{2+} \qquad E^\circ=0.911 \text{ V} \qquad (2\text{-}1)$$

$$Hg^{2+}+2e \rightleftharpoons Hg^0 \qquad E^\circ=0.854 \text{ V} \qquad (2\text{-}2)$$

$$Hg_2^{2+}+2e \rightleftharpoons 2Hg^0 \qquad E^\circ=0.796 \text{ V} \qquad (2\text{-}3)$$

$$Hg_2^{2+} \rightleftharpoons Hg^0 + Hg^{2+} \qquad E^\circ=-0.115 \text{ V} \qquad (2\text{-}4)$$

2.2 汞的化学氧化与生物氧化

汞的氧化在多种环境介质如大气、水体、土壤中普遍存在。这种氧化可能涉及化学或生物微观过程。除了均相反应体系，非均相体系中颗粒物的界面反应可能对汞的氧化有催化作用。整体而言，大气中零价汞的氧化降低了汞的大气驻留时间，增加了汞的大气沉降与地表输入；而水体中溶解性气态汞的氧化则抑制了零价汞的水-气交换，增加了水体中的汞浓度，有助于甲基化、生物累积等后续过程的发生。

2.2.1 大气气相汞化学氧化

大气中气态单质汞氧化是活性气态汞的主要来源之一。中纬度死海地区大气中气态单质汞与活性气态汞浓度的监测显示，活性气态汞浓度的高值对应于单质汞的极低值，表明大气单质汞氧化是活性气态汞的来源（Obrist et al., 2011）。目前关于零价汞大气气相化学氧化的研究主要集中于气相均相体系，对于非均相体系研究较少。

1. 气相均相汞氧化

除植物吸收（Fu et al., 2016；Meng et al., 2018）外，零价汞氧化是大气中汞去除最重要的途径。研究汞的大气氧化途径对于理解汞的全球传输与沉降具有重要的意义。目前，报道的零价汞气相氧化剂包括臭氧、过氧化氢、羟基自由基、$NO_3^\bullet$自由基以及各种卤素物种，其与零价汞的气相反应速率系数如表2-1所示。

表 2-1　汞的气相氧化反应

反应	速率常数 [$cm^3/(mol \cdot s)$]	参考文献
$Hg(0) + O_3 \longrightarrow HgO + O_2$　（2-5）	$(3\pm2)\times10^{-20}$	（Hall，1995）
	8.43×10^{-17}	（Pal and Ariya，2004b）
	$(6.2\pm1.1)\times10^{-19}$	（Snider et al.，2008）
	$(7.4\pm0.5)\times10^{-19}$	（Rutter et al.，2012）
$Hg(0) + HO^{\bullet} \longrightarrow HgO$　（2-6）	$(9.0\pm1.3)\times10^{-14}$	（Pal and Ariya，2004a）
	$(8.7\pm2.8)\times10^{-14}$	（Sommar et al.，2001）
	$<1.2\times10^{-13}$	（Bauer et al.，2003）
$Hg(0) + Cl_2 \longrightarrow HgCl_2$　（2-7）	$(2.6\pm0.2)\times10^{-18}$	（Ariya et al.，2002）
	2.5×10^{-16}	（Menke and Wallis，1980）
	4.8×10^{-15}	（Wang and Anthony，2005）
$Hg(0) + 2Cl^{\bullet} \longrightarrow HgCl_2$　（2-8）	$(1.0\pm0.2)\times10^{-11}$	（Ariya et al.，2002）
	$(1.8\pm0.5)\times10^{-11}$	（Sun et al.，2016）
$Hg(0) + Br_2 \longrightarrow HgBr_2$　（2-9）	$<(0.9\pm0.2)\times10^{-16}$	（Ariya et al.，2002）
$Hg(0) + 2Br^{\bullet} \longrightarrow HgBr_2$　（2-10）	$(3.2\pm0.3)\times10^{-12}$	（Ariya et al.，2002）
	$(1.6\pm0.8)\times10^{-12}$	（Sun et al.，2016）
	2.7×10^{-13}	（Tas et al.，2012）
$Hg(0) + BrO^{\bullet} \longrightarrow$ 产物　（2-11）	$1\times10^{-15} \sim 1\times10^{-13}$	（Obrist et al.，2011）
	1.5×10^{-13}	（Tas et al.，2012）
$Hg(0) + I_2 \longrightarrow HgI_2$　（2-12）	$\leqslant(1.27\pm0.58)\times10^{-19}$	（Raofie et al.，2008）
	$(7.4\pm0.2)\times10^{-17}$	（Chi et al.，2009）
$Hg(0) + H_2O_2 \longrightarrow Hg(OH)_2$　（2-13）	$<8.5\times10^{-19}$	（Tokos et al.，1998）
$Hg(0) + 2ICl \longrightarrow HgCl_2 + I_2$　（2-14）	$(10.5\pm0.3)\times10^{-17}$	（Qu et al.，2010）
$Hg(0) + 2BrCl \longrightarrow HgCl_2 + Br_2$　（2-15）	$(2.3\pm0.2)\times10^{-17}$	（Qu et al.，2009）
$Hg(0) + HCl \longrightarrow$ 产物　（2-16）	1.0×10^{-19}	（Hall，1993）
$Hg(0) + NO_3^{\bullet} \longrightarrow$ 产物　（2-17）	$<4\times10^{-15}$	（Sommar et al.，1997）

1）臭氧

背景大气臭氧浓度通常为 20～30 ppbv（1 ppbv=1 nL/L），但对于重污染空气，其浓度可达数百 ppbv。Hall 首次测定了臭氧对零价汞的气相氧化，其速率常数为 $(3\pm2)\times10^{-20}$ $cm^3/(mol \cdot s)$（Hall，1995）。反应速率常数随表面/体积比（surface-to-volume ratio）的增加而增加，提示反应过程中可能存在一定的非均相界面反应（Hall，1995）。日光照射可将反应速率提高约 6 倍（Hall，1995）。随后，研究者分别给出了 8.43×10^{-17} $cm^3/(mol \cdot s)$（Pal and Ariya，2004b）、$(6.2\pm1.1)\times10^{-19}$ $cm^3/(mol \cdot s)$（Snider et al., 2008)、(7.4±0.5)×10^{-19} cm^3/(mol·s)(Rutter et al., 2012)的速率常数。机理分析则提示这一过程可能涉及形成相对稳定的 HgO$_3$ 中间体（Calvert and Lindberg, 2005; Castro et al., 2009）。模型计算也提示臭氧对零价汞的氧化产物可能为 HgO 固体，而非活性气态汞（reactive gaseous mercury, RGM）（Hedgecock et al., 2005），但这一结果尚存在一定争议。HgO 固体产物的生成得到了高分辨透射电子显微镜与能谱的证实（Snider et al., 2008）。魁北克南部现场观测发现，当臭氧浓度高于 30 ppbv 时，臭氧浓度与气态单质汞浓度呈反比，提示臭氧对大气中零价汞的氧化有所贡献（Poissant, 1997）。需要特别指出的是，虽然许多研究表明臭氧浓度下降与 RGM 形成呈现显著相关性（Ebinghaus et al., 2002; Laurier et al., 2003; Huang et al., 2010; Han et al., 2014），但这种相关性除了可能来自于臭氧直接氧化（Li et al., 2011），也可能与臭氧生成的其他氧化剂（如活性卤物种）对零价汞的氧化有关（Skov et al., 2004）。模型预测显示，臭氧可在零价汞氧化中扮演重要作用：直接氧化零价汞或通过产生其他氧化剂（如羟基自由基）间接氧化零价汞（Pan and Carmichael, 2005; Holmes et al., 2009）。

2）过氧化氢

Tokos 等发现过氧化氢对零价汞的气相氧化速率低于方法检出限，即其速率＜8.5×10^{-19} cm^3/(mol·s)（Tokos et al., 1998）。根据这一研究，在实际大气中，过氧化氢对零价汞的氧化可能并不重要。但在后续研究中应进一步考虑光照（Lahoutifard et al., 2003; Liu et al., 2014）、非均相界面催化（Lahoutifard et al., 2003; Zhou et al., 2018）等对过氧化氢氧化汞的影响及过氧化氢在大气汞氧化中的贡献。

3）氯物种

Cl$_2$ 对零价汞的气相氧化为两步反应：①单质汞与氯原子反应生成 HgCl；②HgCl 进一步与 Cl$_2$ 反应生成 HgCl$_2$（Naruse et al., 2010）。反应产物 HgCl$_2$ 主要吸附在器壁上（Ariya et al., 2002）。目前有关 Cl$_2$ 对零价汞的气相氧化速率常数的报道差异较大（Menke and Wallis, 1980; Ariya et al., 2002; Wang and Anthony, 2005），还存在较大的不确定性。这为现场监测估算 Cl$_2$ 在零价汞氧化中的贡献带来了困难（Stephens et al., 2012）。共存水、SO$_2$、NO 均可抑制 Cl$_2$ 对零价汞的氧化（Agarwal et al., 2006），但＜400 ppmv（1 ppmv=1 mL/L）的 SO$_2$ 对 HCl 氧化零价汞的反应没有影响（Van Otten et al., 2011）。

4）溴物种

分子溴、原子溴以及 BrO· 自由基对零价汞的氧化速率常数见表 2-1。在臭氧存在下 Br$_2$ 与 CH$_2$Br$_2$ 的光解过程中可形成 BrO· 自由基，其对零价汞的氧化产物为 HgBr、HgBrO、HgOBr 以及 HgO（Raofie and Ariya, 2004）。多项研究表明，海洋或沿海地区大气中活性溴物种可能是零价汞的主要氧化剂（Peleg et al.，2007；Holmes et al.，2009；Stephens et al.，2012；Coburn et al.，2016）。

模型计算发现来自于海洋有机溴分解的原子溴是大气零价汞主要的氧化剂，而其反应产物 HgBr 可进一步与其他氧化剂如 NO_2 与 $HO_2^·$ 自由基反应（Horowitz et al.，2017）。根据氧化速率计算，对流层大气原子溴对零价汞氧化导致零价汞的寿命为 0.5～1.7 年，表明原子溴对零价汞的氧化可能是最主要的零价汞大气消除途径（Holmes et al.，2006）。原子溴对零价汞的氧化主要发生在对流层中上层，在这一区域原子溴浓度较高，且较低的温度也可抑制 HgBr 中间体的分解，从而有助于氧化的发生（Holmes et al.，2010）。另一模型分析也表明，原子溴氧化模型可较好地解释极地地区春季汞"亏损"以及夏季总气态汞的再次升高（Holmes et al.，2010）。

现场监测提示，除溴原子外，$BrO^·$ 也可能在零价汞氧化中起着关键作用（Ebinghaus et al.，2002；Moore et al.，2013）。无论南极地区（Ebinghaus et al.，2002）还是中纬度死海地区（Obrist et al.，2011），大气活性气态汞的生成往往与较高 $BrO^·$ 浓度相一致，甚至在温度高至 45 ℃时仍可观测到这一现象。化学箱式模型分析表明，在这些区域活性溴是单质汞的主要氧化剂，溴引发的氧化是海洋地区大气汞氧化的重要途径之一（Obrist et al.，2011）。Tas 等认为，由于死海地区大气较小的 $[Br^·]/[BrO^·]$ 浓度比（0.2～0.5）及其相近的氧化速率，$BrO^·$ 在对流层零价汞的氧化中的作用（可高达 80%～90%）较原子溴更为显著（Tas et al.，2012）。

此外，含碘化合物（I_2/CH_2I_2/CH_2IBr/CH_2ICl/IBr/ICl）也可能通过与臭氧反应，生成 $IO^·$，进而促进原子溴生成，并间接促进溴对单质汞的氧化（Calvert and Lindberg，2004）。

5）碘物种

Raofie 等测定 I_2 对零价汞的气相氧化速率为≤(1.27±0.58)×10^{-19} cm^3/(mol·s)，其主要反应产物为 HgI_2、HgO、HgIO 以及 HgOI（Raofie et al.，2008）。而 Chi 等测定的氧化速率则为 (7.4±0.2)×10^{-17} cm^3/(mol·s)（Chi et al.，2009）。飞灰和粉末活性炭可加速 I_2 对零价汞的氧化（Chi et al.，2009）。理论分析显示，由于 I_2 与单质汞作用较弱，I_2 对零价汞的氧化要弱于其他卤素（Auzmendi-Murua et al.，2014）。现场监测表明，中国东部沿海地区大气悬浮颗粒物的 HCl 溶解颗粒态汞浓度与碘浓度显著正相关（r=0.77，p<0.01）（Duan et al.，2017），提示碘可能在零价汞氧化过程中起着重要作用（Goodsite et al.，2012；Wang et al.，2014），但这一假设需要在实际大气环境中得到验证。

6）混合卤素物种

ICl 对零价汞的氧化速率为(10.5±0.3)×10^{-17} cm^3/(mol·s)，快于 Cl_2（Qu et al.，

2010）。煤飞灰可加速 ICl 对零价汞的氧化（Qu et al.，2010）。这一过程的氧化产物为 $HgCl_2$，而非 HgI_2，提示碘可能主要通过促进反应中间体生成加速了氧化（Qu et al.，2010）。类似地，BrCl 也可导致零价汞的氧化，其氧化速率为 $(2.3±0.2)×10^{-17}$ $cm^3/(mol·s)$（Qu et al.，2009）。BrCl 可能参与了实际大气环境中零价汞的氧化（Lindberg et al.，2002）。

7）羟基自由基

羟基自由基具有极强的氧化活性，可导致零价汞的快速气相氧化。Sommar 等在室温常压下采用相对速率技术研究了羟基自由基对零价汞的氧化：羟基自由基由亚硝酸甲酯光解生成，环己烷为参照化合物，羟基自由基对零价汞氧化的速率常数为 $(8.7±2.8)×10^{-14}$ $cm^3/(mol·s)$（Sommar et al.，2001）。Pal 等之后报道的速率常数 $[(9.0±1.3)×10^{-14}$ $cm^3/(mol·s)]$ 与之相一致（Pal and Ariya，2004a）。这一反应的主要氧化产物为 HgO（分别以气态、悬浮气溶胶或沉积在器壁形式存在）（Pal and Ariya，2004a）。但 Calvert 等认为由于反应体系中 O_3 副产物的存在，以上测定可能高估了羟基自由基与零价汞的反应（Calvert and Lindberg，2005）。羟基自由基对零价汞的氧化可部分地解释亚洲海洋边界层中活性汞和颗粒汞的生成（Chand et al.，2008）。模型计算也显示羟基自由基是大气零价汞氧化的重要途径之一（Bergan and Rodhe，2001；Pal and Ariya，2004a）。

8）$NO_3^·$

硝酸自由基（$NO_3^·$）主要来自于平流层与对流层臭氧对二氧化氮的氧化［式（2-18）］以及五氧化二氮的分解［式（2-19）］（Wayne et al.，1991）。

$$NO_2 + O_3 \longrightarrow NO_3^· + O_2 \qquad (2\text{-}18)$$

$$N_2O_5 \longrightarrow NO_3^· + NO_2 \qquad (2\text{-}19)$$

光照可使硝酸自由基快速分解，因此硝酸自由基主要存在于夜间（Wayne et al.，1991）。经测定，气相中硝酸自由基与零价汞的反应速率系数 $<4×10^{-15}$ $cm^3/(mol·s)$，其产物为氧化汞（Sommar et al.，1997）。但另一项理论计算认为，硝酸自由基与零价汞不能形成强键，因此不能引发零价汞的氧化（Dibble et al.，2012）。2012 年夏季对耶路撒冷城市区域大气连续六周硝酸自由基、汞形态（单质汞、活性气态汞、氧化汞）的监测显示，夜晚 RGM 浓度相对较高（最高达 97 pg/m^3），且夜间 RGM 浓度与硝酸自由基浓度呈正相关，而与臭氧、CO、SO_2 等浓度无关（Peleg et al.，2015）。这一发现提示，硝酸自由基虽然不能直接氧化零价汞，但可能通过与不稳定的 Hg(Ⅰ)的反应间接参与了零价汞氧化。

以上研究表明，多种大气氧化剂均可氧化零价汞。目前，对于大气中零价汞气相均相氧化的主要途径还存在较大争议。不同地区（内陆与沿海）以及不同高度大气氧化剂的时空分布较大差异，导致了这一问题的复杂性。在后续研究中，应进一步提高对大气各氧化性物种浓度以及氧化剂-零价汞氧化动力学测定的准确性，这有助于对零价汞气相均相氧化主要途径的识别。此外，零价汞氧化过程中的同位素分馏也为研究汞的氧化过程提供了新的手段。最近，Sun 等发现，在 Cl· 对零价汞的氧化过程中，汞重同位素主要富集于反应物（Hg^0）；而对于零价汞的 Br· 氧化，汞重同位素则主要富集于氧化产物，$^{202}Hg/^{198}Hg$ 的分馏因子分别为 0.99941 ± 0.00006（Cl·）与 1.00074 ± 0.00014（Br·）（Sun et al.，2016）。Br· 氧化零价汞的 $\Delta^{199}Hg/\Delta^{201}Hg=1.64\pm0.3$，表明汞的非质量同位素分馏主要来自于核体积效应；而 Cl· 氧化零价汞的 $\Delta^{199}Hg/\Delta^{201}Hg=1.89\pm0.18$，表明除了核体积效应，还存在其他非质量同位素分馏效应（Sun et al.，2016）。此外这一氧化过程还存在显著偶数汞同位素的非质量分馏（反应物中 $\Delta^{200}Hg$ 可达 –0.17‰）（Sun et al.，2016）。以上同位素分馏特征的发现为后续识别大气中零价汞氧化的关键过程提供了可能。

2. 气相非均相汞氧化

在环境气压、温度以及光照（290～700 nm）下，以铁氧化物（磁赤铁矿、赤铁矿、磁铁矿、针铁矿）纳米颗粒为大气气溶胶矿物组分替代物模拟了其对气态零价汞的吸收（Kurien et al.，2017）。在实验条件下，光照对磁赤铁矿、赤铁矿、针铁矿的零价汞摄入动力学速率常数提高了 40～900 倍（Kurien et al.，2017）。而磁铁矿对零价汞的摄入动力学不受光照影响（Kurien et al.，2017）。这一摄入过程可能涉及零价汞的氧化，但其具体过程与机制尚有待进一步深入研究。二次有机气溶胶可导致零价汞的气相氧化，这种氧化可能与气溶胶生成的臭氧以及羟基自由基有关（Rutter et al.，2012）。现场检测也表明，大气颗粒物浓度的增加也导致颗粒态汞浓度相应增加，这可能与颗粒物对气态汞的吸附与氧化有关（Kim et al.，2012）。南极冰原汞的总沉降通量与钠离子显著相关（Han et al.，2017a），这可能与海盐颗粒组分如氯化钠、溴化钠对零价汞的氧化有关（Sheu and Mason，2004）。据推测，大气降水中偶数汞同位素的非质量分馏（$\Delta^{200}Hg$）可能与对流顶层气溶胶或固体表面发生的零价汞光氧化有关（Chen et al.，2012）。目前，关于大气中零价汞氧化的非均相过程还知之甚少，这也为大气汞模型带来了极大的不确定性（Subir et al.，2012；Ariya et al.，2015）。

2.2.2　大气水相汞氧化

由于零价汞在大气液滴中的溶解度较低，其极易扩散至气相，因此一般认为大气水相中零价汞的氧化过程并不重要。但据推测，成雨初期大气液滴中可能存在零价汞的氧化（Mason et al.，1997）。目前认为，液滴中汞氧化剂可能包括臭氧、$HOCl/OCl^-$、$HOBr$、H_2O_2 以及 $HO^\bullet$（表 2-2）。

表 2-2 汞的大气水相氧化反应

反应	速率常数 [L/(mol·s)]	参考文献
$Hg(0) + O_3 + 2H^+ \longrightarrow Hg^{2+} + H_2O + O_2$ （2-20）	$(4.7\pm2.2)\times10^7$	（Munthe，1992）
$Hg(0) + HO^\bullet \longrightarrow Hg^+ + OH^-$ （2-21）	2.0×10^9	（Lin and Pehkonen，1997）
	$(2.4\pm0.3)\times10^9$	（Gardfeldt et al.，2001）
$Hg(0) + HOCl \longrightarrow Hg^{2+} + Cl^- + OH^-$ （2-22）	$(2.09\pm0.06)\times10^6$	（Lin and Pehkonen，1998）
$Hg(0) + OCl^- \xrightarrow{H^+} Hg^{2+} + Cl^- + OH^-$ （2-23）	$(1.99\pm0.05)\times10^6$	（Lin and Pehkonen，1998）
$Hg(0) + HOBr \longrightarrow Hg^{2+} + Br^- + OH^-$ （2-24）	0.28 ± 0.02	（Wang and Pehkonen，2004）
$Hg_2^{2+} + O_3 \longrightarrow Hg^{2+} + HgO + O_2$ （2-25）	$(9.2\pm0.9)\times10^6$	（McElroy and Munthe，1991）
$Hg_2^{2+} + H_2O_2 \longrightarrow Hg^{2+} + HgO + H_2O$ （2-26）	<6	（Munthe and McElroy，1992）

在 70～200 ppbv 臭氧的存在下，水相对汞的吸收可增加三个数量级，表明臭氧可显著提高零价汞的水相氧化（Iverfeldt and Lindqvist，1986）。经测定，臭氧对零价汞的氧化速率较快，为 $(4.7\pm2.2)\times10^7$ L/(mol·s)（Munthe，1992）。臭氧也可将 Hg_2^{2+} 氧化为二价汞，其氧化速率为 $(9.2\pm0.9)\times10^6$ L/(mol·s)（McElroy and Munthe，1991）。在 pH 1.0～3.0、温度 283～293 K 范围内，臭氧对零价汞的水相氧化与 pH、温度无关（McElroy and Munthe，1991）。

$HOCl/OCl^-$ 对零价汞的氧化速率常数分别为 $(2.09\pm0.06)\times10^6$ L/(mol·s)、$(1.99\pm0.05)\times10^6$ L/(mol·s)（Lin and Pehkonen，1998）。Hg_2^{2+} 可被 HOCl 快速氧化（Munthe and McElroy，1992）。此外，Cl^- 可促进零价汞的水相氧化，这与 Cl^- 与 Hg^{2+} 形成络合物有关（Yamamoto，1996）。光照也可在一定程度上促进氯离子对零价汞的氧化，这可能与光照生成的活性氯物种有关（Sun et al.，2014）。

溶解 Fe(Ⅲ) 可氧化零价汞，而过氧化氢可进一步促进这种氧化（Ogata et al.，1982）。类似地，芬顿试剂 [Fe(Ⅱ)-H_2O_2] 也可氧化 Hg_2^{2+}（Munthe and McElroy，1992），这可能与 Fe(Ⅲ/Ⅱ)-H_2O_2 体系中生成的羟基自由基有关。随后，研究证实硝酸盐光解生成羟基自由基的确可氧化零价汞（Lin and Pehkonen，1997）。羟基自由基氧化零价汞的速率常数可高达 2.0×10^9 L/(mol·s)（Lin and Pehkonen，1997），这与后续测定 [$(2.4\pm0.3)\times10^9$ L/(mol·s)] 基本一致（Gardfeldt et al.，2001）。相对地，单独的过氧化氢虽然可氧化 Hg(0)（Marek，1997）或 Hg_2^{2+}（Munthe and

McElroy，1992），但这一过程较为缓慢，可能在大气零价汞水相氧化中并不重要。

目前，这些途径在大气零价汞水相氧化中的重要性还不清楚。据模型估计，在日间，臭氧以及HO·途径较为重要（除非大气液滴pH＞5.0）（其中羟基自由基可占整个大气零价汞水相氧化的25%）；而在夜间含氯物种是主要的氧化剂（可占整个大气零价汞水相氧化的90%）（Lin and Pehkonen，1998；1999a）。另一模型也认为大气水相零价汞的主要氧化剂为臭氧与含氯物种（Hedgecock and Pirrone，2001）。

2.2.3 汞氧化与大气汞"亏损"事件

早在20世纪80年代，人们就发现北极地区对流层臭氧破坏，这一过程伴随着对流层大量活性卤素物种的生成（Farman et al.，1985）。由于活性卤物种对零价汞的氧化，这一过程也导致极地大气汞"亏损"现象。1995年4～6月，Schroeder等连续监测了北极加拿大阿勒特（Alert）地区大气中汞浓度的变化，首次发现了北极春季大气汞"亏损"事件，臭氧浓度与大气总气态汞浓度存在显著相关性（r^2=0.8），即在对流层臭氧破坏的同时，大气总气态汞浓度也显著减少（Schroeder et al.，1998）。2000年1月至2001年1月，Ebinghaus等在南极德国诺伊迈尔（Neumayer）研究站连续一年的测试发现，南极春季也存在类似的大气汞"亏损"事件（Ebinghaus et al.，2002）。卫星监测也显示，大气汞"亏损"事件期间大气中BrO·自由基浓度显著增加（Ebinghaus et al.，2002）。之后，在其他纬度地区如中纬度死海地区（Peleg et al.，2007；Obrist et al.，2011）也观测到类似的大气汞"亏损"事件。

大气中零价汞的氧化主要来自于活性卤物种，其中BrO·、Br·的作用最为重要（Lindberg et al.，2002；Ariya et al.，2004；Obrist et al.，2011）。除了BrO·、Br·外，臭氧与卤素反应生成的其他活性卤物种如Cl·、I·、ClO·以及过氧化氢可能也在汞的氧化中起着一定作用（Lindberg et al.，2002；Calvert and Lindberg，2004；Lahoutifard et al.，2006；Hedgecock et al.，2008）。

极地大气汞"亏损"事件导致大气气态零价汞浓度降低，大气RGM与颗粒态汞浓度升高，促进汞的沉降与地表输入。因此，极地大气汞"亏损"事件可使地表雪汞浓度快速增加。如阿拉斯加巴罗（Barrow）地区表层雪（表层1 cm）的测定表明，在9天的大气汞"亏损"事件期间，雪汞浓度由背景值（4.1～15.5 ng/L）增加至147～237 ng/L（Johnson et al.，2008）。

不同学者对极地大气汞"亏损"事件导致北极圈内汞沉降量的估计存在一定差异，分别为100～300 t/a（Lindberg et al.，2002）、89～208 t/a（Skov et al.，2004）以及168～181 t/a（Dastoor et al.，2008）。极地大气汞"亏损"事件中南极地区的汞沉降量低于北极地区，两项估计分别为 50～100 t/a 与＞100 t/a（Ebinghaus et al.，2004）。虽然沉降汞中的一部分可通过还原再次释放到大气中，但仍有一部分汞在极地残留（Ariya et al.，2004）。因此，极地地区是汞重要的汇（Ariya et al.，2004；Bargagli et al.，2005）。南极临近特拉诺瓦（Terra Nova）湾冰沼湖的土壤、地衣、苔藓中汞的浓度高于南极背景值，这被认为与这一地区冰晶促进春季汞"亏损"事件有关（Bargagli et al.，2005），表明极地汞"亏损"事件增加了当地生物对汞的累积。此外，考虑到极地积雪、海冰中甲基化菌的存在（Gionfriddo et al.，2016），极地大气汞"亏损"事件对汞在这一地区的甲基化、生物累积的影响需要引起进一步关注。

2.2.4 表层水汞化学氧化

水相中零价汞氧化在汞的生物地球化学循环中起着重要的作用。氧化降低了水中零价汞的浓度，抑制了汞的挥发，增加了可作为甲基化底物的二价汞的浓度（Batrakova et al.，2014）。除了挥发，光氧化是水中零价汞的另一主要的汇（Lalonde et al.，2001）。在模型中考虑水中零价汞的氧化过程，有助于进一步提高大气-地表水汞交换模型的准确度（Soerensen et al.，2010）。

目前，在多种环境水体如湖水（Garcia et al.，2005）、河水（Whalin and Mason，2006）、湿地水（Krabbenhoft et al.，1998）、海水（Whalin and Mason，2006；Whalin et al.，2007；Ci et al.，2016b）中均观测到了零价汞的氧化。一般认为，零价汞的氧化产物为二价汞，但也有分析认为汞的光氧化与光还原并非简单可逆：溶解性气态汞[Hg(0)]可转化成一些未知的汞氧化形态（Qureshi et al.，2010）。

在黑暗条件下，水中的零价汞可被氧化。在暗氧化过程中，液态汞（汞滴）与溶解气态汞的氧化行为存在差异：在共存氯离子与有氧条件下，液态汞可快速被氧化，其氧化速率与氯离子浓度及液滴表面积呈正比；相比而言，溶解气态汞在水中的氧化较为缓慢（Amyot et al.，2005）。但在黑暗或较弱的光照下，零价汞氧化强于溶解气态汞的生成（Garcia et al.，2005）。光照可以显著促进溶解气态汞的氧化。随着太阳升起，零价汞氧化速率逐渐增加，并在中午达到最大值，之后又逐步下降（Garcia et al.，2005）。日光中的紫外线B（ultraviolet B，UVB）组分在零价汞的光氧化中起着关键作用，而可见光所起作用较弱（Lalonde et al.，2004；Ci et al.，2016b）。光氧化过程不受过滤与热灭菌的影响，表明这种光氧化主要是一种化学作用（Lalonde et al.，2004；Qureshi et al.，2010）。

目前，提出的水中汞的氧化剂主要有羟基自由基、臭氧、活性卤物种、过氧酸（如过氧乙酸和间氯过苯甲酸）（Wigfield and Perkins，1985）、碳酸根自由基、溶解有机质（dissolved organic matter，DOM）等。Lalonde 等发现羟基自由基清除剂甲醇可将半咸水与人工海水中零价汞的氧化降低 25%与 19%，提示羟基自由基在零价汞光氧化中起着一定的作用（Lalonde et al.，2004）。已有报道证实羟基自由基可快速氧化水相中零价汞（Lin and Pehkonen，1999a；Gardfeldt et al.，2001；Zhang and Lindberg，2001）。羟基自由基可来自于硝酸根（Gardfeldt et al.，2001）、亚硝酸根（Mason et al.，2001）、DOM（Vaughan and Blough，1998）的光化学过程。Hines 等则认为明尼苏达北部湖泊水中具有较高的臭氧稳态浓度，其可能是零价汞的主要氧化剂（Hines and Brezonik，2004）。一般海水中零价汞的氧化要显著高于淡水（Yamamoto，1996；Amyot et al.，1997a），这可能与海水中较高的氯离子含量有关。一方面，氯离子可与二价汞形成较为稳定的络合物（Sun et al.，2014），促进氧化反应的发生；另一方面，光照下活性氯（$HOCl/OCl^-$）的生成也可能加速了水相中零价汞氧化（Lin and Pehkonen，1999a；Mason et al.，2001）。

通常认为，氧化性较强的自由基（如羟基自由基）在水相零价汞的光化学氧化中起着主要作用。但最近一项研究提示，氧化活性较弱的碳酸根自由基（$CO_3^{\cdot-}$）也可氧化零价汞，并在水相零价汞光氧化中起着重要作用（He et al.，2014）。电子顺磁共振研究发现，在光照下，碳酸根可与硝酸根生成的羟基自由基进一步反应，生成碳酸根自由基［式（2-27）］。

$$NO_3^- \xrightarrow{h\nu} HO^{\cdot} \xrightarrow{CO_3^{2-}} CO_3^{\cdot-} \qquad (2\text{-}27)$$

在去离子水介质及模拟光照条件下，碳酸根与硝酸根共存体系中零价汞的光氧化速率（$k = 1.44\ h^{-1}$）显著高于碳酸根、硝酸根、DOM 单独体系（$k = 0.1 \sim 0.17\ h^{-1}$）。DOM 可以一定程度清除羟基自由基与碳酸根自由基，从而部分抑制碳酸根与硝酸根共存体系对零价汞的光氧化。作为羟基自由基清除剂的异丙醇可显著抑制零价汞的光氧化，而单线态氧增强剂重水（D_2O）对光氧化影响不大，证明羟基自由基在光氧化中起着重要的直接（直接氧化）或间接作用（通过生成碳酸根自由基氧化零价汞），而单线态氧作用不大。这一结果对于理解汞的自由基反应具有重要的意义：以往研究往往较为重视反应活性较强的自由基（如羟基自由基），认为其在汞的氧化反应中较为重要，而一些反应活性较弱的自由基往往不被重视。但反应活性强的自由基半衰期短，稳态浓度低；一些反应活性较弱的自由基半衰期长，稳态浓度高，反而可能在汞的自由基反应中扮演更为重要的角色。

以往关于 DOM 介导汞的氧化还原反应研究多聚焦于汞的还原，对于 DOM 在零价汞氧化方面的作用知之甚少。1995 年，Yamamoto 等（1995）发现小分子巯基配体谷胱甘肽、L-半胱氨酸、D-青霉胺均可促进去离子水中零价汞的暗氧化，氧化效率为谷胱甘肽＞L-半胱氨酸＞D-青霉胺。有趣的是，L-抗坏血酸的加入进一步促进了 L-半胱氨酸对零价汞的氧化。其他一些不含巯基的小分子有机酸络合剂也可导致零价汞的氧化（赵士波等，2014）。随后，在环境水样、底泥（Melamed et al.，1997）、土壤（do Valle et al.，2006；Soares et al.，2015）中均发现 DOM 这种络合引发的氧化在汞氧化中起着重要的作用。例如，在暗反应条件下，盐水中藻类生成的有机质可促进零价汞的氧化（Poulain et al.，2007）；亚马孙土壤中零价汞的转化率与土壤有机质含量正相关（do Valle et al.，2006）。Gu 等（2011）深入研究了缺氧与暗反应条件下 DOM 对汞的络合氧化，发现 DOM 在汞的氧化还原中起着复杂的双重作用：还原态有机质可通过还原醌有效还原二价汞为零价汞，但进一步增加有机质浓度（>0.2 mg/L），零价汞生成量反而降低，提示有机质中还原性巯基对零价汞的络合氧化。相较于氧化态有机质，还原态有机质对零价汞的氧化更为有效。氧化态与还原态有机质对零价汞可能的氧化途径分别如式（2-28）至式（2-31）所示：

$$R—S—S—R' + Hg(0) \longrightarrow Hg(II) + R—S^- + R'—S^- \quad (2\text{-}28)$$

$$Hg(II) + R—S^- + R'—S^- \longrightarrow R—S—Hg(II)—S—R' \quad (2\text{-}29)$$

$$2R—SH + Hg(0) \longrightarrow R—SH\cdots Hg(0)\cdots HS—R \quad (2\text{-}30)$$

$$R—SH\cdots Hg(0)\cdots HS—R \longrightarrow R—S—Hg(II)—S—R + 2H^+ + 2e^- \quad (2\text{-}31)$$

在 pH 7.0 条件下，有机质与汞的强结合位点丰度（3.5 μmol Hg/g）高于低 pH，因此高 pH 有利于零价汞的氧化。进一步研究表明，不同来源还原态有机质均可有效促进汞的还原与氧化（Zheng et al.，2012）。当有机质/汞浓度比增加到一定程度时（该比值依赖于有机质来源），有机质对汞的氧化进一步增强，汞的还原得到抑制。进一步采用不同结构与硫氧化态的 DOM 小分子模型研究了这一反应，发现：非巯基配体或氧化态硫配体（如二硫）对零价汞的氧化能力较弱，而巯基配体可显著引发零价汞的氧化；零价汞的氧化速率及程度与巯基配体结构、巯基/汞浓度比以及是否存在电子受体有关（Zheng et al.，2013）。低分子量脂肪族巯基对零价汞的氧化强于芳香巯基，高巯基/汞浓度比有助于零价汞的氧化。电子受体（如腐殖酸、蒽醌-2,6-二磺酸钠）的加入提高了零价汞的氧化。低分子量有机酸也在零价汞的氧化中起着一定的作用。酒石酸、柠檬酸及琥珀酸 3 种有机酸在反应初期对水中零价汞均表现出一定的氧化作用，但随后被氧化的零价汞又再次被还原（赵士波等，2014）。3 种有机酸对零价汞的氧化能力顺序为：柠檬酸＞酒石酸＞琥珀酸。

由于汞的还原与氧化同时发生，双同位素示踪有助于同时揭示二价汞的还原与零价汞的氧化速率。双同位素示踪发现，在日光照射下，河水（马里兰 Patuxent 河）与海水（新泽西 Brigantine 岛）中同时存在汞的还原与氧化，且光还原与氧化速率相当（Whalin and Mason，2006）。另一同位素示踪研究也发现，切萨皮克湾（Chesapeake Bay）海水中汞的氧化速率与还原速率相当或高于还原速率（Whalin et al.，2007）。同位素示踪实验结果表明，中国近海海域零价汞的暗氧化远高于暗还原，光氧化略高于光还原（Ci et al.，2016b）。

2.2.5 土壤中汞的氧化

关于土壤中零价汞的氧化研究相对较少。在含汞温度计生产厂周边土壤中零价汞的热解析分析显示，主要的汞形态分别为硫化汞（56%±8%）与有机结合态汞（22%±9%），而零价汞仅占 17%±5%（Boszke et al.，2008）。由于该工厂主要以零价汞形式向土壤排汞，因此推测零价汞在土壤中经历了氧化过程。类似地，巴西废弃金矿区以及美国橡树岭零价汞污染地点的土壤中较高的二价汞及较低的零价汞提示土壤中存在零价汞的氧化（Durao et al.，2009；Miller et al.，2013）。土壤空气中气态单质汞浓度随深度增加而逐渐下降也提示在土壤中存在零价汞的氧化（Obrist et al.，2014）。

热解析分析显示，巴西亚马孙干燥土壤中加入的零价汞（30 mg/kg）可随培育时间缓慢转化为二价汞（28%~68%），零价汞的半衰期约为 148 d。零价汞的氧化动力学可分为两个一级反应阶段：首先是快反应，其次是慢反应，分别对应于零价汞氧化为一价汞以及一价汞继续氧化为二价汞［式（2-32）］：

$$Hg(0) \xrightarrow{4\sim11d} Hg(I) \xrightarrow{133\sim178d} Hg(II) \qquad (2\text{-}32)$$

零价汞的氧化转化率与土壤有机质含量正相关（do Valle et al.，2006），有机质可能通过络合/稳定二价汞的形式促进零价汞的氧化（Soares et al.，2015）。此外，土壤中零价汞的氧化也可能与锰氧化物有关（Windmoller et al.，2015）。在零价汞液滴与土壤或氧化锰作用后，汞液滴表面生成氧化汞，这一氧化过程极大促进了汞的溶解（Miller et al.，2015）。土壤及氧化锰作用分别将汞的溶解提高了 20 倍及 700 倍以上，表明氧化锰对零价汞的氧化更为有效（Miller et al.，2015）。

2.2.6 冰雪中汞的氧化

在极地汞"亏损"事件期间，积雪上方气态单质汞浓度基本不变而积雪表面汞浓度存在日夜变化，表明日间在雪-气界面气态单质汞发生了氧化（Fain et al.，2006；Fain et al.，2008；Ferrari et al.，2008；Dommergue et al.，2013）。这一氧化过程主要在午后光照强度下降时期发生（Poulain et al.，2004b；Ferrari et al.，2008；Dommergue et al.，2012），表明可能涉及光照生成的自由基及有机化合物引发的后续反应（Poulain et al.，2004b）。此外，雪表层活性气态汞占总汞比例的降低也提示活性气态汞逐渐转化为了更为稳定的汞形态（Ferrari et al.，2008）。积雪内的间隙气态汞也可被快速清除，表明存在零价汞的暗氧化过程（Dommergue et al.，2003b；Ferrari et al.，2004a；Fain et al.，2008；Fain et al.，2013）。

积雪表层与深处溴浓度和汞浓度存在明显的正相关，表明溴（$Br^{\cdot}/BrO^{\cdot}$）可能是汞的氧化剂（Ferrari et al.，2004b；Spolaor et al.，2018）。据报道，日光照射可显著催化雪表面生成分子溴，而低pH、臭氧亦可促进分子溴的生成（Pratt et al.，2013）。分子溴可进一步生成活性溴物种（如$Br^{\cdot}/BrO^{\cdot}$）。而夏季$BrO^{\cdot}$浓度降低与地表空气温度升高导致的冰雪融化具有显著的相关性：$BrO^{\cdot}$仅在低于冰点的温度才能检测到（Burd et al.，2017）。这也表明积雪催化了$BrO^{\cdot}$的生成。此外，过氧化氢、氯离子与铁也可促进雪表面零价汞的光氧化（Lahoutifard et al.，2003，2006；Mann et al.，2015b）。

北极沿海地区汞"亏损"事件与海冰动态直接相关，提示冰层有助于零价汞氧化（Moore et al.，2014）。在南极海冰区，冬季亦存在汞的氧化，这可能与海冰表面催化有机溴（$CHBr_3$）与Br_2的生成有关（Mastromonaco et al.，2016b）。

2.2.7 冰冻引发汞的氧化

在寒冷气候条件下，冻结的冰在日间融化，夜间再次冻结，或在夏季融化而冬季再次冻结。这种冻融作用在地表水、土壤以及大气中广泛存在。由于冰冻过程中存在的浓缩效应、电位变化、冰晶表面催化、对流以及温度的不均一性（O'Concubhair and Sodeau，2013），相比于常温，冰冻过程反而可能提高某些反应的速率（Pincock and Kiovsky，1966；Takenaka et al.，1992；Takenaka et al.，1996）。例如，在pH 4.5条件下，冰冻可将亚硝酸盐的氧化速率提高5个数量级（Takenaka et al.，1992；Takenaka et al.，1996）。这种冰冻过程亦可引发水中汞的氧化。冰冻低浓度硫酸、过氧化氢、硫酸/过氧化氢、亚硝酸溶液均可一定程度促进其中溶解气态汞的氧化（O'Concubhair et al.，2012）。而相应的氧化反应在常温下则极为缓慢。冰冻可能导致氧化剂（如溶解氧、过氧化氢、HONO 或 H_2ONO^+）在冰晶包裹水相中的浓缩，从而提高了氧化反应速率［式(2-33)～式(2-34)］。

$$O_{2(aq)} + 4H^+_{(aq)} + 2Hg_{(aq)} \rightleftharpoons 2H_2O + 2Hg^{2+} \qquad (2\text{-}33)$$

$$Hg_{(aq)} + H_2O_{2(aq)} + 2H^+_{(aq)} \rightleftharpoons Hg^{2+}_{(aq)} + 2H_2O_{(aq)} \qquad (2\text{-}34)$$

冰冻过程也可导致卤素离子氧化，生成活性卤物种（如 Cl_2Br^-、Br_2Cl^-、I_2Br^-、IBr_2^-、$BrCl$、Br_2 等）（O'Driscoll et al.，2006a）。这些活性卤物种可能引发零价汞的氧化。目前这方面的研究尚待进一步开展。

2.2.8 汞氧化过程的生物作用

体内或体外实验证实，动物［如人（Halbach and Clarkson，1978；Magos et al.，1978；Ogata，1979；Halbach et al.，1988）、小鼠（Sugata and Clarkson，1979；Dunn et al.，1981；Sichak et al.，1986；Halbach et al.，1988）］、植物（Ogata et al.，1981）、真菌（Ogata et al.，1981）与细菌（Holm and Cox，1975）可将零价汞氧化为二价汞。早在1975年，Holm等（Holm and Cox，1975）发现多种细菌均可氧化零价汞，其中绿脓杆菌（*Pseudomonas aeruginosa*）、大肠杆菌（*Escherichia coli*）、荧光假单胞菌（*Pseudomonas fluorescens*）、柠檬酸杆菌属（*Citrobacter*）等对零价汞氧化能力较弱，而枯草芽孢杆菌（*Bacillus subtilis*）、巨大芽孢杆菌（*Bacillus megaterium*）对零价汞的氧化较为完全。枯草芽孢杆菌与巨大芽孢杆菌均为好氧菌。在黑暗条件下，甲基化厌氧细菌脱硫弧菌（*Desulfovibrio desulfuricans* ND132）亦可将溶解气态汞快速转化为"非吹扫汞"（Colombo et al.，2013）。细胞结合汞的乙基化衍生-原子光谱及X射线吸收近边结构（XANES）分析证实"非吹扫汞"以氧化态的二价汞为主。XANES进一步显示，二价汞共价结合于细胞巯基基团上。同时，*D. desulfuricans* ND132可与零价汞反应生成甲基汞。这一发现表明厌氧微生物可以零价汞为唯一汞源生成甲基汞。进一步考察了三种细菌：专性厌氧菌 *Geothrix fermentans* H5 和兼性厌氧菌 *Shewanella oneidensis* MR-1、*Cupriavidus metallidurans* AE104 对零价汞的氧化（Colombo et al.，2014）。在缺氧条件下，三株细菌均可将溶解气态汞转化为"非吹扫汞"。质量平衡实验显示，"非吹扫汞"的生成对应于挥发汞的降低。细菌对零价汞氧化的速率为 1.6×10^{-4}（*S. oneidensis* MR-1）fg/(cell·min)、2.5×10^{-4}（*C. metallidurans* AE104）fg/(cell·min)、23.1×10^{-4}（*G. fermentans* H5）fg/(cell·min)。另一项研究则进一步揭示了 *D. desulfuricans* ND132对零价汞的氧化与甲基化，其对零价汞的甲基化效率约为二价汞的三分之一（Hu et al.，2013a）。另一株脱硫弧菌（*Desulfovibrio alaskensis* G20）也可氧化零价汞，但不具有甲基化汞的能力（Hu et al.，2013a）。半胱氨酸可显著促进地杆菌 *Geobacter sulphurreducens* PCA对零价汞的氧化与甲基化，其对零价汞的甲基化效率约为二价汞的二分之一。更为重要的是，考虑到不同细菌对零价汞与二价汞甲基化能力的差异，在实际环境中零价汞氧化细菌与甲基化细菌的协同作用可能促进甲基汞的生成（Hu et al.，2013a）。目前，这一重要假设尚有待进一步验证。

目前，关于微生物对零价汞的氧化机制还不是特别明确。现有研究显示微生物对零价汞的氧化可能涉及多种途径与机制。细菌与热灭活细菌（*Geothrix fermentans* H5、*Shewanella oneidensis* MR-1、*Cupriavidus metallidurans* AE104）均可氧化零价汞为二价汞，提示细菌对汞的氧化为一被动过程（Colombo et al., 2014）。汞甲基化基因 *hgcAB* 的敲除可导致 *Geobacter sulfurreducens* PCA 和 *Desulfovibrio desulfuricans* ND132 细胞巯基含量的降低。缺氧条件下，两种细菌对二价汞的还原速率增加，对零价汞的氧化速率则降低（Colombo et al., 2014），这提示零价汞的氧化可能与细菌巯基有关。进一步分析发现，细菌 *Desulfovibrio desulfuricans* ND132 胞内巯基浓度约为细胞壁巯基浓度的 6 倍（Wang et al., 2016）。巯基掩蔽剂可显著抑制细菌对零价汞的氧化活性，而去除细胞壁的细菌原生质体仍可将零价汞快速氧化为二价汞（Wang et al., 2016）。这表明细胞壁组分对零价汞的氧化作用不大。细菌原生质体扩展 X 射线吸收精细结构（EXAFS）分析显示零价汞氧化产生多种不同形式的汞-巯基复合物，且氧化态汞的配位环境随培育时间发生变化：随着培育时间增加，多巯基配位二价汞所占比例增加（Wang et al., 2016）。这一结果意味着 *D. desulfuricans* ND132 对零价汞的氧化为胞内过程，且与胞内巯基分子有关。此外，枯草杆菌过氧化氢酶提取物可导致零价汞氧化，且过氧化氢对这一过程有显著促进（Ogata et al., 1981；Ogata and Aikoh, 1983）。过氧化氢对淡水汞氧化酶活性的促进作用也提示这一零价汞氧化过程可能涉及细菌的过氧化氢酶（Siciliano et al., 2002）。这表明，过氧化氢酶可能也在细菌零价汞氧化中起着重要作用。另一项研究显示，大肠杆菌 *katG* 基因突变或过表达显著影响零价汞的氧化：*katG* 基因过表达增强零价汞的氧化，*katG* 基因缺失则降低零价汞的氧化，表明过氧化氢酶 katE 自身难以实现对零价汞的氧化（Smith et al., 1998）。但是，*katE* 与 *katG* 基因同时缺失的大肠杆菌仍可部分氧化零价汞，提示存在过氧化氢酶之外的其他零价汞氧化途径。

哺乳动物体内也存在零价汞的氧化过程。早在二十世纪六七十年代，研究人员就猜测哺乳动物血红细胞中的过氧化氢酶参与了吸入零价汞的氧化，从而阻止其外排，增加了血红细胞对汞的净摄入（Kudsk, 1965；Magos et al., 1974）。血红细胞对零价汞的体外氧化研究表明，零价汞的氧化依赖于过氧化氢的生成速率与血红细胞过氧化氢酶活性（Halbach and Clarkson, 1978）。谷胱甘肽过氧化物酶不参与零价汞的氧化，但与过氧化氢酶竞争过氧化氢底物，从而抑制过氧化氢酶对零价汞的氧化（Halbach and Clarkson, 1978；Halbach et al., 1988）。因此，过氧化氢酶失活的血红细胞对零价汞的摄入显著低于正常血红细胞（Ogata, 1979）。类似地，暴露零价汞蒸气后，过氧化氢酶失活小鼠通过呼吸排出零价汞量高于正常小鼠（Sugata and Clarkson, 1979）。在溶液体系中进一步证实了过氧化氢酶对零价汞的氧化，外源性过氧化氢的加入可促进零价汞的液相氧化（Wigfield and Perkins, 1983）。在血红细胞、大鼠脑匀浆（Sichak et al., 1986）中均可观测到该过氧化氢酶氧化途径。除了哺乳动物，其他来源的过氧化氢酶（如蘑菇、枯草杆菌和菠菜来源）也可氧化零价汞（Ogata and Aikoh，1983）。其他过氧化物酶，如辣根过氧化物酶（Ogata and Aikoh，1984）、乳过氧化物酶（Ogata et al.，1982；Ogata and Aikoh，1984）亦可氧化零价汞。辣根过氧化物酶对零价汞的氧化反应对于汞与辣根过氧化物酶均为一级反应，该反应涉及零价汞向过氧化物酶 Fe-heme 的双电子转移以及过氧化氢对 Fe-heme 的氧化再生（Wigfield and Tse，1985；1986）。

微生物可能在土壤（Smith et al.，1998）、湖泊底泥与水（Siciliano et al.，2002）中零价汞的氧化中起着重要作用。向安大略湖湖水中加入过氧化氢，30 min 后汞氧化酶活性可提高 250%（Siciliano et al.，2002）。在这一水环境中，光照生成的过氧化氢可能引发湖水中微生物对溶解气态汞的氧化（Siciliano et al.，2002）。

2.3　汞的化学与生物还原

与汞的氧化类似，汞的还原在多种环境介质如大气、水、土壤、冰雪中均可发生。大气气相/水相中汞的还原增加了零价汞在大气中的半衰期与长距离迁移（Horowitz et al.，2017）；表层水与土壤中汞的还原增加了零价汞向大气中的释放，降低了水/土壤中的汞浓度。

2.3.1　大气气相汞化学还原

大气中氧化态汞可通过均相或非均相界面过程还原。目前关于大气中汞的气相还原的研究相对较少。

1. 气相均相汞还原

大气中汞可能的均相还原反应如表 2-3 所示。在光照下，RGM 如气态 $HgCl_2$、$Hg(OH)_2$ 均可还原为气态零价汞（Lin and Pehkonen，1999b；Pongprueksa et al.，2008）。一氧化碳也可介导 RGM 的还原（Pongprueksa et al.，2008）。模拟显示，RGM 直接光还原或一氧化碳介导的还原可能在气相汞还原中较为重要（Pongprueksa et al.，2008）。此外，在燃煤电厂烟羽中二氧化硫可能导致 RGM 的气相还原（Seigneur et al.，2006；Landis et al.，2014；Castro and Sherwell，2015）。燃煤电厂下风向汞形态现场测试分析显示，二价气态汞占总汞比例随距离点源的增加而下降，这一现象不足以用二价汞的干沉降来解释。据此推测在烟羽中存在二氧化硫对二价气态汞的还原（Lohman et al.，2006）。考虑二氧化硫对二价气态汞的还原可进一步提高汞大气模型的精度（Seigneur et al.，2006；Vijayaraghavan et al.，2008）。除了均相还原，以上过程也可能导致固相二价汞的非均相还原。

表 2-3　大气气相均相汞还原

反应	反应速率	参考文献
$Hg(OH)_2 \xrightarrow{h\nu} Hg(0)$ （2-35）	—	（Lin and Pehkonen，1999b）
$RGM(HgCl_2) \xrightarrow{h\nu} Hg(0)$ （2-36）	$1\times10^5\ s^{-1}$	（Schroeder et al.，1991；Pongprueksa et al.，2008）
$HgO + CO \longrightarrow Hg(0) + CO_2$ （2-37）	$5\times10^{18}\ cm^3/(mol\cdot s)$	（Pongprueksa et al.，2008）
$RGM + SO_2 \longrightarrow Hg(0)$ （2-38）	$8\times10^{18}\ cm^3/(mol\cdot s)$	（Seigneur et al.，2006）

2. 气相非均相汞还原

以往多认为大气中二价汞的还原主要在大气液滴中进行，因此对二价汞在大气颗粒物表面的非均相还原研究较少。燃煤电厂烟羽中零价汞约占总汞的 84%，而根据燃煤及 EPRI-ICR 汞形态分析模型，零价汞仅约占 42%，这提示在烟羽中可能存在二价汞的还原（Edgerton et al.，2006；Lohman et al.，2006）。随后，通过对燃煤电厂羽流的进一步分析发现，烟羽中存在二价汞的还原，这一还原过程受燃煤类型和用量的影响，其还原比例为 0～55%（Landis et al.，2014）。烟羽中二价汞的还原可能与二氧化硫等还原性气氛有关。固定床流通反应器模拟显示，紫外、可见光与模拟日光光照均可显著增加氯化汞在氯化钠干气溶胶表面的还原，汞还原比率随光照时间延长而增加（Tong et al.，2013）。辐照强度归一化后，三种光源对汞的还原能力相当。干气溶胶中铁可抑制二价汞的光还原，其中三价铁的抑制作用显著强于二价铁。三价铁络合剂甲磺酸去铁胺可减弱三价铁对还原的抑制作用。实验室模拟显示，高碳、低碳/低硫、高硫飞灰中二价汞的光还原较为快速，在 1000 W/m² 光照强度下，二价汞的平均半衰期仅为 1.6 h（Tong et al.，2014）。低碳/低硫飞灰中汞的还原速率约为其他飞灰样品的 1.5 倍。炭黑与左旋葡聚糖模拟气溶胶中二价汞光还原速率与煤飞灰样品相当，己二酸模拟气溶胶中二价汞光还原速率约为煤飞灰样品的 3～5 倍。煤飞灰水提取物气溶胶中二价汞还原速率略高于或相当于煤飞灰，表明煤飞灰中溶解性组分可能在还原中起着重要作用。密度泛函理论计算进一步揭示了 $HgCl_2$、$HgBr_2$、$Hg(NO_3)_2$、$HgSO_4$ 在洁净 Fe(110)、NaCl(100)、NaCl(111)^{Na} 表面的还原途径，发现 Fe(110)、NaCl(111)^{Na} 表面二价汞更易于还原（Tacey et al.，2016）。此外，Fe(110)表面零价汞的解吸需要 0.5 eV 的外部能量；而 NaCl(111)^{Na} 表面则无须外部能量。

2.3.2　大气水相汞化学还原

大气中氧化态汞的还原是大气汞循环的重要组成部分，据估计大气二价汞还原主要发生在大气液滴中（Shia et al.，1999）。由于零价汞较高的蒸气压与低的水溶性，液相还原生成的零价汞可快速扩散至气相，导致大气中零价汞浓度的增加与大气汞半衰期的延长（Horowitz et al.，2017）。

1. 大气水相均相汞还原

加拿大南魁北克 St. Anicet 及美国纽约州多个采样点大气中水蒸气混合比与总气态汞呈线性正相关（Poissant，1997；Han et al.，2004），表明大气液相中存在二价汞的再还原与零价汞产物的液气再分配，也说明相较于汞氧化，大气液相中汞的还原占主导。在较低臭氧浓度（<30 ppbv）下，总气态汞对水蒸气混合比的线性相关斜率是高臭氧浓度（>30 ppbv）下的两倍，表明臭氧对总气态汞浓度存在抑制（Poissant，1997），这可能来自于臭氧对零价汞的氧化或臭氧对大气液相中具有汞还原能力的物质的消耗。此外，气温升高总气态汞对水蒸气混合比的线性相关斜率增加，表明增加温度有助于液相中汞的还原生成（Han et al.，2004）。汞的大气模型也显示，需要大气液相中汞的还原才可得到与观测相符的模拟（Ryaboshapko et al.，2002；Han et al.，2004；Selin et al.，2007；Shia et al.，1999；Bash et al.，2014）。纳木错大气汞形态观测数据也提示，在大气中存在日光引发活性气态汞还原（de Foy et al.，2016）。另一研究也证实大气降雨中二价汞可被还原为零价汞（Lamborg et al.，1999）。

表 2-4 列出了大气水相中二价汞可能的一些还原途径，主要包括：①氢氧化汞的直接光解：波长大于 290 nm 的模拟日光可将水中的 $Hg(OH)_2$ 还原为零价汞，这一过程可能在汞的大气化学中是重要的（Munthe and McElroy，1992）。相对地，氯化汞较为稳定，未见光还原（Munthe and McElroy，1992）。②亚硫酸对汞的还原：研究显示，水相中 SO_3^{2-} 可以与二价汞络合，生成不稳定的 $HgSO_3$ 中间体。$HgSO_3$ 降解生成一价汞，并进一步快速还原为零价汞（Munthe et al.，1991）。该反应可在黑暗条件下进行，不受日光照射影响（Munthe and McElroy，1992）。反应速率随温度升高而增大，随 pH 降低而减小（Solis et al.，2017）。此外，该还原反应速率在较低亚硫酸浓度下较为显著，随亚硫酸浓度增加而降低，这可能与亚硫酸导致 pH 的降低有关（Munthe et al.，1991）。一项模型研究提示，大气 SO_2 可能在雾滴二价汞还原中起着重要作用（Pleijel and Munthe，1995）。但另一项两相箱式模型研究显示，大气中亚硫酸盐可快速氧化，其对汞的还原作用可忽略（Pan and Carmichael，2005）。③过氧羟基自由基途径：有研究认为在环境条件下 $HO_2^{\bullet}$ 与 $O_2^{\bullet-}$ 不能还原二价汞，大气汞还原无须考虑这一过程（Gardfeldt and Jonsson，2003），但也有一些研究认为草酸光解生成的 $HO_2^{\bullet}$ 可引起二价汞的还原（Pehkonen and Lin，1998）。模型研究显示，过氧羟基所起还原作用可能对大气汞转化较为重要

(Pan and Carmichael，2005）。④二元羧酸对汞的光还原：在光照下，大气液相中的二元羧酸如草酸、丙二酸和琥珀酸均可将二价汞还原为零价汞。草酸对二价汞的光还原较一元羧酸如甲酸与乙酸更为显著（Pehkonen and Lin，1998），而与其他二元羧酸如丙二酸、琥珀酸相当（Si and Ariya，2008）。这一还原过程可能涉及二元羧酸-二价汞络合物的分子内电子转移（Gardfeldt and Jonsson，2003；Si and Ariya，2008）或二元羧酸光解生成的高还原活性的二氧化碳自由基（$CO_2^{\cdot-}$）（Ababneh et al.，2006；Berkovic et al.，2010）。紫外-可见光谱提示二元羧酸可与二价汞形成络合物（Gardfeldt and Jonsson，2003），基质辅助激光解吸电离飞行时间质谱可检测到二元羧酸与二价汞络合物的形成，其进一步的光解产物为零价汞、羟基羧酸与一元羧酸（Si and Ariya，2008）。溶解氧在羧酸汞还原中的作用尚存在争议。有研究显示，溶解氧可在一定程度上促进草酸对汞的光还原（Pehkonen and Lin，1998）。但另外的研究表明，通氧抑制了草酸对汞的光还原（Gardfeldt and Jonsson，2003；Si and Ariya，2008），这可能与溶解氧消耗 $CO_2^{\cdot-}$（Ababneh et al.，2006）或溶解氧导致还原汞形态的再氧化有关（Si and Ariya，2008）。氯离子可部分抑制草酸、丙二酸和琥珀酸对汞的光还原，这可能与氯离子与二元羧酸竞争结合汞离子有关（Si and Ariya，2008）。此外，Fe(III)可显著促进草酸-Hg(II)体系零价汞的光生成（Ababneh et al.，2006）。模型计算显示，考虑二元羧酸对二价汞的水相光还原可将汞湿沉降的预测准确度提高13%（Bash et al.，2014）。

表2-4　大气水相中汞的还原

反应	平衡或速率因子	参考文献
$Hg(OH)_2 \xrightarrow{h\nu} Hg(0)$ （2-39）	$3\times10^{-7}\ s^{-1}$	（Lin and Pehkonen，1999b）
$HgSO_3 \longrightarrow Hg(0)$ （2-40）	$0.6\ s^{-1}$	（Munthe et al.，1991）
$Hg(II) + 有机酸 \xrightarrow{h\nu} Hg(0)$ （2-41）	$1.2\times10^4\ cm^3/(mol\cdot s)$（草酸）；$4.9\times10^3\ cm^3/(mol\cdot s)$（丙二酸）；$2.8\times10^3\ cm^3/(mol\cdot s)$（琥珀酸）	（Si and Ariya，2008）
$Hg(II) + HO_2^{\cdot} \longrightarrow Hg(I) + O_2 + H(I)$ （2-42）	$1.7\times10^4\ cm^3/(mol\cdot s)$	（Pehkonen and Lin，1998）
$Hg(I) + HO_2^{\cdot} \longrightarrow Hg(0) + O_2 + H(I)$ （2-43）	Fast	（Pehkonen and Lin，1998）

2. 大气水相非均相汞还原

水相中的固相汞形态（如HgO等）可被光还原。而其他固相基质如铁矿物也可影响汞的水相还原。水相中的针铁矿（α-FeOOH）、赤铁矿（α-Fe₂O₃）、磁赤铁矿（γ-Fe$_2$O$_3$）与采集的大气气溶胶可促进草酸对二价汞的光还原，其中针铁矿、磁赤铁矿对光还原的促进最为显著（Lin and Pehkonen，1997）。反应初期，赤铁矿也能促进草酸体系汞的还原，但生成的羟基自由基进一步导致零价汞的再氧化。据推测，这一对光还原的促进作用机制涉及颗粒物-水微界面草酸自由基的生成以及草酸自由基与溶解氧反应生成的氢过氧自由基[如式（2-44）~式（2-50）所示]。但光反应过程中生成的溶解性二价铁也可能在汞的还原中起着关键作用。

$$>\text{Fe(III)}—\text{OH} + \text{C}_2\text{O}_4^{2-} + \text{H}^+ \rightleftharpoons >\text{Fe(III)}—\text{C}_2\text{O}_4^- + \text{H}_2\text{O} \quad (2\text{-}44)$$

$$>\text{Fe(III)}—\text{C}_2\text{O}_4^- + h\nu \longrightarrow \text{Fe(II)} + \text{C}_2\text{O}_4^{\cdot-} \quad (2\text{-}45)$$

$$>\text{Fe(II)} \longrightarrow > + \text{Fe}^{2+} \quad (2\text{-}46)$$

$$\text{C}_2\text{O}_4^{\cdot-} + \text{O}_2 \longrightarrow \text{O}_2^{\cdot-} + 2\text{CO}_2 \quad (2\text{-}47)$$

$$\text{O}_2^{\cdot-} + \text{H}^+ \rightleftharpoons \text{HO}_2^{\cdot} \quad (2\text{-}48)$$

$$\text{HO}_2^{\cdot} + \text{Hg}^{2+} \longrightarrow \text{Hg}^+ + \text{O}_2 + \text{H}^+ \quad (2\text{-}49)$$

$$\text{HO}_2^{\cdot} + \text{Hg}^+ \longrightarrow \text{Hg}^0 + \text{O}_2 + \text{H}^+ \quad (2\text{-}50)$$

与以上光反应相反的是，煤飞灰可将亚硫酸钠对二价汞的暗还原速率降低40%~90%（Feinberg et al.，2015）。

2.3.3 表层水均相汞还原

地表水中二价汞还原生成挥发性零价汞是汞生物地球化学循环的重要过程之一。多种环境参数如光照强度与波长、DOM 浓度与组成、汞形态、溶解氧、pH、共存离子等均可对这一过程产生影响。

1. 光照强度与波长

通常光照可显著促进二价汞的还原。暗反应对比实验显示，湖水中溶解气态汞的生成主要由光照引发（Amyot et al.，1994）。对于淡水（O'Driscoll et al.，2003；Amyot et al.，2004；O'Driscoll et al.，2007；Poulain et al.，2007；Zhang and Dill，2008；Amyot et al.，1994）与海水（Lanzillotta and Ferrara，2001；Fantozzi et al.，2007；Poulain et al.，2007）体系，溶解气态汞生成量与日光照射强度显著正相关。模拟光照实验也显示，溶解气态汞的生成随光照强度增加而增加（Costa and Liss，2000；李希嘉等，2014；Sun et al.，2015）。因此，环境水体溶解气态汞的生成呈现出显著的日变化（O'Driscoll et al.，2003；Zhang and Dill，2008；O'Driscoll et al.，2007）（Lanzillotta and Ferrara，2001；Fantozzi et al.，2007）及季节变化（Zhang and

Dill，2008）。光照可导致汞的快速还原，但溶解气态汞生成与挥发之间存在一定的滞后（O'Driscoll et al.，2007），湖水体系这种滞后性较河水更为明显（O'Driscoll et al.，2006b）。

模拟实验显示，光照波长越短，溶解气态汞生成比率越高（Costa and Liss，2000；李希嘉等，2014）。在太阳光谱中，紫外线 A（ultraviolet A，UVA）、UVB、可见光所占能量比例为可见光＞UVA＞UVB。UVB 遮光膜（Mylar 膜）对加拿大湖泊中溶解气态汞生成影响较小，说明日光中 UVB 波段的贡献较小（Amyot et al.，1994）。现场测试数据显示，水库水溶解气态汞生成速率与早晨 UVA 辐照呈正比（$r=0.9582$，$p<0.01$）（Zhang and Dill，2008）。但对于东亚海域海水，UVB 辐照占总光致还原贡献的约 30%～55%，其余光还原的贡献主要来自于可见光，仅在部分样品中观测到 UVA 辐照对光还原的贡献（Ci et al.，2016b）。这说明，日光不同波段的贡献还可能与当地的日光光谱组成及环境水样的水化学条件有关。

由于随水深存在光的衰减，汞的光还原主要发生在表水层（Amyot et al.，1994；Monperrus et al.，2007）。不同波长光照在水中具有不同的衰减系数，在考虑深度梯度下汞的光还原时需要考虑这一问题（O'Driscoll et al.，2007）。因此，水体深处光照对还原的作用可以忽略。水体深处汞的还原可能归结于 DOM 对汞的暗还原（Alberts et al.，1974）或汞的微生物还原（Peretyazhko et al.，2006b；Monperrus et al.，2007）。

2. 溶解有机质浓度与组成

在表层水中，DOM 是控制 Hg(Ⅱ)还原的主要因素之一。DOM 可通过光生水合电子、超氧阴离子或有机自由基引发汞的还原，也可能通过 Hg(Ⅱ)-DOM 络合物内的电子转移导致汞还原。DOM 或 Hg(Ⅱ)-DOM 络合物对光的吸收一方面可引发汞的还原；另一方面，"过剩"的 DOM 吸收太阳光可导致光照随水深衰减，降低了光照引发还原的有效性，从而抑制汞的还原。此外，DOM 也可能在零价汞的氧化中起着一定作用。大部分研究表明，在较低浓度 DOM 条件下，DOM 可有效促进汞的还原（Costa and Liss，2000；Lanzillotta et al.，2002；O'Driscoll et al.，2004；Fantozzi et al.，2007）。如加拿大魁北克地区湖泊中溶解气态汞的生成量与 DOM 浓度呈显著正相关（O'Driscoll et al.，2004）。在阿拉斯加湖泊中，低浓度 DOM 促进汞的光还原，但高浓度显著抑制汞的光还原，这主要与 DOM 浓度增加导致光的衰减有关（Tseng et al.，2004）。另一项关于水库汞还原的研究也发现，在水体浊度小于 10 NTU 情况下，溶解气态汞与 DOM 浓度呈正比，而当水体浊度大于 10 NTU，溶解气态汞与 DOM 浓度呈反比（Ahn et al.，2010）。类似地，随着腐殖酸浓度的增加，汞的暗还原先显著增加后缓慢下降（Jiang et al.，2014）。在暗反应及较高的腐殖质浓度条件下（＞1000 mg/L），汞还原效率随腐殖质浓度增加而下降（Rocha et al.，2000）。由于暗反应不涉及光的衰减，这一现象可能与 DOM 导致零价汞的氧化有关。

DOM 分子结构是影响汞还原的关键因素之一。不同来源 DOM 对汞的还原速率和还原容量均存在差异，可能与 DOM 中涉及氧化还原基团如酚羟基、羧基等的差异有关（Zhang et al.，2011；Jiang et al.，2014）。例如，受森林砍伐影响的湖泊中 DOM 对汞的光还原效率低于未受森林砍伐影响湖泊，这可能与两类湖泊中 DOM 结构的差异有关（O'Driscoll et al.，2004）。水相汞结合形态计算和零价汞分析显示，不同来源腐殖酸中 $Hg(OR)_2$ 结构对汞暗还原的贡献高于 RSHgOR 与 $Hg(SR)_2$（Jiang et al.，2015）。天然水体汞还原过程中两种奇数同位素非质量分馏的程度与 Hg/DOM 比例有关，提示不同 Hg/DOM 比例导致的 Hg 结合位点差异可能导致了不同的汞还原机制（Zheng and Hintelmann，2009）。不同官能团 DOM 小分子模型对汞的光还原进一步显示，具有还原性硫基团的小分子（如半胱氨酸、谷胱甘肽）对汞的还原低于非还原硫基小分子（如丝氨酸、草酸）。这一过程的非质量分馏也存在较大差异：对于含硫配体，零价汞产物富集奇数同位素（^{199}Hg 与 ^{201}Hg）；而非含硫配体奇数同位素（^{199}Hg 与 ^{201}Hg）则富集在反应物 [Hg(Ⅱ)] 中，表明两类不同配体产生了相反的磁同位素效应（Zheng and Hintelmann，2010）。这也提示，DOM 中的巯基与羧基在汞还原中所起作用是不同的。

从目前的实验研究结果看，DOM 中的巯基可抑制汞的还原。例如，在腐殖质对汞的暗还原中，溶解性腐殖质中总硫的增加可降低汞还原（Rocha et al.，2003；Chakraborty et al.，2015；Vudamala et al.，2017）。以小分子硫醇与巯基乙酸为有机质模型，发现汞的光还原可能受 Hg(Ⅱ)-巯基络合物介导（Si and Ariya，2011；2015），除了还原之外，巯基乙酸还可介导硫化汞生成（Si and Ariya，2015）。硫化汞的生成可能使汞的还原得到抑制。DOM 中的氨基、酚羟基与羧基可能均在汞的还原中起着重要作用。采用多种表征和主成分分析研究巴西亚马孙地区里奥内格罗（Rio Negro）盆地土壤腐殖质对汞的还原过程，发现与汞还原相关的主要官能团是酚羟基、羧基和氨基（Serudo et al.，2007）。一些研究认为提高—COOH/—OH 官能团比例可降低汞还原（Rocha et al.，2003；Vudamala et al.，2017），但也有研究认为高—COOH/—OH 官能团比例有助于汞的暗还原（Chakraborty et al.，2015）。天然水 DOM 中羧基的甲基化掩蔽可抑制零价汞生成，表明这一还原过程可能涉及羧基介导的 Hg(Ⅱ)-DOM 络合物分子内电子转移过程（Allard and Arsenie，1991）。模型研究发现，光照下 1,4-萘醌三线态可介导甲酸生成二氧化碳自由基（Berkovic et al.，2012），二氧化碳自由基可还原二价汞为零价汞（Berkovic et al.，2010；Berkovic et al.，2012）。以水杨酸、邻羟基苯甲酸、邻氨基苯甲酸、邻苯二甲酸、对羟基苯甲酸、对氨基苯甲酸为 DOM 模型分子对汞的光还原研究发现，光还原速率不仅受苯环上的羟基、氨基、羧基取代基影响，还与取代基位置有关（He et al.，2012）。DOM 模型分子对汞的光还原速率为：邻氨基苯甲酸＞邻羟基苯甲酸＞邻苯二甲酸，邻氨基苯甲酸＞对氨基苯甲酸，邻羟基苯甲酸＞对羟基苯甲酸。还原速率与这些化合物的浓度及紫外吸收光谱有关（He et al.，2012）。这一结果也确证了氨基、酚羟基、羧基等基团在汞还原中的作用。

3. 无机汞形态

光照之前向湖水或海水中加入二价汞可显著提高溶解气态汞的生成速率，这表明可见光还原 Hg(Ⅱ)络合物是限制溶解气态汞生成的因素之一（Amyot et al.，1997a；Amyot et al.，1997b；Amyot et al.，1997c；O'Driscoll et al.，2004）。例如，北极地区 Amituk 湖水中总汞与其他湖泊相当，但却未观测到溶解气态汞生成，这可能与湖水中可见光还原 Hg(Ⅱ)络合物浓度较低有关（Amyot et al.，1997b）。研究显示阿拉斯加湖水中溶解气态汞的生成与溶解活性汞（dissolved labile Hg）（定义为水中可被 0.45 μm 滤膜过滤的且可被 $SnCl_2$ 还原的汞）的浓度线性正相关（r^2=0.82，$p<0.0001$），也表明溶解气态汞主要来自于水溶相中汞络合物（以汞-有机络合物为主）（Tseng et al.，2004）。

据估计，淡水体系可还原汞（$[Hg_{reducible}]_0$）约占过滤水样总汞的 37.8%（O'Driscoll et al.，2006b）。Ci 等研究发现，对于东亚海域海水中可还原汞占总汞的比例［%Hg_r(Ⅱ)/THg］为 5.8%～48.5%（平均 24.9%±10.6%），随总汞浓度升高呈指数降低，即在开阔海洋水域总汞较低，%Hg_r(Ⅱ)/THg 较高，在沿海水域总汞较高，%Hg_r(II)/THg 较低（Ci et al.，2016b）。这些现象提示在水体中一部分汞形态不可还原或还原能力较差。最近研究发现光照 Hg(Ⅱ)-DOM 络合物除了生成零价汞之外，也导致 Sn(II)还原汞（或称之为反应活性汞）浓度的下降，这可能与二价汞和 DOM 反应生成还原惰性的硫化汞有关（Manceau et al.，2015；Luo et al.，2017）。与之相一致的是，硫化汞体系中零价汞的生成挥发远低于二价汞-有机质络合物（Zhu et al.，2018）。

4. 溶解氧

多项研究显示，除氧可在一定程度上促进环境水样或模拟水样二价汞的暗/光还原。对于东亚地区海水，相对于空气吹扫，氩气吹扫可显著提高二价汞的暗还原（Ci et al.，2016b）。圭亚那河水通氮脱氧可显著促进二价汞的光还原（300～450 nm），氮气饱和条件下总溶解气态汞的生成量较空气饱和条件高 122%（Beucher et al.，2002）。对于青岛近岸海水，无论在 UVB（297 nm、15.6 μW/cm^2）还是在

UVA（365 nm、78.9 μW/cm^2）光照下，氩气吹脱时汞的还原均高于空气吹脱（马学琳等，2015）。除氧也可在一定程度上促进富里酸溶液中零价汞的光还原生成（Allard and Arsenie，1991）。对于 365 nm 紫外光照，去离子水中硝酸汞的光还原为：通氮气＞通空气＞通氧气；但对于 254 nm 紫外光照，去离子水中硝酸汞的光还原能力为：通氮气＞通氧气＞通空气（Byrne et al.，2009），说明溶解氧的作用还与光照波长有关。水中溶解氧对巯基丙醇对二价汞的光还原影响不大（Si and Ariya，2011）。但在 pH 3.9 条件下，通氧（溶解氧浓度 6.2 mg/L）却可显著促进草酸溶液中氯化汞的光还原（320～800 nm）；光照下检测到过氧化氢生成，表明这一过程可能涉及溶解氧生成的超氧阴离子或氢过氧自由基中间体（Lin and Pehkonen，1997）。也有研究认为，氧气可显著抑制光照下丙二酸（Si and Ariya，2008）或草酸铁（Ababneh et al.，2006）对二价汞的光还原。

5. pH

早在 1974 年，Alberts 等就观察到暗反应条件下 pH 6.5～8.2 范围内腐殖酸对汞的还原速率随 pH 降低而升高（Alberts et al.，1974）。在 pH 4.0～8.0 范围内，多种国际腐殖酸协会的标准腐殖质对汞的暗还原亦随 pH 降低而升高（Chakraborty et al.，2015）。在 pH 6.0～8.5 范围内，水库水溶解气态汞生成与 pH 负相关（$p<0.01$），据推测这与 pH 导致二价汞与 DOM 结合位点（由巯基结合变为羧基结合）的变化有关（Ahn et al.，2010）。pH 2.9～8.8 范围内，腐殖质对汞的暗还原在 pH 4.5 处还原最为显著（Allard and Arsenie，1991）。在 pH 2.0～11.0 范围内，巴西里奥内格罗地区河流腐殖质对汞的还原在 pH 8.0 达到最大（Rocha et al.，2000）。pH 3.5～7.8 范围内淡水体系测试显示，溶解气态汞生成量在 pH 6.0 条件下最高（O'Driscoll et al.，2006b）。也有研究发现腐殖质对汞的还原随 pH（5.0～9.0）升高而增加（Matthiessen，1998）。

6. 共存离子

共存离子可通过间接途径影响汞的还原。东亚海域海水（Ci et al.，2016b）、北极地区水体（Poulain et al.，2007）、腐殖质溶液（Chakraborty et al.，2015）中溶解气态汞的生成随氯离子浓度的增加显著下降，这可能与光还原活性较弱的 $HgCl_x^{(x-2)}$ 组分比例增加或者氯离子促进零价汞氧化有关（Sun et al.，2014）。去离子水中汞形态计算与光还原实验显示，$Hg(OH)_2$ 较 $HgCl_x^{(x-2)}$ 更易光还原为零价汞（Sun et al.，2014），因此在 pH 7.0 的去离子水中，在 280～800 nm 光照下，氯离子浓度增加降低了 $Hg(OH)_2$ 的比例与二价汞的还原速率（Sun et al.，2014）。竞争离子 Eu^{3+} 的存在可显著降低腐殖质对汞的暗还原。类似地，高浓度钙离子可降低

DOM 向二价汞的电子转移，从而抑制零价汞的生成（Vudamala et al., 2017）。但提高离子强度对硫醇对汞的光还原速率没有影响（Si and Ariya, 2011）。在北极海水体系中，溶解气态汞的生成与铁浓度显著正相关（r=0.96, p=0.0017）（Poulain et al., 2007）。5 μmol/L 或 10 μmol/L 三价铁的加入可显著增加淡水中溶解气态汞的生成，这可能与 Fe(III)-有机酸络合物光解生成还原性的自由基有关（Zhang and Lindberg, 2001）。

7. 悬浮颗粒物

悬浮颗粒物对汞还原的影响较为复杂。阿拉斯加地区淡水湖泊中悬浮颗粒物浓度与溶解气态汞浓度呈反比（Tseng et al., 2004）；0.2 μm 筛孔过滤的青岛近岸海水中汞的光还原高于未过滤海水（马学琳等, 2015）；悬浮颗粒物的过滤去除将圣弗朗索瓦湾湿地水中溶解气态汞的光生成提高了 30%（Garcia et al., 2005）。对于含有高浓度悬浮颗粒物的东亚海域海水，过滤降低了水中总汞及可还原二价汞 [$Hg_r(II)$] 库，显著降低了暗还原零价汞的生成量（53.5%~77.5%），但对光还原的影响（增加或基本不变）与水化学参数有关（Ci et al., 2016b）。在法国吉伦特（Gironde）河口低浊度区，溶解气态汞生成主要受日光辐照所控制；但在高浊度区，溶解气态汞生成受到抑制，其浓度受日光照射与悬浮颗粒物浓度比（rad/SPM）的控制（Castelle et al., 2009）。说明在这些环境水体中悬浮颗粒物对汞的还原主要起抑制作用。对于圣劳伦斯河/安大略湖（Amyot et al., 2000）、大西洋海水（Qureshi et al., 2010）中汞的光还原，悬浮颗粒物的去除对其影响不大。过滤去除悬浮颗粒物对法属圭亚那河水中汞的光还原没有影响，但却降低了暗反应溶解气态汞的生成（Beucher et al., 2002），提示暗条件下悬浮颗粒物参加的非均相反应在汞的还原中起着一定作用。也有报道指出过滤可以在一定程度上降低湖水与海水中汞的光还原（Amyot et al., 1997a; Poulain et al., 2007）；燃煤电厂脱硫海水中汞的光还原与悬浮颗粒物浓度显著正相关（Sun et al., 2015），表明悬浮颗粒物在汞的光还原中起着一定作用。模拟实验显示，胡敏酸对不同矿物结合态汞的还原作用存在差异，无论在暗反应还是光照条件下，不同来源胡敏酸对矿物结合态汞的还原顺序为：碳酸钙结合态汞、赤铁矿结合态汞、氧化锰结合态汞（刘丽琼等, 2011）。这表明悬浮颗粒物中无机物的组成可显著影响汞的还原。

悬浮颗粒物对汞还原影响的可能原因如下：①悬浮颗粒物对光的吸收与散射降低了水体透光率，并影响入射光光谱，这可在一定程度上降低二价汞的光还原（O'Driscoll et al., 2007; Castelle et al., 2009）；②悬浮颗粒物存在或结合的一些还原性物种可在黑暗或光照条件下一定程度促进二价汞还原（Nriagu, 1994; Beucher et al., 2002），其中的微生物可能在暗还原中也起着一定作用（Fantozzi et al., 2009）（参见 2.3.8 节微生物还原部分）；③悬浮颗粒物对二价汞的吸附降低了水均相体系中二价汞与 DOM 等的结合，抑制了 DOM 对汞的光还原（Tseng et al.，2004）；④悬浮颗粒物可与还原生成的零价汞结合，这部分结合态零价汞不可吹扫，从而低估了零价汞的生成（Wang et al.，2015）。

2.3.4 表层水非均相汞还原

在水相非均相体系中，自然界中存在的多种含铁矿物颗粒均可将二价汞还原为零价汞。这一还原过程对于地表水、底泥/土壤间隙水以及大气液相中汞的还原可能具有重要的意义。但目前的研究尚局限于实验室模拟，真实环境中含铁矿物颗粒在汞的水相非均相还原中的作用尚待进一步证实。

多种含铁矿物（如菱铁矿、绿锈、磁铁矿、黄铁矿等）中含有二价铁。这些固相二价铁可引发二价汞的还原。

绿锈（green rust）为 Fe(Ⅱ)和 Fe(Ⅲ)的氢氧化物混合物，广泛存在于低氧环境。Hg LⅢ 边 XANES 以及 EXAFS 分析研究表明，碱式绿锈可将 $HgCl_2$ 还原为零价汞，X 射线衍射谱（XRD）显示二价汞的还原过程伴随着绿锈向磁铁矿的氧化转化（O'Loughlin et al.，2003）。这可能与绿锈中二价铁对二价汞的直接还原有关［式（2-51）和式（2-52）］。

$$>Fe(Ⅱ) + Hg(Ⅱ) \longrightarrow >Fe(Ⅲ) + Hg(Ⅰ) \quad (2\text{-}51)$$

$$>Fe(Ⅱ) + Hg(Ⅰ) \longrightarrow >Fe(Ⅲ) + Hg(0) \quad (2\text{-}52)$$

菱铁矿（$FeCO_3$）可将二价汞还原为零价汞（Ha et al.，2017）。二价汞还原可在数分钟内完成，且反应速率随菱铁矿表面积增加而增加（Ha et al.，2017）。汞 LⅢ 边 XANES 分析显示零价汞主要吸附在菱铁矿表面，表明电子转移反应发生在菱铁矿-水界面（Ha et al.，2017）。

磁铁矿（Fe_3O_4）也可将二价汞还原为零价汞（Wiatrowski et al.，2009）。动力学分析显示还原反应可在数分钟内发生，反应速率随磁铁矿表面积及溶液 pH 增加而增加，随氯离子浓度增加而降低（Wiatrowski et al.，2009）。X 射线光电子能谱分析检测到二价汞在磁铁矿表面的吸附，提示电子转移过程可能涉及二价汞的吸附及界面过程（Wiatrowski et al.，2009）。穆斯堡尔谱分析显示在汞还原过程，磁铁矿中的二价铁部分氧化为三价铁（Wiatrowski et al.，2009）。磁铁矿可具有不同的二价铁与三价铁比例［Fe(Ⅱ)/Fe(Ⅲ)］。Hg LⅢ 边 XANES 以及吹扫捕集-电感耦合等离子体质谱分析显示，在不存在氯离子条件下，Fe(Ⅱ)/Fe(Ⅲ) 为 0.29、0.38、0.50 的磁铁矿均可还原二价汞为汞离子（Pasakarnis et al.，2013）。但在氯离子存在情况下，当磁铁矿中二价铁比例较低［Fe(Ⅱ)/Fe(Ⅲ)<0.42］时，除零价汞外二价汞部分还原生成氯化亚汞；而当磁铁矿中二价铁比例较高［Fe(Ⅱ)/Fe(Ⅲ)≥0.42］

时，即使在高浓度氯离子情况下，也仅观测到零价汞生成。此外，磁铁矿中二价铁比例也对二价汞还原速率有一定影响（Pasakarnis et al.，2013）。类似地，微生物源磁铁矿亦可将二价汞还原为零价汞（Yee et al.，2010）。

在环境水体（尤其是底泥）中，二价汞通常以与有机质或生物质结合的形式存在。这种结合可在一定程度上影响含铁矿物对二价汞的还原。在高汞：生物质（来自于枯草芽孢杆菌）比例（50 μmol/L：2 g/L）条件下，汞主要通过羧基与生物质结合，在 pH 6.5 2 h 及 pH 5.0 2 d 条件下，观测到磁铁矿对二价汞的还原（Mishra et al.，2011）。在低汞：生物质比例（5 μmol/L：2 g/L）条件下，汞主要通过巯基与生物质结合，未观测到磁铁矿对二价汞的还原（Mishra et al.，2011）。而另一种还原性较强的含铁矿物绿锈则仍可在两天内将 20%的二价汞还原为零价汞（Mishra et al.，2011）。这一结果表明，羧基结合态汞未明显抑制磁铁矿对汞的还原，但巯基结合态汞显著抑制磁铁矿与绿锈对汞的还原。

溶解性 S^{2-} 与汞的络合通常可抑制零价汞的生成。但含硫及二价铁的铁矿物如黄铁矿、四方硫铁矿却可将二价汞还原为零价汞。在碱性 pH 条件下，无论无氧或有氧条件下，经过 37 个月的反应，氧化汞均可部分被黄铁矿或四方硫铁矿部分还原为零价汞（Svensson et al.，2006）。在含水及中性/酸性 pH 条件下，有氧/黄铁矿或无氧/四方硫铁矿条件下，氧化汞也可部分被还原为零价汞（Svensson et al.，2006）。Hg LIII 边 EXAFS 在二价汞与四方硫铁矿培育过程中除形成 β-HgS 与表面络合态二价汞外，部分二价汞还原为零价汞（Bone et al.，2014）。吹扫捕集-原子荧光光谱对零价汞的测定与 EXAFS 结果相符（Bone et al.，2014）。这与之前研究（Svensson et al.，2006）相一致。但该研究亦显示，四方硫铁矿对二价汞的还原反应在 1 小时内即可发生（Bone et al.，2014）。

如前所述，合成铁氧化物［针铁矿（α-FeOOH）、赤铁矿（α-Fe_2O_3）、磁赤铁矿（γ-Fe_2O_3）］可促进有机酸（草酸、甲酸、乙酸）对二价汞的光还原（Lin and Pehkonen，1997）。此外，在黑暗缺氧条件下，与均相体系相比，溶解二价铁在三价铁矿物（针铁矿、赤铁矿）上的吸附也可促进二价汞的还原（Amirbahman et al.，2013）。零价汞的非均相生成与二价铁平衡吸附、溶解/吸附二价铁对二价汞的还原（速控步）、二价汞的吸附（速控步）等步骤有关（Amirbahman et al.，2013）。零价汞的非均相生成速率可用下式描述（Amirbahman et al.，2013）：

$$r_{het} = k_{het}[>Fe(II)][Hg(OH)_2]$$

式中，[>Fe(II)] 为总吸附二价铁浓度，k_{het} 为 5.36 × 10^3 L/(mol·min)（赤铁矿）、4.69 × 10^3 L/(mol min)（针铁矿）。

2.3.5 土壤表面汞的还原

土壤气态汞的释放是汞自然排放一次源与二次源（沉降汞的再释放）的重要组成部分之一，对汞的生物地球化学循环具有重要的意义。近些年来，对土壤释汞通量与影响因素进行了较多研究。据估计，全世界每年陆地表面汞释放量为607 t/a（Agnan et al.，2016）。来自132项研究结果（包括来自20万次测定的1290项通量）的综合分析显示，释汞通量的时空分布极不均一（Agnan et al.，2016）。土壤中汞的还原主要可分为以下三个步骤（Zhang and Lindberg, 1999）：①二价汞的还原；②土壤中零价汞向土壤表层的扩散；③土壤-大气界面零价汞的释放。因此，任何影响以上三个步骤的环境因素都可能对土壤释汞通量产生影响。通常，土壤中二价汞的生物/非生物还原是决速步（Schluter, 2000）。研究显示，土壤释汞通量与土壤汞浓度与形态、气象因素（光照、风速等）、土壤物理化学（土壤温度、湿度、孔隙度等）等有关，这些因素之间也存在一定的复合作用（高锦玉等，2016）。

1. 土壤汞浓度

综合分析表明，当土壤中汞浓度较低时，土壤释汞通量与汞浓度不存在显著相关性（Agnan et al.，2016）。例如对于广东省不同土地利用情况的土壤（裸土、森林土、稻田土）（汞浓度 0.03~1.88 μg/g）研究显示，释汞通量与土壤汞浓度之间相关性并不显著（Fu et al., 2012）。但高汞地区释汞通量与汞浓度通常呈正相关（Agnan et al.，2016；Eckley et al., 2016）。例如，美国加州矿山废弃物及其周边区域土壤基质（汞浓度 2.7~9060.0 μg/g）中汞与释汞通量呈正相关：平均汞浓度为254 μg/g 的土壤对应于释汞通量约为 2000 ng/(m^2·h)（Gustin et al.，2003）。两项综合研究进一步表明，高汞地区土壤汞浓度与释汞通量存在下列类似的关系：

$logHg_{flux} = 0.427logHg_{soil}+1.432$（$r^2 = 0.26$，$n = 381$）（Agnan et al.，2016）

$logHg_{flux} = 0.54logHg_{soil}+1.3$（$r^2 = 0.59$，$n = 445$）（Eckley et al.，2016）

2. 无机汞形态

关于土壤中汞形态对释汞通量的研究较少。对于以 Hg(0)、Hg(Ⅱ)或硫化汞为主的土壤，其土壤汞挥发活化能（E_a）分别为（12.8±2.5）kcal/mol、（17.3±7.7）kcal/mol、（25.8±2.6）kcal/mol（Schluter, 2000）。活化能越大，土壤中汞越难还原释放。随着土壤汞浓度的升高，土壤汞挥发活化能下降，但相对于其他汞形态，硫化汞的挥发活化能降低较少（Kocman and Horvat, 2010）。天然及人工汞富集固体基质中汞的释放表明，在日光照射下，辰砂（α-HgS）的光释放显著低于氯硫汞矿（corderoite，$Hg_3S_2Cl_2$）、黑辰砂（β-HgS）及基质结合态汞（有机质或无机相结合汞)(Gustin et al.,2002)。实验室模拟表明,加入土壤中 Hg^{2+} 的释汞通量显著高于 Hg_2^{2+},这可能与 Hg_2^{2+} 易与氯离子反应生成难溶性的 Hg_2Cl_2 有关(Yang et al.,2007)。

3. 光照强度与波长

日光照射是影响土壤释汞通量的最关键因素之一。一方面,日光照射可引发土壤中二价汞的还原;另一方面,光照赋予的能量可促进土壤中零价汞的解吸附(Gustin et al.,2003)。由于光照的穿透有限,光致还原主要发生在土壤表层(<0.5 cm)(Carpi and Lindberg,1997)。多项研究显示,土壤释汞通量与总辐射强度呈显著正相关(Lindberg et al.,1999;Gustin et al.,2003;李仲根等,2006;Kocman and Horvat,2010;Ci et al.,2016a;Haynes et al.,2017)。短时间(20 min)内将铝箔遮盖日光照射(但温度基本不变)可显著降低土壤释汞通量,表明日光照射而非温度是影响土壤释汞的最关键因素(Zhang et al.,2001)。

也有研究显示,在日间光照强度达到最高之前,土壤释汞通量即达到最大值(Han et al.,2017b),这说明土壤基质中零价汞生成与释放的活化能较低或土壤表层可供还原释放的汞浓度有限。土壤覆盖物可降低其受光强度与面积。因此,裸露土壤汞的释放通量最高(Agnan et al.,2016;Eckley et al.,2016),而植被、凋落物等对日光的遮挡可抑制土壤气态汞的释放(Choi and Holsen,2009;Bagnato et al.,2014;Eckley et al.,2016;Zhou et al.,2017)。

日光照射中不同波段在土壤释汞中所起作用可能是不同的。实验室模拟光照显示,紫外光在土壤释汞中起着主要作用(Moore and Carpi,2005)。相较暗反应而言,可见光与 UVA(365 nm、1.147 W/m^2)对土壤释汞通量的影响较小,而 UVB(302 nm、0.778 W/m^2)与 UVC(254 nm、1.070 W/m^2)对汞释放的促进作用更为明显(Xin et al.,2007)。另一项研究也显示,UVA(365 nm)辐照对固体基质汞释放无显著作用(Choi and Holsen,2009),UVB(302 nm)辐照可显著促进固体基质的汞释放(Choi and Holsen,2009),UVC 可引发气相零价汞的氧化,从而促进汞的沉降(Choi and Holsen,2009)。也有研究表明,UVA(320~380 nm、0.368 mW/cm^2)可促进土壤气汞释放,而 UVB(280~320 nm、0.259 mW/cm^2)对汞释放的促进作用较 UVA 更为显著。但也有模拟光照实验发现,光照中的 UVA 波段在光还原中起着最主要的作用(Mauclair et al.,2008)。在采用天然日光辐照时,紫外遮光膜(380 nm 紫外光透过率 40%,340 nm 以下紫外光透过率<0.3%,可见光透过率为 98%)可将释汞通量降低 24%,这一结果提示日光中的紫外光与可见光均可显著提高土壤的汞释放(Zhang et al.,2001)。以上不同光照波段在土壤释汞中的作用的研究结果存在一定矛盾,这可能与所采用光源的波长与强度以

及固相基质中汞形态的差异有关。在未来研究中，所用模拟光源应尽量与日光照射的各波段的光谱分布与照射强度保持一致，才能真正反映实际环境土壤释汞中不同波长光照的真实作用。

4. 大气零价汞浓度

土壤向大气汞的扩散取决于土壤空气与大气零价汞的浓度梯度。因此，大气中零价汞可抑制土壤向大气的释汞通量。对于背景点，大气汞浓度与释汞通量通常呈负相关（Gillis and Miller，2000；Xin and Gustin，2007；Zhou et al.，2017）。统计数据显示，大气汞浓度与释汞通量存在以下关系（Agnan et al.，2016）：

$$Hg(0)_{flux} = -0.900Hg(0)_{air} + 2.477 \ (r^2 = 0.11, n = 266)$$

上式显示，随大气零价汞浓度增加，零价汞释放通量减少。大气零价汞浓度过高时（超过 2.75 ng/m³），土壤释汞通量变为负值（即大气向土壤沉降）（Agnan et al.，2016）。另一方面，随着零价汞排放点源（如金属冶炼厂）的关闭，土壤释汞通量可由负值（沉降）变为正值（挥发）（Eckley et al.，2015）。这提示我们，随着大气零价汞排放控制技术的改进以及汞排放点源的关闭，其临近区域土壤可由汞的汇转变为汞排放源。在点源减少后较长时间内，土壤中沉积的汞仍将持续在环境中循环。

但是，在汞污染区域，大气零价汞浓度与释汞通量呈现正相关（Agnan et al.，2016）：

$$\log Hg(0)_{flux} = 0.815\log Hg(0)_{air} + 0.700 \ (r^2 = 0.36, n = 262)$$

这主要归因于汞污染区域较高土壤释汞通量导致零价汞在近地表大气累积，而非大气零价汞浓度对释汞通量的影响（Agnan et al.，2016）。

5. 降水与灌溉

多项研究表明，降水与灌溉初期，土壤释汞通量显著增加（Lindberg et al.，1999；Wallschlager et al.，2000；Gustin et al.，2003；Ci et al.，2016a；Zhou et al.，2017；Howard and Edwards，2018）。降水与灌溉初期，水对土壤间隙中空气的排出作用及土壤颗粒上零价汞的解吸附可能增加了零价汞的释放（Zhang and Lindberg，1999；Zhou et al.，2017）。此外，水可促进二价汞在土壤中的移动，并进而促进间隙水中二价汞的还原（Zhang and Lindberg，1999）。进一步的降水或水淹可导致表层土壤水的饱和，堵塞土壤中零价汞的迁移孔道，从而抑制零价汞的释放（Zhou et al.，2017）。

6. 土壤温度

温度升高有利于二价汞的还原、零价汞的解吸附与迁移，因此土壤温度通常与其释汞量呈正相关，可由下式表示（Lindberg et al.，1995；Gustin et al.，2003）：

$$k = Ae^{\frac{E_a}{RT}}$$

式中，k 为土壤释汞速率常数，E_a 为汞释放的活化能，T 为热力学温度，R 为气体常数，A 为常数。因此，土壤释汞通量往往与温度呈现指数相关性（Gustin et al.，1997；王少峰等，2004；Wang et al.，2006；Haynes et al.，2017）。由于温度的影响，通常夏季土壤向大气的释汞通量远高于冬季（王少锋等，2004；Ci et al.，2016a）。

在寒冷地区，土壤存在冻融循环以及零度以下的温度循环。冻融条件下，土壤释汞通量呈现出与土壤冻融温度相似的正相关变化趋势，即在土壤冻融温度最高时，土壤释汞通量也最高（Walters et al.，2016）。随着土壤含水率的增加（4.5%～18%），冻融过程中汞释放通量随冻融温度的变化更为显著，最大汞释放通量也相应增加。但当土壤含水率达 20%时，汞释放通量随冻融温度的变化趋势被抑制，这与高含水率条件下土壤孔隙中充满水或冰，阻断了零价汞传输通道有关。据推测，冻融条件下土壤释汞通量与冻融过程中能量的输入与输出有关。零度以下温度循环时，土壤释汞通量与空气温度循环趋势相反，即空气温度最低时，土壤释汞通量最高，这一现象在土壤含水率为 20%时更为明显（Walters et al.，2016）。这表明，零度以下温度循环与冻融循环零价汞的释放机制是不同的。即使在较低的温度下，土壤中水分仍有一部分以液态形式存在。当冷空气通过土壤表面时，冷却土壤基质，降低残存液态水。随着水的冰冻，零价汞从冰晶中浓缩累积至液相，从而促进了零价汞向大气的释放。当空气温度增加时，停止了进一步冰冻，零价汞释放得到抑制。这一现象主要局限于近表层土壤。

7. 其他土壤性质

土壤释放零价汞的累积量及还原速率均随孔隙水比率先上升后下降（Pannu et al.，2014）。这表明，湿润土壤可在一定程度上促进土壤释汞通量（Park et al.，2014），但过高的土壤湿度（39%）可降低汞释放通量（Zhou et al.，2017）。

土壤有机质含量对其释汞通量的影响较为复杂。一些研究显示，模拟固相基质中零价汞的释放与基质中有机质（添加腐殖酸与富里酸）含量负相关（Mauclair et al.，2008），表明有机质可能抑制汞的还原。但也有研究显示，对于森林土壤，土壤有机质与土壤总气态汞呈现显著正相关，提示有机质对汞还原的促进作用；对于草地土壤，土壤有机质对土壤总气态汞的影响并不明显（Moore and Castro，

2012）。森林土壤汞同位素指纹分析也显示，零价汞的生成释放可能与土壤天然有机质介导的化学还原有关（Jiskra et al., 2015）。

土壤氧化还原电位降至 400 mV 以下时，可促进二价汞的还原（Moore and Castro, 2012; Hindersmann et al., 2014）。但在较低的氧化还原电位情况下，土壤中还原硫浓度也可能增加，导致硫化汞的生成，从而抑制汞的还原（Gabriel and Williamson, 2004）。

土壤渗透性可影响零价汞在土壤中的迁移。例如，渗透性较低的黏土可在很大程度上限制零价汞的迁移，从而抑制零价汞的土壤释放（Gabriel and Williamson, 2004）。

土壤中加入草酸铁与赤铁矿可显著促进零价汞的释放，表明土壤中的含铁矿物可能对汞的还原有重要的促进作用（Peretyazhko et al., 2006a）。

8. 风速

风可将土壤上空大气的零价汞迁移，从而增大土壤表层与大气中零价汞的浓度梯度，提高土壤向大气的零价汞迁移系数（Kim and Kim, 1999）。研究显示，风速由 0.2 m/s 增加至 0.8 m/s 时，对于汞浓度大于 150 μg/g 土壤，土壤释汞通量增加 2 倍；但对汞浓度小于 150 μg/g 的土壤，风速的影响并不显著（Gustin et al., 1997）。

9. 臭氧

当空气中的臭氧浓度为约 50 ppb 时，可将含二价汞的固体基质中的汞释放通量提高 1.7~51 倍，将含零价汞的固体基质中的汞释放通量降低 75%以上（Engle et al., 2005）。臭氧影响释汞通量的机制还不清楚，但可能涉及其与固相基质、汞之间的非均相反应。臭氧对含二价汞固体基质中气汞释放的增强作用与光照相当。基质高湿度可降低臭氧对汞释放的促进作用。据估计，随着未来 100 年内大气臭氧浓度的翻倍，陆地基质中零价汞的释放将增加 65%~72%（Engle et al., 2005）。

10. 生物作用

通常认为，土壤汞还原中微生物因素较为次要。放射性辐照灭菌实验显示，美国 Adirondacks 山脉落叶与针叶林土壤中气态汞释放以非生物过程为主，生物过程仅起较小的作用（Choi and Holsen, 2009）。但抗汞菌 *Acidithiobacillus ferrooxidans* SUG 2-2 的加入可显著促进土壤中氯化汞的挥发（可达 50%以上），提示在自然土壤中汞的微生物挥发仍可能占有一定比例（Takeuchi et al., 2001）。放射性辐照灭菌后，加拿大新斯科舍（Nova Scotia）森林土壤中零价汞生成累积量的对数和还原速率常数均显著低于未灭菌土壤，表明土壤微生物在汞还原中的作用不容忽视（Pannu et al.，2014）。避光条件下汞污染水滨土壤在模拟淹水条件下零价汞的生成与 Fe/Mn 的微生物异化还原过程正相关，也提示微生物在汞还原中的关键作用（Poulin et al.，2016）。目前，土壤微生物在汞还原中的重要性与普遍性尚需进一步证实，土壤中还原微生物的种类也需进一步识别。

2.3.6　底泥中汞的还原

底泥中零价汞的存在对于汞的迁移与甲基化等具有重要意义，目前关于底泥中汞的还原过程研究极少。加拿大两个湖泊底泥的热解析形态分析显示，底泥中零价汞浓度为 6.3~60.3 pg/g ww（wet weight，湿重），占总汞的 7.4%~28.4%（Bouffard and Amyot，2009）。两个湖泊底泥零价汞的浓度接近。吸附实验显示零价汞可被底泥快速吸附，20 min 零价汞的吸附可接近 100%。相较而言，夏季表层（1 cm）底泥零价汞的吸附较秋季底泥与深层（10 cm）底泥更强（Bouffard and Amyot，2009）。零价汞的吸附与底泥有机质含量正相关，与底泥粒度、上覆水 pH 及氧气浓度负相关（Bouffard and Amyot，2009）。这些底泥中的零价汞可能来自于底泥原位生成或水中零价汞的吸附。

放射性汞同位素 ^{197}Hg 示踪显示，经 20 h 培养，底泥中 ^{197}Hg(Ⅱ)可被还原，还原率为 0.13%~1.6%/g(湿底泥)（Guevara et al.，2007）。加拿大哈得孙湾（Hundson Bay）低地淡水湖泊底泥的有氧培养实验显示，底泥在有氧环境下汞的还原较低（<1%）（Brazeau et al.，2013）。生物碳源与盐（尤其是钠离子）的加入可使零价汞的生成提高 10 倍以上（Brazeau et al.，2013）。甲醛灭菌可完全抑制零价汞的生成，表明这一还原与生物作用有关。底泥中指示微生物丰度的谷氨酰胺合成酶基因的拷贝数与零价汞的生成正相关，也提示这一还原过程与微生物有关（Brazeau et al.，2013）。葡萄牙塔古斯（Tagus）河口自然公园潮滩沉积物在水淹 20~30 min 内观测到溶解气态汞的快速增加（最大 40 pg/L）（Cesario et al.，2017）。据推测，除了汞的光化学还原，溶解气态汞生成可能还与底泥间隙水中汞的微生物还原有关（Cesario et al.，2017）。

2.3.7　冰雪中汞的还原

极地地区大气汞"亏损"事件可使冰雪表层汞浓度快速增加。但加拿大北极地区与格陵兰岛冰川冰芯、雪冰浅层的总汞浓度和大气沉降通量数据显示其汞浓度低于相关其他环境介质（Zheng，2015），据推测这可能与冰雪表面沉降二价汞的还原与再释放有关（Zheng，2015）。模型分析也表明，雪中汞的还原是冰雪区大气中气态单质汞的重要来源之一，区域气态单质汞模型的构建必须考虑雪中汞的还原问题（Toyota et al.，2014）。

积雪或冰晶中汞的同位素指纹指示了汞的再还原过程。融化海冰再冰冻过程形成霜花（frost flower）中汞的同位素分析显示在这一过程中 Δ^{199}Hg 值降低，表明这一过程中冰雪表面存在沉降二价汞的光还原再释放（Sherman et al.，2012）。北极积雪中的汞同位素呈现较负的非质量分馏，结合通量箱实验，发现这一非质量分馏与日光照射引发汞的还原与再释放有关（Sherman et al.，2010）。

积雪中汞的再还原导致大气汞"亏损"事件中积累在雪中的汞浓度快速降低（Ariya et al.，2004；Durnford and Dastoor，2011）。阿拉斯加巴罗（Barrow）地区表层雪（表层1 cm）的测定则表明，在大气汞"亏损"事件停止后表层雪中汞的浓度迅速下降（Johnson et al.，2008）。通量箱实验分析显示，1天雪中汞释放量为总汞的4%～7%，这可能代表了雪中较易光还原汞的部分（Johnson et al.，2008）。加拿大安大略实验湖区积雪中的汞在沉降后24 h内即可损失超过40%（Lalonde et al.，2003）。在大气汞"亏损"事件之后的几天内，加拿大马尼托巴（Manitoba）地区积雪中总汞即可由（67.8±97.7）ng/L 降至（4.25±1.85）ng/L（Kirk et al.，2006）。随着表层雪总汞浓度的下降，积雪间隙空气中总气态汞浓度由约1.4～3.4 ng/m^{3}升高至约 20～150 ng/m^3，表明沉降二价汞还原为气态单质汞并从积雪中扩散出去（Kirk et al.，2006）。加拿大努纳武特（Nunavut）地区积雪总汞在大气汞"亏损"事件之后的2天内快速降低92%而达到背景浓度，积雪表层3 cm的光反应是导致汞还原的主要因素（Poulain et al.，2004b）。加拿大魁北克地区冰冻湖泊积雪中沉降的汞在24 h内降低54%（Lalonde et al.，2002）。据估计，北极圈内冰雪表面汞还原导致的再释放可达249～262 t/a（Dastoor et al.，2008）。

类似地，南极海冰中总汞浓度由冬季的9.7 ng/L降至春季的4.7 ng/L，而零价汞浓度由冬季的0.07 ng/L增加至夏季的0.105 ng/L，这表明在冰中可能存在汞的还原释放（Mastromonaco et al.，2016a）。

光照是引发雪表层二价汞还原的主要驱动力（Dommergue et al.，2003a；Dommergue et al.，2003b；Mann et al.，2014）。-10℃日光照射可导致溶解气态汞的显著增加，但暗反应下溶解气态汞基本不变（Lalonde et al.，2002）。这一过程可能涉及雪中的有机质、Cl$^-$及铁等［如式(2-53)～式(2-62)所示］。

$$\text{HgCl}_n^{(n-2)-} \xrightarrow{h\nu} \text{HgCl}_{n-1}^{(n-2)-} + \text{Cl}^\cdot \quad (2\text{-}53)$$

$$2\text{HgCl}_{n-1}^{(n-2)-} \longrightarrow \text{Hg}(0) + \text{HgCl}_n^{(n-2)-} + (n-2)\text{Cl}^- \quad (2\text{-}54)$$

$$\text{HCHO} \xrightarrow{h\nu} \text{H} + \text{HCO} \quad (2\text{-}55)$$

$$\text{HCO} + \text{O}_2 \longrightarrow \text{HO}_2^{\cdot-} + \text{CO} \quad (2\text{-}56)$$

$$Hg(II) + HO_2^{\bullet-} \longrightarrow Hg(0) + 产物 \qquad (2\text{-}57)$$

$$DOM—COO—Hg(II) \xrightarrow{h\nu} Hg(0) \qquad (2\text{-}58)$$

$$DOM + O_2 \xrightarrow{h\nu} O_2^{\bullet-} \qquad (2\text{-}59)$$

$$DOM + Fe(III) \xrightarrow{h\nu} Fe(II) \qquad (2\text{-}60)$$

$$Hg(II) + 2O_2^{\bullet-} \longrightarrow Hg(0) + 2O_2 \qquad (2\text{-}61)$$

$$Hg(II) + 2Fe(II) \longrightarrow Hg(0) + 2Fe(III) \qquad (2\text{-}62)$$

由于光还原的影响，加拿大魁北克 Kuujjuarapik/Whapmagoostui 与美国科罗拉多州 Niwot Ridge 不同深度积雪间隙的气态汞均存在显著的日变化特征：午后气态汞浓度达到最大值，而凌晨之后浓度最低（Dommergue et al.，2003b；Fain et al.，2013）。此外，由于夏季极地地区光照较强，这一地区夏季气态零价汞浓度也相应较高（Aspmo et al.，2006）。在积雪融化期间，积雪表层零价汞的生成也存在类似的日变化特征（Dommergue et al.，2003a）。对来自于高山、城市和极地积雪的零价汞释放通量的实验室测试显示，零价汞的释放主要受日光照射所驱动，二价汞在雪中的自然寿命仅为 4~6 h（Dommergue et al.，2007）。采用不同波长滤光片研究了日光辐照波长的作用，发现日光中的 UVB 组分在还原中的作用十分显著（Dommergue et al.，2007）。类似地，紫外遮光膜实验也证实，雪中汞的光还原主要来自于日光的 UVB 组分，而非可见光与 UVA（Lalonde et al.，2003）。紫外光照（280~400 nm）模拟实验显示，紫外光强度越强，在冰冻雪及融化雪中汞的光还原量越高（Mann et al.，2015a；Mann et al.，2015b），零价汞通量与紫外光强呈正相关（Mann et al.，2015a）。高浓度氯离子与铁可同时促进汞的光还原与光氧化，导致零价汞的生成总量较低（Mann et al.，2015b）。但在格陵兰岛 Station Nord 站点，三月份白天雪中气态单质汞浓度低于大气，这可能与活性溴对汞的光氧化有关（Ferrari et al.，2004b）。这一地区不同深度雪连续测定显示，零价气态汞的浓度在日落之后开始增加，在凌晨 2:00 左右达到最大值，在日出之前再次降低至最小值（Ferrari et al.，2004b）。美国科罗拉多州 Niwot Ridge 雪汞观测也发现，在风速较低的情况下，即使在多云的天气或夜间，雪表层气态单质汞浓度仍高于大气，表明也存在较弱的气态单质汞生成（Fain et al.，2013）。在光照下，积雪温度与雪龄可间接影响汞的光还原。实验室模拟显示，当温度增加至接近雪熔点（−2℃）时，准一级还原速率常数逐步下降，但光还原汞的总量增加 10 倍，这可能与雪：水比例的变化有关（Mann et al.，2015a）。此外，积雪的雪龄越长，午间最大汞通量越小（Mann et al.，2015a）。

积雪中汞的再还原导致在垂直方向上存在气态零价汞的浓度梯度。加拿大努纳武特地区春季大气汞"亏损"事件中沉降于雪表面汞的光致再还原过程导致雪表层气态单质汞浓度的升高（Steffen et al.，2002）。在低风速条件下，非汞"亏损"事件期间，雪表层与雪面之上 1～2 m 气态单质汞浓度基本保持不变；而在汞"亏损"事件期间，雪表层气态单质汞浓度增加至大气气态单质汞浓度的两倍以上（Steffen et al.，2002）。雪间隙空气中气态单质汞浓度较表层之上 2 m 气态单质汞浓度高 41%，表明大气中一部分气态单质汞来自于雪中汞的还原释放（Steffen et al.，2002）。美国科罗拉多州 Niwot Ridge 雪汞观测则发现，在整个冬季期间，气态单质汞均可通过光化学过程在雪-气界面生成，造成间隙空气中气态单质汞浓度升高（可达 8 ng/m^3）（Fain et al.，2013）。积雪中气态单质汞的垂直梯度分析显示表层生成的气态单质汞可向积雪下部传输，同时雪层下部也存在气态单质汞的暗氧化（Fain et al.，2013）。

除了光照因素之外，微生物也可能在冰雪汞还原中起一定的作用。北极积雪中抗汞菌占可培养细菌的 31%，包括 α-、β-和 γ-变形杆菌、厚壁菌、放线菌和拟杆菌（Moller et al.，2011）。这些分离的抗汞菌中，约 25%具有还原二价汞的能力（Moller et al.，2011）。这表明，在日光无法到达的积雪深处，这些微生物可能在雪中汞还原过程中起着一定的作用（Moller et al.，2011）。基因分析显示，*merA* 基因的水平转移是北极微生物群落适应汞浓度变化的可能机制（Moller et al.，2014）。

2.3.8 汞还原过程的生物作用

1. 微生物介导汞的还原

在二十世纪六七十年代，人们就发现多种细菌具有汞抗性（Smith，1967；Summers and Sugarman，1974；Izaki，1977）。后续研究发现，MER 位点或转座子 501（Tn501）中结构基因编码汞还原酶（merA）。这种汞还原酶是一种在多种好氧/厌氧菌的胞浆中广泛存在的可将二价汞催化还原为零价汞的氧化还原酶。汞还原酶的活性形式是同型二聚体，单体由两个结构域组成。类似金属伴侣 N 末端结构域 NmerA 含有两个半胱氨酸残基，可将其他转运蛋白（如 merT 和 MerP）或无机配体结合汞转运至 merA 的活性中心（Hong et al.，2014）。活性中心的两个半胱氨酸残基与汞离子结合，NADPH 将活性位点中的 FAD 还原为 FADH$^-$，FADH$^-$进而将半胱氨酸残基结合汞还原为零价汞，进而从细胞中挥发至胞外［式（2-63）至式（2-65）］（Osborn et al.，1997）。

$$Hg^{2+} + 2Cys\text{—}S^- \longrightarrow Cys\text{—}S\text{—}Hg\text{—}S\text{—}Cys \qquad (2\text{-}63)$$

$$FAD + NADPH \longrightarrow FADH^- + NADP^+ \qquad (2\text{-}64)$$

$$\text{Cys—S—Hg—S—Cys} + \text{FADH}^- \longrightarrow \text{H}^+ + \text{Hg}^0 + \text{FAD} + 2\text{Cys}-\text{S}^- \quad (2\text{-}65)$$

需要特别指出的是，merA 还原途径仅是微生物还原汞的机制之一，微生物还存在许多其他汞还原途径，如细胞色素 c 氧化酶途径、光合细菌电子供体途径等。

来自抗汞氧化亚铁硫杆菌（*Acidithiobacillus ferrooxidans*）质膜的细胞色素 c 氧化酶具有还原汞的能力（Sugio et al., 2001; Sugio et al., 2003）。细胞色素 c 氧化酶以 Fe^{2+} 作为底物，可将 39%的加入的二价汞（7 nmol）转化为零价汞，而加热灭活的细胞色素 c 氧化酶不具汞还原能力（Sugio et al., 2001）。在磷酸缓冲液（pH 7.4）及厌氧条件下，*G. sulfurreducens* PCA（10^{11} cell/L）可将二价汞（50 nmol/L）还原为零价汞，还原反应服从准一级动力学，二价汞半衰期小于 2 h（Hu et al., 2013b）。加热灭活细菌及细菌过滤液不能还原二价汞，外膜细胞色素酶基因敲除后的细胞可显著降低二价汞的还原速率（Hu et al., 2013b），表明这一还原过程涉及细胞色素酶。细胞表面对二价汞的吸附与汞的还原过程相竞争，因此较低二价汞：细菌比例（<10～20 mol Hg/cell）可抑制汞的还原。

光合兼养微生物为维持体内的氧化还原平衡，需将胞内累积的辅酶因子的电子传递至胞外（Gregoire and Poulain, 2014）。在光合营养条件及胞内还原辅酶过量条件下，以丁酸为碳源的紫细菌（*Rhodobacter capsulatus* SB1003、*Rhodopseudomonas palustris* TIE-1、*Rhodobacter sphaeroides*）可采用二价汞作为胞外电子受体维持体内氧化还原平衡，将二价汞还原为零价汞（Gregoire and Poulain, 2016）。而在单纯化学营养条件下生长的紫细菌不能还原二价汞。加热灭活紫细菌对二价汞的还原极低。光合营养条件下细菌生长速率随二价汞浓度增加而增加，当向系统中加入其他电子受体（如碳酸氢盐、二甲基亚砜）时，零价汞的生成受到抑制。进一步分析显示，在碳酸氢盐存在下，紫细菌胞内[NADH]/[NAD$^+$]比例下降。由于二价汞的电子受体作用，一定浓度的二价汞（200 nmol/L）可显著促进紫细菌生长。

在厌氧条件下，异化金属还原菌 *Shewanella oneidensis* MR-1 也可通过 Fe(Ⅱ)依赖的途径还原二价汞，这一过程不涉及 merA（Wiatrowski et al., 2006）。在以氧气或富马酸作为电子受体情况下，二价汞可被还原；但羟基氧化铁作为电子受体时还原速率更高。*Geobacter sulfurreducens* PCA 与 *Geobacter metallireducens* GS-15 还原二价汞的活性与 *Shewanella oneidensis* MR-1 相当，但其他异化金属还原菌 *Anaeromyxobacter dehalogenans* 2CP-C 与硝酸盐还原菌 *Pseudomonas stutzeri* OX-1 不能还原二价汞（Wiatrowski et al., 2006）。

此外一些微生物的氧化还原酶可在胞外生成超氧阴离子自由基（Diaz et al., 2013），而超氧阴离子自由基可将二价汞还原为单质汞（Aikoh, 2002），因此微生物还原汞的超氧阴离子途径也可能存在。这一途径可能在好氧与兼氧微生物还原汞的过程中起着一定的作用。

一些微生物如 *Geobacter bemidjiensis* Bem 也可将甲基汞降解为零价汞，这一降解过程可能涉及有机汞裂解酶（merB）将甲基汞降解为二价汞，汞还原酶进一步还原二价汞为零价汞（Lu et al.，2016）。

汞还原微生物在多种环境介质如水（Barkay et al.，1989；Mason et al.，1994；Vandal et al.，1995；Moller et al.，2014）、底泥（Nelson and Colwell，1974；Chadhain et al.，2006；Kannan et al.，2006；Oyetibo et al.，2015；Figueiredo et al.，2016）、土壤（Petrus et al.，2015）、冰雪（Moller et al.，2014）中广泛存在。例如海洋水体混合层存在二价汞的还原，这主要归因于水中的原核微生物等（Mason et al.，1995）；对美国威斯康星 Pallette 湖水样品加热灭活，可显著降低汞的还原，也提示这一过程与生物相关（Vandal et al.，1995）。淡水与海水微宇宙实验显示，大于 1 μm 微生物的去除对汞还原影响不大，提示汞的还原主要来自于细菌（Barkay et al.，1989）。淡水体系抗汞菌中约 50%含有 mer（Tn21）类似 DNA 序列，而海水体系抗汞菌仅约 12%含有 mer（Tn21）类似 DNA 序列，说明微生物还原汞机制的多样性（Barkay et al.，1989）。

微生物对汞的还原还可被用以去除废水、土壤中的汞污染（Hansen et al.，1984；von Canstein et al.，1999；Dey and Patke，2000；Wagner-Dobler et al.，2000；Takeuchi et al.，2001；Wagner-Dobler，2003；Leonhauser et al.，2006；Velasquez-Riano and Benavides-Otaya，2016）。转基因技术可将汞还原酶表达至其他微生物、藻类、植物等中（Chang et al.，1998）。如表达汞还原酶的真核微藻 *Chlorella* sp. DT 可将水中的汞有效挥发去除（Huang et al.，2006）。含 *merA* 转基因的多种植物在汞污染土壤的挥发去除方面也有极大的潜力（Rugh et al.，1996；Rugh et al.，1998；Che et al.，2003；Hussein et al.，2007；Haque et al.，2010）。类似地，固定化的工程菌生产的汞还原酶也可用于水中汞的去除（Chang et al.，1999）。

2. 微藻介导汞的还原

早在 1972 年，人们就观测到衣藻随二价汞暴露的适应性，从而推测藻类可将二价汞还原为挥发性的零价汞（Benbassat et al.，1972）。随后，发现小球藻也可显著降低培养介质中的 $HgCl_2$ 浓度，这一降低作用与小球藻接种量有关（Benbassat and Mayer，1975）。小球藻对二价汞的还原在黑暗条件下较弱，但光照可显著提高零价汞的生成与挥发，且随光照时间延长而增加（Yamamoto et al.，1999）。在以 $HgBr_2$ 形态为主的溴离子浓度下，零价汞的挥发最为显著（Yamamoto et al.，1999）。类似地，在金卤灯光照（$\lambda>365$ nm）下，柱胞鱼腥藻、小球藻与菱形硅藻对二价汞的光还原产物溶解气态汞随光照时间逐渐增加,而后达到稳态(Deng et al.,2008；Deng et al.,2009；Deng et al.,2010)。对柱胞鱼腥藻、小球藻、菱形硅藻,这一还原过程对二价汞的反应级数分别为1、0.39与0.76。三种微藻对二价汞的还原能力为：柱胞鱼腥藻＞菱形硅藻＞小球藻(Deng et al.,2015)。这一光还原过程随藻浓度与溶液pH增加而增加。藻悬浮液中三价铁与腐殖质的加入可进一步促进二价汞的还原。据推测,汞的还原与光化学过程中生成的有机自由基以及超氧自由基有关［式(2-66)～式(2-72)］。

$$Fe(III) + 天然有机酸 \longrightarrow Fe(III)-Org \qquad (2\text{-}66)$$

$$Fe(III)-Org + h\nu \longrightarrow Fe(II) + 有机自由基 \qquad (2\text{-}67)$$

$$有机自由基 + Hg(II) \longrightarrow Hg(0) + 产物 + CO_2 \qquad (2\text{-}68)$$

$$有机自由基 + O_2 \longrightarrow O_2^{\cdot -} + 产物 + CO_2 \qquad (2\text{-}69)$$

$$DOM + h\nu \longrightarrow DOM^{\cdot} \qquad (2\text{-}70)$$

$$DOM^{\cdot} + O_2 \longrightarrow DOM^{\cdot +} + O_2^{-} \qquad (2\text{-}71)$$

$$O_2^{\cdot -} + Hg(II) \longrightarrow Hg(0) + O_2 \qquad (2\text{-}72)$$

现场实验显示,湖水与海水中的溶解气态汞的生成与微藻水华的强度与持续时间有关(Poulain et al.,2004a；Kuss et al.,2015)。原位培养实验显示,暗反应、过滤处理或加入光合成抑制剂均可抑制溶解气态汞的生成(Poulain et al.,2004a)。波罗的海海水中零价汞的生成遵循准一级动力学,来自于光化学过程、光依赖的生物过程以及低光过程(主要来自于生物过程)的贡献分别占总零价汞生成的30%、30%与40%(Kuss et al.,2015)。现场分析显示,蓝藻中的聚球藻与束丝藻在汞的生物还原中起着重要作用(Kuss et al.,2015)。

通常认为,微藻对二价汞的还原是其重要的脱毒机制之一。但关于微藻还原二价汞的过程与机制尚不清楚。一些研究认为,微藻对二价汞还原主要来自于微藻生成的一些生物源有机物,如小球藻中的低分子量热稳定性组分(Benbassat and Mayer,1977)以及角毛藻产生的生物源有机物(Lanzillotta et al.,2004)。但也有一些研究显示甲醛灭活的藻细胞不能还原生成溶解气态汞,提示这一过程可能与微藻的氧化还原活性酶有关(Morelli et al.,2009)。此外,低光照下汞的还原(Kuss et al.,2015)以及光合成抑制剂(Poulain et al.,2004a)对汞还原的抑制也显示光能营养型微藻生物活性对汞还原的关键作用(Gregoire and Poulain,2014)。微藻对汞的还原可能涉及两步单电子转移过程。扫描电镜与能谱分析显示未处理与酸处理海黍子马尾藻吸附二价汞后,在藻表面均可观测到零价汞与Hg_2Cl_2颗粒(Carro et al., 2011)。

2.4 展　　望

　　近些年来，环境中汞的氧化还原过程研究取得了长足的进展，但仍有一系列关键问题有待进一步研究与解决。①非均相微界面氧化还原：大气与水体中汞的氧化还原反应并非局限于均相反应。目前，人们对大气与水体中非均相微界面参与的汞氧化还原还知之甚少，研究有待加强。②大气氧化性物种的生成与时空分布：大气中氧化性物种的时空分布特征研究对于理解其在汞氧化中的作用至关重要。尚需对大气中氧化性物种（尤其是活性卤素）的生成与时空分布进行深入研究。③实验室模拟应充分反映环境大气与水体条件：接近真实条件下实验室模拟获得的汞氧化还原参数才能进一步用于模型模拟与现场数据的阐释。④环境中"新"的汞形态：以往对另一环境中的汞形态"一价汞"研究不足。一价汞可能是汞氧化还原过程中的重要中间体。应发展多种环境介质中一价汞的分析方法，阐明环境中一价汞的赋存及其在汞氧化还原转化中扮演的角色。新近也发现环境水体中颗粒结合态零价汞的存在。这一"新"的汞形态对汞的氧化还原、挥发可能具有重要影响，其环境意义需要进一步阐明。⑤同位素技术的运用：由于汞的氧化还原反应通常同时发生，多同位素示踪技术的运用可解析同一体系中汞氧化/还原反应的发生程度，有助于对其过程与机制的深入理解。此外，汞同位素组成精确测定与汞同位素分馏特征分析也为汞氧化还原提供了新的工具与研究视角。在相关研究中，应注意将实验室模拟与现场测试紧密结合，以识别不同氧化还原反应过程在汞转化中的重要性。

参 考 文 献

高锦玉，王昊，蔡武，吴靖霆，何云峰，2016. 土壤释汞通量影响因素研究进展. 地球与环境，44: 261-269.

李希嘉，钟紫旋，孙荣国，杨鲲，王定勇，2014. 不同波长和强度光照对水体汞还原的影响. 环境科学，35: 1788-1792.

李仲根，冯新斌，汤顺林，王少峰，2006. 封闭式城市生活垃圾填埋场向大气释放汞的途径. 环境科学，27: 19-23.

刘丽琼，江韬，魏世强，张艳敏，李雪梅，2011. 胡敏酸对不同矿物结合汞的还原作用. 中国环境科学，31: 1998-2004.

马学琳，刘汝海，魏莱，王艳，刘诗璇，2015. 紫外光对冬季近岸海水中溶解性气态汞产生的影响. 中国环境科学，35: 3462-3467.

王少锋，冯新斌，仇广乐，付学吾，2004. 贵州红枫湖地区冷暖两季土壤/大气界面间汞交换通量的对比. 环境科学, 25: 123-127.
赵士波, 孙荣国, 王定勇, 王小文, 张成, 2014. 低分子有机酸对汞氧化还原反应的影响. 环境科学, 35: 2193-2200.
Ababneh F A, Scott S L, Al-Reasi H A, Lean D R S, 2006. Photochemical reduction and reoxidation of aqueous mercuric chloride in the presence of ferrioxalate and air. Science of the Total Environment, 367: 831-839.
Agarwal H, Stenger H G, Wu S, Fan Z, 2006. Effects of H_2O, SO_2, and NO on homogeneous Hg oxidation by Cl_2. Energy & Fuels, 20: 1068-1075.
Agnan Y, Le Dantec T, Moore C W, Edwards G C, Obrist D, 2016. New constraints on terrestrial surface atmosphere fluxes of gaseous elemental mercury using a global database. Environmental Science & Technology, 50: 507-524.
Ahn M C, Kim B, Holsen T M, Yi S M, Han Y J, 2010. Factors influencing concentrations of dissolved gaseous mercury (DGM) and total mercury (TM) in an artificial reservoir. Environmental Pollution, 158: 347-355.
Aikoh H, 2002. Reduction of mercuric ion *in vitro* by superoxide anion. Physiological Chemistry and Physics and Medical NMR, 34: 185-189.
Alberts J J, Schindler J E, Miller R W, Nutter D E, 1974. Elemental mercury evolution mediated by humic acid. Science, 184: 895-896.
Allard B, Arsenie I, 1991. Abiotic reduction of mercury by humic substances in Aquatic system: An important process for the mercury cycle. Water Air and Soil Pollution, 56: 457-464.
Amirbahman A, Kent D B, Curtis G P, Marvin-Dipasquale M C, 2013. Kinetics of homogeneous and surface-catalyzed mercury(II) reduction by iron(II). Environmental Science & Technology, 47: 7204-7213.
Amyot M, Gill G A, Morel F M M, 1997a. Production and loss of dissolved gaseous mercury in coastal seawater. Environmental Science & Technology, 31: 3606-3611.
Amyot M, Lean D, Mierle G, 1997b. Photochemical formation of volatile mercury in high Arctic lakes. Environmental Toxicology and Chemistry, 16: 2054-2063.
Amyot M, Lean D R S, Poissant L, Doyon M R, 2000. Distribution and transformation of elemental mercury in the St. Lawrence River and Lake Ontario. Canadian Journal of Fisheries and Aquatic Sciences, 57: 155-163.
Amyot M, Mierle G, Lean D, Mcqueen D J, 1997c. Effect of solar radiation on the formation of dissolved gaseous mercury in temperate lakes. Geochimica Et Cosmochimica Acta, 61: 975-987.
Amyot M, Mierle G, Lean D R S, Mcqueen D J, 1994. Sunlight-induced formation of dissolved gaseous mercury in lake waters. Environmental Science & Technology, 28: 2366-2371.
Amyot M, Morel F M M, Ariya P A, 2005. Dark oxidation of dissolved and liquid elemental mercury in aquatic environments. Environmental Science & Technology, 39: 110-114.
Amyot M, Southworth G, Lindberg S E, Hintelmann H, Lalonde J D, Ogrinc N, Poulain A J, Sandilands K A, 2004. Formation and evasion of dissolved gaseous mercury in large enclosures amended with ($HgCl_2$)-Hg-200. Atmospheric Environment, 38: 4279-4289.
Andersson M E, Gardfeldt K, Wangberg I, Stromberg D, 2008. Determination of Henry's law constant for elemental mercury. Chemosphere, 73: 587-592.
Ariya P A, Amyot M, Dastoor A, Deeds D, Feinberg A, Kos G, Poulain A, Ryjkov A, Semeniuk K, Subir M, Toyota K, 2015. Mercury physicochemical and biogeochemical transformation in the atmosphere and at atmospheric interfaces: A review and future directions. Chemical Reviews,

115: 3760-3802.

Ariya P A, Dastoor A P, Amyot M, Schroeder W H, Barrie L, Anlauf K, Raofie F, Ryzhkov A, Davignon D, Lalonde J, Steffen A, 2004. The Arctic: A sink for mercury. Tellus Series B-Chemical and Physical Meteorology, 56: 397-403.

Ariya P A, Khalizov A, Gidas A, 2002. Reactions of gaseous mercury with atomic and molecular halogens: Kinetics, product studies, and atmospheric implications. Journal of Physical Chemistry A, 106: 7310-7320.

Aspmo K, Temme C, Berg T, Ferrari C, Gauchard P A, Fain X, Wibetoe G, 2006. Mercury in the atmosphere, snow and melt water ponds in the North Atlantic Ocean during Arctic summer. Environmental Science & Technology, 40: 4083-4089.

Auzmendi-Murua I, Castillo A, Bozzelli J W, 2014. Mercury oxidation via chlorine, bromine, and iodine under atmospheric conditions: Thermochemistry and kinetics. Journal of Physical Chemistry A, 118: 2959-2975.

Bagnato E, Barra M, Cardellini C, Chiodini G, Parello F, Sprovieri M, 2014. First combined flux chamber survey of mercury and CO_2 emissions from soil diffuse degassing at Solfatara of Pozzuoli crater, Campi Flegrei (Italy): Mapping and quantification of gas release. Journal of Volcanology and Geothermal Research, 289: 26-40.

Bargagli R, Agnorelli C, Borghini F, Monaci F, 2005. Enhanced deposition and bioaccumulation of mercury in Antarctic terrestrial ecosystems facing a coastal polynya. Environmental Science & Technology, 39: 8150-8155.

Barkay T, Liebert C, Gillman M, 1989. Environmental significance of the potential for mer(Tn21)-mediated reduction of Hg^{2+} to Hg^0 in natural waters. Applied and Environmental Microbiology, 55: 1196-1202.

Bash J O, Carlton A G, Hutzell W T, Bullock O R, 2014. Regional air quality model application of the aqueous-phase photo reduction of atmospheric oxidized mercury by dicarboxylic acids. Atmosphere, 5: 1-15.

Batrakova N, Travnikov O, Rozovskaya O, 2014. Chemical and physical transformations of mercury in the ocean: A review. Ocean Science, 10: 1047-1063.

Bauer D, D'ottone L, Campuzano-Jost P, Hynes A J, 2003. Gas phase elemental mercury: A comparison of LIF detection techniques and study of the kinetics of reaction with the hydroxyl radical. Journal of Photochemistry and Photobiology A-Chemistry, 157: 247-256.

Benbassat D, Gruner N, Shuval H I, Shelef G, 1972. Growth of chlamydomonas in a medium containing mercury. Nature, 240: 43-44.

Benbassat D, Mayer A M, 1977. Reduction of mercury-chloride by chlorella: Evidence for a reducing factor. Physiologia Plantarum, 40: 157-162.

Benbassat D, Mayer A M, 1975. Volatilization of mercury by algae. Physiologia Plantarum, 33: 128-131.

Bergan T, Rodhe H, 2001. Oxidation of elemental mercury in the atmosphere; Constraints imposed by global scale modelling. Journal of Atmospheric Chemistry, 40: 191-212.

Berkovic A M, Bertolotti S G, Villata L S, Gonzalez M C, Diez R P, Martire D O, 2012. Photoinduced reduction of divalent mercury by quinones in the presence of formic acid under anaerobic conditions. Chemosphere, 89: 1189-1194.

Berkovic A M, Gonzalez M C, Russo N, Michelini M D C, Pis Diez R, Martire D O, 2010. Reduction of mercury (II) by the carbon dioxide radical anion: A theoretical and experimental investigation.

Journal of Physical Chemistry A, 114: 12845-12850.

Beucher C, Wong-Wah-Chung P, Richard C, Mailhot G, Bolte M, Cossa D, 2002. Dissolved gaseous mercury formation under UV irradiation of unamended tropical waters from French Guyana. Science of the Total Environment, 290: 131-138.

Bone S E, Bargar J R, Sposito G, 2014. Mackinawite (FeS) reduces mercury (II) under sulfidic conditions. Environmental Science & Technology, 48: 10681-10689.

Boszke L, Kowalski A, Astel A, Baranski A, Gworek B, Siepak J, 2008. Mercury mobility and bioavailability in soil from contaminated area. Environmental Geology, 55: 1075-1087.

Bouffard A, Amyot M, 2009. Importance of elemental mercury in lake sediments. Chemosphere, 74: 1098-1103.

Brazeau M L, Blais J M, Paterson A M, Keller W, Poulain A J, 2013. Evidence for microbially mediated production of elemental mercury (Hg^0) in subarctic lake sediments. Applied Geochemistry, 37: 142-148.

Burd J A, Peterson P K, Nghiem S V, Perovich D K, Simpson W R, 2017. Snowmelt onset hinders bromine monoxide heterogeneous recycling in the Arctic. Journal of Geophysical Research-Atmospheres, 122: 8297-8309.

Byrne H E, Borello A, Bonzongo J-C, Mazyck D W, 2009. Investigations of photochemical transformations of aqueous mercury: Implications for water effluent treatment technologies. Water Research, 43: 4278-4284.

Calvert J G, Lindberg S E, 2005. Mechanisms of mercury removal by O_3 and OH in the atmosphere. Atmospheric Environment, 39: 3355-3367.

Calvert J G, Lindberg S E, 2004. The potential influence of iodine-containing compounds on the chemistry of the troposphere in the polar spring. II. Mercury depletion. Atmospheric Environment, 38: 5105-5116.

Carpi A, Lindberg S E, 1997. Sunlight-mediated emission of elemental mercury from soil amended with municipal sewage sludge. Environmental Science & Technology, 31: 2085-2091.

Carro L, Barriada J L, Herrero R, De Vicente M E S, 2011. Adsorptive behaviour of mercury on algal biomass: Competition with divalent cations and organic compounds. Journal of Hazardous Materials, 192: 284-291.

Castelle S, Schaefer J, Blanc G, Dabrin A, Lanceleur L, Masson M, 2009. Gaseous mercury at the air-water interface of a highly turbid estuary (Gironde Estuary, France). Marine Chemistry, 117: 42-51.

Castro L, Dommergue A, Ferrari C, Maron L, 2009. A DFT study of the reactions of O_3 with Hg degrees or Br. Atmospheric Environment, 43: 5708-5711.

Castro M S, Sherwell J, 2015. Effectiveness of emission controls to reduce the atmospheric concentrations of mercury. Environmental Science & Technology, 49: 14000-14007.

Cesario R, Poissant L, Pilote M, O'driscoll N J, Mota A M, Canario J, 2017. Dissolved gaseous mercury formation and mercury volatilization in intertidal sediments. Science of the Total Environment, 603: 279-289.

Chadhain S M N, Schaefer J K, Crane S, Zylstra G J, Barkay T, 2006. Analysis of mercuric reductase (merA) gene diversity in an anaerobic mercury-contaminated sediment enrichment. Environmental Microbiology, 8: 1746-1752.

Chakraborty P, Vudamala K, Coulibaly M, Ramteke D, Chennuri K, Lean D, 2015. Reduction of mercury(II) by humic substances-influence of pH, salinity of aquatic system. Environmental

Science and Pollution Research, 22: 10529-10538.

Chand D, Jaffe D, Prestbo E, Swartzendruber P C, Hafner W, Weiss-Penzias P, Kato S, Takami A, Hatakeyama S, Kajii Y Z, 2008. Reactive and particulate mercury in the Asian marine boundary layer. Atmospheric Environment, 42: 7988-7996.

Chang J S, Chao Y P, Fong Y M, Hwang Y P, Lin P J, 1998. Cloning of mercury resistance determinants in *Escherichia coli* and analysis of mercury reduction activity *in vivo* and *in vitro*. Journal of the Chinese Institute of Chemical Engineers, 29: 265-274.

Chang J S, Hwang Y P, Fong Y M, Lin P J, 1999. Detoxification of mercury by immobilized mercuric reductase. Journal of Chemical Technology and Biotechnology, 74: 965-973.

Che D S, Meagher R B, Heaton A C P, Lima A, Rugh C L, Merkle S A, 2003. Expression of mercuric ion reductase in Eastern cottonwood (*Populus deltoides*) confers mercuric ion reduction and resistance. Plant Biotechnology Journal, 1: 311-319.

Chen J B, Hintelmann H, Feng X B, Dimock B, 2012. Unusual fractionation of both odd and even mercury isotopes in precipitation from Peterborough, ON, Canada. Geochimica Et Cosmochimica Acta, 90: 33-46.

Chi Y, Yan N Q, Qu Z, Qiao S H, Jia J P, 2009. The performance of iodine on the removal of elemental mercury from the simulated coal-fired flue gas. Journal of Hazardous Materials, 166: 776-781.

Choi H D, Holsen T M, 2009. Gaseous mercury emissions from unsterilized and sterilized soils: The effect of temperature and UV radiation. Environmental Pollution, 157: 1673-1678.

Ci Z J, Peng F, Xue X A, Zhang X S, 2016a. Air-surface exchange of gaseous mercury over permafrost soil: An investigation at a high-altitude (4700 m a.s.l.) and remote site in the central Qinghai-Tibet Plateau. Atmospheric Chemistry and Physics, 16: 14741-14754.

Ci Z J, Zhang X S, Yin Y G, Chen J S, Wang S W, 2016b. Mercury redox chemistry in waters of the Eastern Asian Seas: From polluted coast to clean open ocean. Environmental Science & Technology, 50: 2371-2380.

Coburn S, Dix B, Edgerton E, Holmes C D, Kinnison D, Liang Q, Ter Schure A, Wang S Y, Volkamer R, 2016. Mercury oxidation from bromine chemistry in the free troposphere over the southeastern US. Atmospheric Chemistry and Physics, 16: 3743-3760.

Colombo M J, Ha J, Reinfelder J R, Barkay T, Yee N, 2013. Anaerobic oxidation of Hg(0) and methylmercury formation by *Desulfovibrio desulfuricans* ND132. Geochimica Et Cosmochimica Acta, 112: 166-177.

Colombo M J, Ha J, Reinfelder J R, Barkay T, Yee N, 2014. Oxidation of Hg(0) to Hg(II) by diverse anaerobic bacteria. Chemical Geology, 363: 334-340.

Costa M, Liss P, 2000. Photoreduction and evolution of mercury from seawater. Science of the Total Environment, 261: 125-135.

Dastoor A P, Davignon D, Theys N, Van Roozendael M, Steffen A, Ariya P A, 2008. Modeling dynamic exchange of gaseous elemental mercury at polar sunrise. Environmental Science & Technology, 42: 5183-5188.

De Foy B, Tong Y D, Yin X F, Zhang W, Kang S C, Zhang Q G, Zhang G S, Wang X J, Schauer J J, 2016. First field-based atmospheric observation of the reduction of reactive mercury driven by sunlight. Atmospheric Environment, 134: 27-39.

Deng L, Deng N S, Mou L W, Zhu F T, 2010. Photo-induced transformations of Hg(II) species in the presence of *Nitzschia hantzschiana*, ferric ion, and humic acid. Journal of Environmental

Sciences, 22: 76-83.
Deng L, Fu D F, Deng N S, 2009. Photo-induced transformations of mercury(II) species in the presence of algae, Chlorella vulgaris. Journal of Hazardous Materials, 164: 798-805.
Deng L, Shi J C, Yang C Q, Deng N S, 2015. Photoreduction of mercury(II) in aqueous suspensions of different algae. Fresenius Environmental Bulletin, 24: 324-334.
Deng L, Wu F, Deng N S, Zuo Y G, 2008. Photoreduction of mercury(II) in the presence of algae, *Anabaena cylindrical*. Journal of Photochemistry and Photobiology B-Biology, 91: 117-124.
Dey S, Patke D S, 2000. Mercury biotransformation and its potential for remediation of mercury contamination in water. Journal of Environmental Biology, 21: 47-54.
Diaz J M, Hansel C M, Voelker B M, Mendes C M, Andeer P F, Zhang T, 2013. Widespread production of extracellular superoxide by heterotrophic bacteria. Science, 340: 1223-1226.
Dibble T S, Zelie M J, Mao H, 2012. Thermodynamics of reactions of ClHg and BrHg radicals with atmospherically abundant free radicals. Atmospheric Chemistry and Physics, 12: 10271-10279.
Do Valle C M, Santana G P, Windmoller C C, 2006. Mercury conversion processes in Amazon soils evaluated by thermodesorption analysis. Chemosphere, 65: 1966-1975.
Dommergue A, Bahlmann E, Ebinghaus R, Ferrari C, Boutron C, 2007. Laboratory simulation of Hg(0) emissions from a snowpack. Analytical and Bioanalytical Chemistry, 388: 319-327.
Dommergue A, Barret M, Courteaud J, Cristofanelli P, Ferrari C P, Gallee H, 2012. Dynamic recycling of gaseous elemental mercury in the boundary layer of the Antarctic Plateau. Atmospheric Chemistry and Physics, 12: 11027-11036.
Dommergue A, Ferrari C P, Gauchard P A, Boutron C F, Poissant L, Pilote M, Jitaru P, Adams F C, 2003a. The fate of mercury species in a Sub-Arctic snowpack during snowmelt. Geophysical Research Letters, 30: 1621.
Dommergue A, Ferrari C P, Magand O, Barret M, Gratz L E, Pirrone N, Sprovieri F, 2013. Monitoring of gaseous elemental mercury in central Antarctica at Dome Concordia. Proceedings of the 16th International Conference on Heavy Metals in the Environment, 1: 17003.
Dommergue A, Ferrari C P, Poissant L, Gauchard P A, Boutron C F, 2003b. Diurnal cycles of gaseous mercury within the snowpack at Kuujjuarapik/Whapmagoostui, Quebec, Canada. Environmental Science & Technology, 37: 3289-3297.
Duan L, Cheng N, Xiu G L, Wang F J, Chen Y, 2017. Characteristics and source appointment of atmospheric particulate mercury over East China Sea: Implication on the deposition of atmospheric particulate mercury in marine environment. Environmental Pollution, 224: 26-34.
Dunn J D, Clarkson T W, Magos L, 1981. Ethanol reveals novel mercury detoxification step in tissues. Science, 213: 1123-1125.
Durao W A, Palmieri H E L, Trindade M C, Branco O E D, Carvalho C A, Fleming P M, Da Silva J B B, Windmoller C C, 2009. Speciation, distribution, and transport of mercury in contaminated soils from Descoberto, Minas Gerais, Brazil. Journal of Environmental Monitoring, 11: 1056-1063.
Durnford D, Dastoor A, 2011. The behavior of mercury in the cryosphere: A review of what we know from observations. Journal of Geophysical Research-Atmospheres, 116: D06305.
Ebinghaus R, Kock H H, Temme C, Einax J W, Lowe A G, Richter A, Burrows J P, Schroeder W H, 2002. Antarctic springtime depletion of atmospheric mercury. Environmental Science & Technology, 36: 1238-1244.
Ebinghaus R, Temme C, Lindberg S E, Scott K J, 2004. Springtime accumulation of atmospheric

mercury in polar ecosystems. Journal De Physique Iv, 121: 195-208.

Eckley C S, Blanchard P, Mclennan D, Mintz R, Sekela M, 2015. Soil-air mercury flux near a large industrial emission source before and after closure (Flin Flon, Manitoba, Canada). Environmental Science & Technology, 49: 9750-9757.

Eckley C S, Tate M T, Lin C J, Gustin M, Dent S, Eagles-Smith C, Lutz M A, Wickland K P, Wang B, Gray J E, Edwards G C, Krabbenhoft D P, Smith D B, 2016. Surface-air mercury fluxes across Western North America: A synthesis of spatial trends and controlling variables. Science of the Total Environment, 568: 651-665.

Edgerton E S, Hartsell B E, Jansen J J, 2006. Mercury speciation in coal-fired power plant plumes observed at three surface sites in the southeastern US. Environmental Science & Technology, 40: 4563-4570.

Engle M A, Gustin M S, Lindberg S E, Gertler A W, Ariya P A, 2005. The influence of ozone on atmospheric emissions of gaseous elemental mercury and reactive gaseous mercury from substrates. Atmospheric Environment, 39: 7506-7517.

Fain X, Ferrari C P, Dommergue A, Albert M, Battle M, Arnaud L, Barnola J M, Cairns W, Barbante C, Boutron C, 2008. Mercury in the snow and firn at Summit Station, Central Greenland, and implications for the study of past atmospheric mercury levels. Atmospheric Chemistry and Physics, 8: 3441-3457.

Fain X, Ferrari C P, Gauchard P A, Magand O, Boutron C, 2006. Fast depletion of gaseous elemental mercury in the Kongsvegen glacier snowpack in Svalbard. Geophysical Research Letters, 33: L06826.

Fain X, Helmig D, Hueber J, Obrist D, Williams M W, 2013. Mercury dynamics in the Rocky Mountain, Colorado, snowpack. Biogeosciences, 10: 3793-3807.

Fantozzi L, Ferrara R, Frontini F P, Dini F, 2007. Factors influencing the daily behaviour of dissolved gaseous mercury concentration in the Mediterranean Sea. Marine Chemistry, 107: 4-12.

Fantozzi L, Ferrarac R, Frontini F P, Dini F, 2009. Dissolved gaseous mercury production in the dark: Evidence for the fundamental role of bacteria in different types of Mediterranean water bodies. Science of the Total Environment, 407: 917-924.

Farman J C, Gardiner B G, Shanklin J D, 1985. Large losses of total ozone in Antarctica reveal seasonal ClO_x/NO_x interaction. Nature, 315: 207-210.

Feinberg A I, Kurien U, Ariya P A, 2015. The kinetics of aqueous mercury(II) reduction by sulfite over an array of environmental conditions. Water Air and Soil Pollution, 226.

Ferrari C P, Dommergue A, Boutron C F, Jitaru P, Adams F C, 2004a. Profiles of mercury in the snow pack at Station Nord, Greenland shortly after polar sunrise. Geophysical Research Letters, 31: L03401.

Ferrari C P, Dommergue A, Boutron C F, Skov H, Goodsite M, Jensen B, 2004b. Nighttime production of elemental gaseous mercury in interstitial air of snow at Station Nord, Greenland. Atmospheric Environment, 38: 2727-2735.

Ferrari C P, Padova C, Fain X, Gauchard P-A, Dommergue A, Aspmo K, Berg T, Cairns W, Barbante C, Cescon P, Kaleschke L, Richter A, Wittrock F, Boutron C, 2008. Atmospheric mercury depletion event study in Ny-Alesund (Svalbard) in spring 2005. Deposition and transformation of Hg in surface snow during springtime. Science of the Total Environment, 397: 167-177.

Figueiredo N L, Canario J, O'driscoll N J, Duarte A, Carvalho C, 2016. Aerobic mercury-resistant bacteria alter mercury speciation and retention in the Tagus Estuary(Portugal). Ecotoxicology and

Environmental Safety, 124: 60-67.

Fu X, Feng X, Zhang H, Yu B, Chen L, 2012. Mercury emissions from natural surfaces highly impacted by human activities in Guangzhou province, South China. Atmospheric Environment, 54: 185-193.

Fu X W, Zhu W, Zhang H, Sommar J, Yu B, Yang X, Wang X, Lin C J, Feng X B, 2016. Depletion of atmospheric gaseous elemental mercury by plant uptake at Mt. Changbai, Northeast China. Atmospheric Chemistry and Physics, 16: 12861-12873.

Gabriel M C, Williamson D G, 2004. Principal biogeochemical factors affecting the speciation and transport of mercury through the terrestrial environment. Environmental Geochemistry and Health, 26: 421-434.

Gai K, Hoelen T P, Hsu-Kim H, Lowry G V, 2016. Mobility of four common mercury species in model and natural unsaturated soils. Environmental Science & Technology, 50: 3342-3351.

Garcia E, Poulain A J, Amyot M, Ariya P A, 2005. Diel variations in photoinduced oxidation of Hg-0 in freshwater. Chemosphere, 59: 977-981.

Gardfeldt K, Jonsson M, 2003. Is bimolecular reduction of Hg(II) complexes possible in aqueous systems of environmental importance. Journal of Physical Chemistry A, 107: 4478-4482.

Gardfeldt K, Sommar J, Stromberg D, Feng X B, 2001. Oxidation of atomic mercury by hydroxyl radicals and photoinduced decomposition of methylmercury in the aqueous phase. Atmospheric Environment, 35: 3039-3047.

Gillis A A, Miller D R, 2000. Some local environmental effects on mercury emission and absorption at a soil surface. Science of the Total Environment, 260: 191-200.

Gionfriddo C M, Tate M T, Wick R R, Schultz M B, Zemla A, Thelen M P, Schofield R, Krabbenhoft D P, Holt K E, Moreau J W, 2016. Microbial mercury methylation in Antarctic sea ice. Nature Microbiology, 1: 16127.

Gonzalez-Raymat H, Liu G L, Liriano C, Li Y B, Yin Y G, Shi J B, Jiang G B, Cai Y, 2017. Elemental mercury: Its unique properties affect its behavior and fate in the environment. Environmental Pollution, 229: 69-86.

Goodsite M E, Plane J M C, Skov H, 2004. A theoretical study of the oxidation of Hg(0) to $HgBr_2$ in the troposphere. Environmental Science & Technology, 38: 1772-1776.

Gregoire D S, Poulain A J, 2014. A little bit of light goes a long way: The role of phototrophs on mercury cycling. Metallomics, 6: 396-407.

Gregoire D S, Poulain A J, 2016. A physiological role for Hg^{II} during phototrophic growth. Nature Geoscience, 9: 121-125.

Gu B H, Bian Y R, Miller C L, Dong W M, Jiang X, Liang L Y, 2011. Mercury reduction and complexation by natural organic matter in anoxic environments. Proceedings of the National Academy of Sciences of the United States of America, 108: 1479-1483.

Guevara S R, Zizek S, Repinc U, Catan S P, Jacimovic R, Horvat M, 2007. Novel methodology for the study of mercury methylation and reduction in sediments and water using Hg-197 radiotracer. Analytical and Bioanalytical Chemistry, 387: 2185-2197.

Gustin M S, Biester H, Kim C S, 2002. Investigation of the light-enhanced emission of mercury from naturally enriched substrates. Atmospheric Environment, 36: 3241-3254.

Gustin M S, Coolbaugh M F, Engle M A, Fitzgerald B C, Keislar R E, Lindberg S E, Nacht D M, Quashnick J, Rytuba J J, Sladek C, Zhang H, Zehner R E, 2003. Atmospheric mercury emissions from mine wastes and surrounding geologically enriched terrains. Environmental Geology, 43:

339-351.

Gustin M S, Taylor G E, Maxey R A, 1997. Effect of temperature and air movement on the flux of elemental mercury from substrate to the atmosphere. Journal of Geophysical Research-Atmospheres, 102: 3891-3898.

Ha J Y, Zhao X H, Yu R Q, Barkay T, Yee N, 2017. Hg(II) reduction by siderite ($FeCO_3$). Applied Geochemistry, 78: 211-218.

Halbach S, Ballatori N, Clarkson T W, 1988. Mercury-vapor uptake and hydrogen-peroxide detoxification in human and mouse red blood-cells. Toxicology and Applied Pharmacology, 96: 517-524.

Halbach S, Clarkson T W, 1978. Enzymatic oxidation of mercury-vapor by erythrocytes. Biochimica Et Biophysica Acta, 523: 522-531.

Hall B, 1995. The gas-phase oxidation of elemental mercury by ozone. Water Air and Soil Pollution, 80: 301-315.

Hall B B, Bloom N S, 1993. Annual Report to the Electric Power Research Institute. Palo Alto, CA.

Han Y, Huh Y, Hur S D, Hong S, Chung J W, Motoyama H, 2017a. Net deposition of mercury to the Antarctic Plateau enhanced by sea salt. Science of the Total Environment, 583: 81-87.

Han Y J, Holsen T M, Lai S O, Hopke P K, Yi S M, Liu W, Pagano J, Falanga L, Milligan M, Andolina C, 2004. Atmospheric gaseous mercury concentrations in New York State: Relationships with meteorological data and other pollutants. Atmospheric Environment, 38: 6431-6446.

Han Y J, Kim J E, Kim P R, Kim W J, Yi S M, Seo Y S, Kim S H, 2014. General trends of atmospheric mercury concentrations in urban and rural areas in Korea and characteristics of high-concentration events. Atmospheric Environment, 94: 754-764.

Han Y J, Kim P R, Lee G S, Lee J I, Noh S, Yu S M, Park K S, Seok K S, Kim H, Kim Y H, 2017b. Mercury concentrations in environmental media at a hazardous solid waste landfill site and mercury emissions from the site. Environmental Earth Sciences, 76: 361.

Hansen C L, Zwolinski G, Martin D, Williams J W, 1984. Bacterial removal of mercury from sewage. Biotechnology and Bioengineering, 26: 1330-1333.

Haque S, Zeyaullah M, Nabi G, Srivastava P S, Ali A, 2010. Transgenic tobacco plant expressing environmental *E. coli merA* gene for enhanced volatilization of ionic mercury. Journal of Microbiology and Biotechnology, 20: 917-924.

Haynes K M, Kane E S, Potvin L, Lilleskov E A, Kolka R K, Mitchell C P J, 2017. Gaseous mercury fluxes in peatlands and the potential influence of climate change. Atmospheric Environment, 154: 247-259.

He F, Zhao W R, Liang L Y, Gu B H, 2014. Photochemical oxidation of dissolved elemental mercury by carbonate radicals in water. Environmental Science & Technology Letters, 1: 499-503.

He F, Zheng W, Liang L Y, Gu B H, 2012. Mercury photolytic transformation affected by low-molecular-weight natural organics in water. Science of the Total Environment, 416: 429-435.

Hedgecock I M, Pirrone N, 2001. Mercury and photochemistry in the marine boundary layer-modelling studies suggest the *in situ* production of reactive gas phase mercury. Atmospheric Environment, 35: 3055-3062.

Hedgecock I M, Pirrone N, Sprovieri F, 2008. Chasing quicksilver northward: Mercury chemistry in the Arctic troposphere. Environmental Chemistry, 5: 131-134.

Hedgecock I M, Trunfio G A, Pirrone N, Sprovieri F, 2005. Mercury chemistry in the MBL:

Mediterranean case and sensitivity studies using the AMCOTS (Atmospheric Mercury Chemistry over the Sea) model. Atmospheric Environment, 39: 7217-7230.

Hindersmann I, Hippler J, Hirner A V, Mansfeldt T, 2014. Mercury volatilization from a floodplain soil during a simulated flooding event. Journal of Soils and Sediments, 14: 1549-1558.

Hines N A, Brezonik P L, 2004. Mercury dynamics in a small Northern Minnesota lake: Water to air exchange and photoreactions of mercury. Marine Chemistry, 90: 137-149.

Holm H W, Cox M F, 1975. Transformation of elemental mercury by bacteria. Applied Microbiology, 29: 491-494.

Holmes C D, Jacob D J, Corbitt E S, Mao J, Yang X, Talbot R, Slemr F, 2010. Global atmospheric model for mercury including oxidation by bromine atoms. Atmospheric Chemistry and Physics, 10: 12037-12057.

Holmes C D, Jacob D J, Mason R P, Jaffe D A, 2009. Sources and deposition of reactive gaseous mercury in the marine atmosphere. Atmospheric Environment, 43: 2278-2285.

Holmes C D, Jacob D J, Yang X, 2006. Global lifetime of elemental mercury against oxidation by atomic bromine in the free troposphere. Geophysical Research Letters, 33: L20808.

Hong L, Sharp M A, Poblete S, Bieh R, Zamponi M, Szekely N, Appavou M S, Winkler R G, Nauss R E, Johs A, Parks J M, Yi Z, Cheng X L, Liang L Y, Ohl M, Miller S M, Richter D, Gompper G, Smith J C, 2014. Structure and dynamics of a compact state of a multidomain protein, the mercuric ion reductase. Biophysical Journal, 107: 393-400.

Horowitz H M, Jacob D J, Zhang Y X, Dibble T S, Slemr F, Amos H M, Schmidt J A, Corbitt E S, Marais E A, Sunderland E M, 2017. A new mechanism for atmospheric mercury redox chemistry: Implications for the global mercury budget. Atmospheric Chemistry and Physics, 17: 6353-6371.

Howard D, Edwards G C, 2018. Mercury fluxes over an Australian alpine grassland and observation of nocturnal atmospheric mercury depletion events. Atmospheric Chemistry and Physics, 18: 129-142.

Hu H, Lin H, Zheng W, Tomanicek S J, Johs A, Feng X, Elias D A, Liang L, Gu B, 2013a. Oxidation and methylation of dissolved elemental mercury by anaerobic bacteria. Nature Geoscience, 6: 751-754.

Hu H Y, Lin H, Zheng W, Rao B, Feng X B, Liang L Y, Elias D A, Gu B H, 2013b. Mercury reduction and cell-surface adsorption by *Geobacter sulfurreducens* PCA. Environmental Science & Technology, 47: 10922-10930.

Huang C C, Chen M W, Hsieh J L, Lin W H, Chen P C, Chien L F, 2006. Expression of mercuric reductase from *Bacillus megaterium* MB1 in eukaryotic microalga *Chlorella* sp. DT: An approach for mercury phytoremediation. Applied Microbiology and Biotechnology, 72: 197-205.

Huang J Y, Choi H D, Hopke P K, Holsen T M, 2010. Ambient Mercury Sources in Rochester, NY: Results from principle components analysis (PCA) of mercury monitoring network data. Environmental Science & Technology, 44: 8441-8445.

Hussein S, Ruiz O N, Terry N, Daniell H, 2007. Phytoremediation of mercury and organomercurials in chloroplast transgenic plants: Enhanced root uptake, translocation to shoots, and volatilization. Environmental Science & Technology, 41: 8439-8446.

Iverfeldt A, Lindqvist O, 1986. Atmospheric oxidation of elemental mercury by ozone in the aqueous phase. Atmospheric Environment, 20: 1567-1573.

Izaki K, 1977. Enzymatic reduction of mercurous ions in *Escherichia-coli* bearing *R* factor. Journal of Bacteriology, 131: 696-698.

Jiang T, Skyllberg U, Wei S Q, Wang D Y, Lu S, Jiang Z M, Flanagan D C, 2015. Modeling of the structure-specific kinetics of abiotic, dark reduction of Hg(II) complexed by O/N and S functional groups in humic acids while accounting for time-dependent structural rearrangement. Geochimica et Cosmochimica Acta, 154: 151-167.

Jiang T, Wei S Q, Flanagan D C, Li M J, Li X M, Wang Q, Luo C, 2014. Effect of abiotic factors on the mercury reduction process by humic acids in aqueous systems. Pedosphere, 24: 125-136.

Jiskra M, Wiederhold J G, Skyllberg U, Kronberg R M, Hajdas I, Kretzschmar R, 2015. Mercury deposition and re-emission pathways in boreal forest Soils investigated with Hg isotope signatures. Environmental Science & Technology, 49: 7188-7196.

Johnson K P, Blum J D, Keeler G J, Douglas T A, 2008. Investigation of the deposition and emission of mercury in Arctic snow during an atmospheric mercury depletion event. Journal of Geophysical Research-Atmospheres, 113: D17304.

Kannan S K, Mahadevan S, Krishnamoorthy R, 2006. Characterization of a mercury-reducing *Bacillus cereus* strain isolated from the Pulicat Lake sediments, south east coast of India. Archives of Microbiology, 185: 202-211.

Kim K H, Kim M Y, 1999. The exchange of gaseous mercury across soil-air interface in a residential area of Seoul, Korea. Atmospheric Environment, 33: 3153-3165.

Kim P R, Han Y J, Holsen T M, Yi S M, 2012. Atmospheric particulate mercury: Concentrations and size distributions. Atmospheric Environment, 61: 94-102.

Kirk J L, Louis V L S, Sharp M J, 2006. Rapid reduction and reemission of mercury deposited into snowpacks during atmospheric mercury depletion events at Churchill, Manitoba, Canada. Environmental Science & Technology, 40: 7590-7596.

Kocman D, Horvat M, 2010. A laboratory based experimental study of mercury emission from contaminated soils in the River Idrijca catchment. Atmospheric Chemistry and Physics, 10: 1417-1426.

Krabbenhoft D P, Hurley J P, Olson M L, Cleckner L B, 1998. Diel variability of mercury phase and species distributions in the Florida Everglades. Biogeochemistry, 40: 311-325.

Kudsk F N, 1965. The influence of ethyl alcohol on the absorption of mercury vapour from the lungs in man. Acta Pharmacologica et Toxicologica, 23: 263.

Kurien U, Hu Z Z, Lee H, Dastoor A P, Ariya P A, 2017. Radiation enhanced uptake of $Hg^0_{(g)}$ on iron(oxyhydr)oxide nanoparticles. RSC Advances, 7: 45010-45021.

Kuss J, Wasmund N, Nausch G, Labrenz M, 2015. Mercury emission by the Baltic Sea: A consequence of cyanobacterial activity, photochemistry, and low-light mercury transformation. Environmental Science & Technology, 49: 11449-11457.

Lahoutifard N, Poissant L, Scott S L, 2003. Heterogeneous scavenging of atmospheric mercury by snow spiked with hydrogen peroxide. Journal De Physique Iv, 107: 711-714.

Lahoutifard N, Poissant L, Scott S L, 2006. Scavenging of gaseous mercury by acidic snow at Kuujjuarapik, Northern Quebec. Science of the Total Environment, 355: 118-126.

Lalonde J D, Amyot M, Doyon M R, Auclair J C, 2003. Photo-induced Hg(II) reduction in snow from the remote and temperate Experimental Lakes Area (Ontario, Canada). Journal of Geophysical Research-Atmospheres, 108: 4200.

Lalonde J D, Amyot M, Kraepiel A M L, Morel F M M, 2001. Photooxidation of Hg^0 in artificial and natural waters. Environmental Science & Technology, 35: 1367-1372.

Lalonde J D, Amyot M, Orvoine J, Morel F M M, Auclair J C, Ariya P A, 2004. Photoinduced oxidation of Hg^0(aq) in the waters from the St. Lawrence estuary. Environmental Science & Technology, 38: 508-514.

Lalonde J D, Poulain A J, Amyot M, 2002. The role of mercury redox reactions in snow on snow-to-air mercury transfer. Environmental Science & Technology, 36: 174-178.

Lamborg C H, Rolfhus K R, Fitzgerald W F, Kim G, 1999. The atmospheric cycling and air-sea exchange of mercury species in the South and equatorial Atlantic Ocean. Deep-Sea Research Part Ii-Topical Studies in Oceanography, 46: 957-977.

Landis M S, Ryan J V, Ter Schure A F H, Laudal D, 2014. Behavior of mercury emissions from a commercial coal-fired power plant: The relationship between stack speciation and near-field plume measurements. Environmental Science & Technology, 48: 13540-13548.

Lanzillotta E, Ceccarini C, Ferrara R, 2002. Photo-induced formation of dissolved gaseous mercury in coastal and offshore seawater of the Mediterranean basin. Science of the Total Environment, 300: 179-187.

Lanzillotta E, Ceccarini C, Ferrara R, Dini F, Frontini E, Banchetti R, 2004. Importance of the biogenic organic matter in photo-formation of dissolved gaseous mercury in a culture of the marine diatom *Chaetoceros* sp. Science of the Total Environment, 318: 211-221.

Lanzillotta E, Ferrara R, 2001. Daily trend of dissolved gaseous mercury concentration in coastal seawater of the Mediterranean basin. Chemosphere, 45: 935-940.

Laszlo Macos S H, Clarksons T W, 1978. Role of catalase in the oxidation of mercury vapor. Biochemical Pharmacology, 27: 1373-1377.

Laurier F J G, Mason R P, Whalin L, Kato S, 2003. Reactive gaseous mercury formation in the North Pacific Ocean's marine boundary layer: A potential role of halogen chemistry. Journal of Geophysical Research-Atmospheres, 67: A245.

Leonhauser J, Rohricht M, Wagner-Dobler I, Deckwer W D, 2006. Reaction engineering aspects of microbial mercury removal. Engineering in Life Sciences, 6: 139-148.

Li Z, Xia C H, Wang X M, Xiang Y R, Xie Z Q, 2011. Total gaseous mercury in Pearl River Delta region, China during 2008 winter period. Atmospheric Environment, 45: 834-838.

Lin C J, Pehkonen S O, 1997. Aqueous free radical chemistry of mercury in the presence of iron oxides and ambient aerosol. Atmospheric Environment, 31: 4125-4137.

Lin C J, Pehkonen S O, 1999a. Aqueous phase reactions of mercury with free radicals and chlorine: Implications for atmospheric mercury chemistry. Chemosphere, 38: 1253-1263.

Lin C J, Pehkonen S O, 1999b. The chemistry of atmospheric mercury: A review. Atmospheric Environment, 33: 2067-2079.

Lin C J, Pehkonen S O, 1998. Oxidation of elemental mercury by aqueous chlorine (HOCl/OCl$^-$): Implications for tropospheric mercury chemistry. Journal of Geophysical Research-Atmospheres, 103: 28093-28102.

Lindberg S E, Brooks S, Lin C J, Scott K J, Landis M S, Stevens R K, Goodsite M, Richter A, 2002. Dynamic oxidation of gaseous mercury in the Arctic troposphere at polar sunrise. Environmental Science & Technology, 36: 1245-1256.

Lindberg S E, Kim K H, Meyers T P, Owens J G, 1995. Micrometeorological gradient approach for quantifying air-surface exchange of mercury-vapor - tests over contaminated soils. Environmental Science & Technology, 29: 126-135.

Lindberg S E, Zhang H, Gustin M, Vette A, Marsik F, Owens J, Casimir A, Ebinghaus R, Edwards G, Fitzgerald C, Kemp J, Kock H H, London J, Majewski M, Poissant L, Pilote M, Rasmussen P,

Schaedlich F, Schneeberger D, Sommar J, Turner R, Wallschlager D, Xiao Z, 1999. Increases in mercury emissions from desert soils in response to rainfall and irrigation. Journal of Geophysical Research-Atmospheres, 104: 21879-21888.

Liu Y X, Zhang J, Pan J F, 2014. Photochemical oxidation removal of Hg^0 from flue gas containing SO_2/NO by an ultraviolet irradiation/hydrogen peroxide (UV/H_2O_2) process. Energy & Fuels, 28: 2135-2143.

Lohman K, Seigneur C, Edgerton E, Jansen J, 2006. Modeling mercury in power plant plumes. Environmental Science & Technology, 40: 3848-3854.

Lu X, Liu Y R, Johs A, Zhao L D, Wang T S, Yang Z M, Lin H, Elias D A, Pierce E M, Liang L Y, Barkay T, Gu B H, 2016. Anaerobic mercury methylation and demethylation by *Geobacter bemidjiensis* Bem. Environmental Science & Technology, 50: 4366-4373.

Luo H W, Yin X P, Jubb A M, Chen H M, Lu X, Zhang W H, Lin H, Yu H Q, Liang L Y, Sheng G P, Gu B H, 2017. Photochemical reactions between mercury (Hg) and dissolved organic matter decrease Hg bioavailability and methylation. Environmental Pollution, 220: 1359-1365.

Magos L, Sugata Y, Clarkson T W, 1974. Effects of 3-amino-1, 2, 4-triazole on mercury uptake by *in vitro* human blood samples and by whole rats. Toxicology and Applied Pharmacology, 28: 367-373.

Manceau A, Lemouchi C, Enescu M, Gaillot A C, Lanson M, Magnin V, Glatzel P, Poulin B A, Ryan J N, Aiken G R, Gautier-Luneau I, Nagy K L, 2015. Formation of mercury Sulfide from Hg(II)-thiolate complexes in natural organic matter. Environmental Science & Technology, 49: 9787-9796.

Mann E, Ziegler S, Mallory M, O'driscoll N, 2014. Mercury photochemistry in snow and implications for Arctic ecosystems. Environmental Reviews, 22: 331-345.

Mann E A, Mallory M L, Ziegler S E, Avery T S, Tordon R, O'driscoll N J, 2015a. Photoreducible mercury loss from Arctic snow is influenced by temperature and snow age. Environmental Science & Technology, 49: 12120-12126.

Mann E A, Mallory M L, Ziegler S E, Tordon R, O'driscoll N J, 2015b. Mercury in Arctic snow: Quantifying the kinetics of photochemical oxidation and reduction. Science of the Total Environment, 509: 115-132.

Marek M, 1997. Dissolution of mercury vapor in simulated oral environments. Dental Materials, 13: 312-315.

Mason R P, Fitzgerald W F, Morel F M M, 1994. The biogeochemical cycling of elemental mercury-anthropogenic influences. Geochimica Et Cosmochimica Acta, 58: 3191-3198.

Mason R P, Lawson N M, Sheu G R, 2001. Mercury in the Atlantic Ocean: Factors controlling air-sea exchange of mercury and its distribution in the upper waters. Deep-Sea Research Part Ii-Topical Studies in Oceanography, 48: 2829-2853.

Mason R P, Lawson N M, Sullivan K A, 1997. The concentration, speciation and sources of mercury in Chesapeake Bay precipitation. Atmospheric Environment, 31: 3541-3550.

Mason R P, Morel F M M, Hemond H F, 1995. The role of microorganisms in elemental mercury formation in natural waters. Water Air and Soil Pollution, 80: 775-787.

Mastromonaco M G N, Gardfeldt K, Langer S, Dommergue A, 2016a. Seasonal study of mercury species in the Antarctic sea ice environment. Environmental Science & Technology, 50: 12705-12712.

Mastromonaco M N, Gardfeldt K, Jourdain B, Abrahamsson K, Granfors A, Ahnoff M, Dommergue A,

Mejean G, Jacobi H W, 2016b. Antarctic winter mercury and ozone depletion events over sea ice. Atmospheric Environment, 129: 125-132.

Matthiessen A, 1998. Reduction of divalent mercury by humic substances - kinetic and quantitative aspects. Science of the Total Environment, 213: 177-183.

Mauclair C, Layshock J, Carpi A., 2008. Quantifying the effect of humic matter on the emission of mercury from artificial soil surfaces. Applied Geochemistry, 23: 594-601.

Mcelroy W J, Munthe J, 1991. The oxidation of mercury(I) by ozone in acidic aqueous solutions. Acta Chemica Scandinavica, 45: 254-257.

Melamed R, Boas R C V, Goncalves C O, Paiva E C, 1997. Mechanisms of physico-chemical interaction of mercury with river sediments from a gold mining region in Brazil: Relative mobility of mercury species. Journal of Geochemical Exploration, 58: 119-124.

Meng B, Li Y B, Cui W B, Jiang P, Liu G L, Wang Y M, Richards J, Feng X B, Cai Y, 2018. Tracing the uptake, transport, and fate of mercury in sawgrass (*Cladium jamaicense*) in the Florida Everglades using a multi-isotope technique. Environmental Science & Technology, 52: 3384-3391.

Menke R, Wallis G, 1980. Detection of mercury in air in the presence of chlorine and water-vapor. American Industrial Hygiene Association Journal, 41: 120-124.

Miller C L, Watson D B, Lester B P, Howe J Y, Phillips D H, He F, Liang L Y, Pierce E M, 2015. Formation of soluble mercury oxide coatings: Transformation of elemental mercury in soils. Environmental Science & Technology, 49: 12105-12111.

Miller C L, Watson D B, Lester B P, Lowe K A, Pierce E M, Liang L Y, 2013. Characterization of soils from an industrial complex contaminated with elemental mercury. Environmental Research, 125: 20-29.

Mishra B, O'loughlin E J, Boyanov M I, Kemner K M, 2011. Binding of Hg^{II} to high-affinity sites on bacteria inhibits reduction to Hg^0 by mixed $Fe^{II/III}$ phases. Environmental Science & Technology, 45: 9597-9603.

Moller A K, Barkay T, Abu Al-Soud W, Sorensen S J, Skov H, Kroer N, 2011. Diversity and characterization of mercury-resistant bacteria in snow, freshwater and sea-ice brine from the High Arctic. Fems Microbiology Ecology, 75: 390-401.

Moller A K, Barkay T, Hansen M A, Norman A, Hansen L H, Sorensen S J, Boyd E S, Kroer N, 2014. Mercuric reductase genes (*merA*) and mercury resistance plasmids in High Arctic snow, freshwater and sea-ice brine. Fems Microbiology Ecology, 87: 52-63.

Monperrus M, Tessier E, Amouroux D, Leynaert A, Huonnic P, Donard O F X, 2007. Mercury methylation, demethylation and reduction rates in coastal and marine surface waters of the Mediterranean Sea. Marine Chemistry, 107: 49-63.

Moore C, Carpi A, 2005. Mechanisms of the emission of mercury from soil: Role of UV radiation. Journal of Geophysical Research-Atmospheres, 110: D24302.

Moore C W, Castro M S, 2012. Investigation of factors affecting gaseous mercury concentrations in soils. Science of the Total Environment, 419: 136-143.

Moore C W, Obrist D, Luria M, 2013. Atmospheric mercury depletion events at the Dead Sea: Spatial and temporal aspects. Atmospheric Environment, 69: 231-239.

Moore C W, Obrist D, Steffen A, Staebler R M, Douglas T A, Richter A, Nghiem S V, 2014. Convective forcing of mercury and ozone in the Arctic boundary layer induced by leads in sea ice. Nature, 506: 81-84.

Morelli E, Ferrara R, Bellini B, Dini F, Di Giuseppe G, Fantozzi L, 2009. Changes in the non-protein thiol pool and production of dissolved gaseous mercury in the marine diatom *Thalassiosira weissflogii* under mercury exposure. Science of the Total Environment, 408: 286-293.

Munthe J, 1992. The aqueous oxidation of elemental mercury by ozone. Atmospheric Environment Part A-General Topics, 26: 1461-1468.

Munthe J, Mcelroy W J, 1992. Some aqueous reactions of potential importance in the atmospheric chemistry of mercury. Atmospheric Environment Part A-General Topics, 26: 553-557.

Munthe J, Xiao Z F, Lindqvist O, 1991. The aqueous reduction of divalent mercury by sulfite. Water Air & Soil Pollution, 56: 621-630.

Naruse I, Yoshiie R, Kameshima T, Takuwa T, 2010. Gaseous mercury oxidation behavior in homogeneous reaction with chlorine compounds. Journal of Material Cycles and Waste Management, 12: 154-160.

Nelson J D Jr., Colwell R R, 1974. The ecology of mercury-resistant bacteria in Chesapeake Bay. Microbial Ecology, 1: 191-218.

Nriagu J O, 1994. Mechanistic steps in the photoreduction of mercury in natural waters. Science of the Total Environment, 154: 1-8.

O'Driscoll P, Lang K, Minogue N, Sodeau J, 2006a. Freezing halide ion solutions and the release of interhalogens to the atmosphere. Journal of Physical Chemistry A, 110: 4615-4618.

O'Concubhair R, O'sullivan D, Sodeau J R, 2012. Dark oxidation of dissolved gaseous mercury in polar ice mimics. Environmental Science & Technology, 46: 4829-4836.

O'Concubhair R, Sodeau J R, 2013. The effect of freezing on reactions with environmental impact. Accounts of Chemical Research, 46: 2716-2724.

O'Driscoll N J, Lean D R S, Loseto L L, Carignan R, Siciliano S D, 2004. Effect of dissolved organic carbon on the photoproduction of dissolved gaseous mercury in lakes: Potential impacts of forestry. Environmental Science & Technology, 38: 2664-2672.

O'Driscoll N J, Poissant L, Canario J, Ridal J, Lean D R S, 2007. Continuous analysis of dissolved gaseous mercury and mercury volatilization in the upper St. Lawrence River: Exploring temporal relationships and UV attenuation. Environmental Science & Technology, 41: 5342-5348.

O'Driscoll N J, Siciliano S D, Lean D R S, 2003. Continuous analysis of dissolved gaseous mercury in freshwater lakes. Science of the Total Environment, 304: 285-294.

O'Driscoll N J, Siciliano S D, Lean D R S, Amyot M, 2006b. Gross photoreduction kinetics of mercury in temperate freshwater lakes and rivers: Application to a general model of DGM dynamics. Environmental Science & Technology, 40: 837-843.

O'Loughlin E J, Kelly S D, Kemner K M, Csencsits R, Cook R E, 2003. Reduction of Ag-I, Au-III, Cu-II, and Hg-II by Fe-II/Fe-III hydroxysulfate green rust. Chemosphere, 53: 437-446.

Obrist D, Pokharel A K, Moore C, 2014. Vertical profile measurements of soil air suggest immobilization of gaseous elemental mercury in mineral soil. Environmental Science & Technology, 48: 2242-2252.

Obrist D, Tas E, Peleg M, Matveev V, Fain X, Asaf D, Luria M, 2011. Bromine-induced oxidation of mercury in the mid-latitude atmosphere. Nature Geoscience, 4: 22-26.

Ogata M, Aikoh H, 1984. Mechanism of metallic mercury oxidation in vitro by catalase and peroxidase. Biochemical Pharmacology, 33: 490-493.

Ogata M, Aikoh H, 1983. The oxidation of metallic mercury by catalase in relation to acatalasemia. Industrial Health, 21: 219-230.

Ogata M, Ikeda, M. And Sugata Y, 1979. *In vitro* mercury uptake by human acatalasemic erythrocytes. Arch. Environ. Health, 34: 218-221.

Ogata M, Kenmotsu K, Hirota N, Aikoh H, 1982. Mercury oxidation *in vitro* by ferric compounds. Archives of Toxicology, 50: 93-95.

Ogata M, Kenmotsu K, Hirota N, Naito M, 1981. Relationship between uptake of mercury-vapor by mushrooms and its catalase activity. Bulletin of Environmental Contamination and Toxicology, 27: 816-820.

Onat E, 1974. Solubility studies of metallic mercury in pure water at various temperatures. Journal of Inorganic & Nuclear Chemistry, 36: 2029-2032.

Osborn A M, Bruce K D, Strike P, Ritchie D A, 1997. Distribution, diversity and evolution of the bacterial mercury resistance (mer) operon. FEMS Microbiology Reviews, 19: 239-262.

Oyetibo G O, Ishola S T, Ikeda-Ohtsubo W, Miyauchi K, Ilori M O, Endo G, 2015. Mercury bioremoval by *Yarrowia strains* isolated from sediments of mercury-polluted estuarine water. Applied Microbiology and Biotechnology, 99: 3651-3657.

Pal B, Ariya P A, 2004a. Gas-phase HO center dot-Initiated reactions of elemental mercury: Kinetics, product studies, and atmospheric implications. Environmental Science & Technology, 38: 5555-5566.

Pal B, Ariya P A, 2004b. Studies of ozone initiated reactions of gaseous mercury: Kinetics, product studies, and atmospheric implications. Physical Chemistry Chemical Physics, 6: 572-579.

Pan L, Carmichael G R, 2005. A two-phase box model to study mercury atmospheric mechanisms. Environmental Chemistry, 2: 205-214.

Pannu R, Siciliano S D, O'driscoll N J, 2014. Quantifying the effects of soil temperature, moisture and sterilization on elemental mercury formation in boreal soils. Environmental Pollution, 193: 138-146.

Park S Y, Holsen T M, Kim P R, Han Y J, 2014. Laboratory investigation of factors affecting mercury emissions from soils. Environmental Earth Sciences, 72: 2711-2721.

Pasakarnis T S, Boyanov M I, Kemner K M, Mishra B, O'loughlin E J, Parkin G, Scherer M M, 2013. Influence of chloride and Fe(II) content on the reduction of Hg(II) by magnetite. Environmental Science & Technology, 47: 6987-6994.

Pehkonen S O, Lin C J, 1998. Aqueous photochemistry of mercury with organic acids. Journal of the Air & Waste Management Association, 48: 144-150.

Peleg M, Matveev V, Tas E, Luria M, Valente R J, Obrist D, 2007. Mercury depletion events in the troposphere in mid-latitudes at the Dead Sea, Israel. Environmental Science & Technology, 41: 7280-7285.

Peleg M, Tas E, Obrist D, Matveev V, Moore C, Gabay M, Luria M, 2015. Observational evidence for involvement of nitrate radicals in nighttime oxidation of mercury. Environmental Science & Technology, 49: 14008-14018.

Peretyazhko T, Charlet L, Grimaldi M, 2006a. Production of gaseous mercury in tropical hydromorphic soils in the presence of ferrous iron: A laboratory study. European Journal of Soil Science, 57: 190-199.

Peretyazhko T, Charlet L, Muresan B, Kazimirov V, Cossa D, 2006b. Formation of dissolved gaseous mercury in a tropical lake (Petit-Saut reservoir, French Guiana). Science of the Total Environment, 364: 260-271.

Petrus A K, Rutner C, Liu S N, Wang Y J, Wiatrowski H A, 2015. Mercury reduction and methyl mercury degradation by the soil bacterium *Xanthobacter autotrophicus* Py2. Applied and Environmental Microbiology, 81: 7833-7838.

Pincock R E, Kiovsky T E, 1966. Kinetics of reactions in frozen solutions. Journal of Chemical Education, 43: 358-360.

Pirrone N, Cinnirella S, Feng X, Finkelman R B, Friedli H R, Leaner J, Mason R, Mukherjee A B, Stracher G B, Streets D G, Telmer K, 2010. Global mercury emissions to the atmosphere from anthropogenic and natural sources. Atmospheric Chemistry and Physics, 10: 5951-5964.

Pleijel K, Munthe J, 1995. Modeling the atmospheric mercury cycle-chemistry in fog droplets. Atmospheric Environment, 29: 1441-1457.

Poissant L, 1997. Field observations of total gaseous mercury behaviour: Interactions with ozone concentration and water vapour mixing ratio in air at a rural site. Water Air and Soil Pollution, 97: 341-353.

Pongprueksa P, Lin C J, Lindberg S E, Jang C, Braverman T, Bullock O R, Ho T C, Chu H W, 2008. Scientific uncertainties in atmospheric mercury models III: Boundary and initial conditions, model grid resolution, and Hg(II) reduction mechanism. Atmospheric Environment, 42: 1828-1845.

Poulain A J, Amyot M, Findlay D, Telor S, Barkay T, Hintelmann H, 2004a. Biological and photochemical production of dissolved gaseous mercury in a boreal lake. Limnology and Oceanography, 49: 2265-2275.

Poulain A J, Garcia E, Amyot M, Campbell P G C, Raofie F, Ariya P A, 2007. Biological and chemical redox transformations of mercury in fresh and salt waters of the high arctic during spring and summer. Environmental Science & Technology, 41: 1883-1888.

Poulain A J, Lalonde J D, Amyot M, Shead J A, Raofie F, Ariya P A, 2004b. Redox transformations of mercury in an Arctic snowpack at springtime. Atmospheric Environment, 38: 6763-6774.

Poulin B A, Aiken G R, Nagy K L, Manceau A, Krabbenhoft D P, Ryan J N, 2016. Mercury transformation and release differs with depth and time in a contaminated riparian soil during simulated flooding. Geochimica Et Cosmochimica Acta, 176: 118-138.

Pratt K A, Custard K D, Shepson P B, Douglas T A, Pohler D, General S, Zielcke J, Simpson W R, Platt U, Tanner D J, Huey L G, Carlsen M, Stirm B H, 2013. Photochemical production of molecular bromine in Arctic surface snowpacks. Nature Geoscience, 6: 351-356.

Qu Z, Yan N, Liu P, Jia J, Yang S, 2010. The role of iodine monochloride for the oxidation of elemental mercury. Journal of Hazardous Materials, 183: 132-137.

Qu Z, Yan N Q, Liu P, Chi Y P, Jia J, 2009. Bromine chloride as an oxidant to improve elemental mercury removal from coal-fired flue gas. Environmental Science & Technology, 43: 8610-8615.

Qureshi A, O'driscoll N J, Macleod M, Neuhold Y M, Hungerbuhler K, 2010. Photoreactions of mercury in surface ocean water: Gross reaction kinetics and possible pathways. Environmental Science & Technology, 44: 644-649.

Raofie F, Ariya P A, 2004. Product study of the gas-phase BrO$^-$ initiated oxidation of Hg^0: Evidence for stable Hg^{1+} compounds. Environmental Science & Technology, 38: 4319-4326.

Raofie F, Snider G, Ariya P A, 2008. Reaction of gaseous mercury with molecular iodine, atomic iodine, and iodine oxide radicals: Kinetics, product studies, and atmospheric implications. Canadian Journal of Chemistry-Revue Canadienne De Chimie, 86: 811-820.

Rocha J C, Junior E S, Zara L F, Rosa A H, Dos Santos A, Burba P, 2000. Reduction of mercury(II)by tropical river humic substances (Rio Negro): A possible process of the mercury cycle in Brazil.

Talanta, 53: 551-559.

Rocha J C, Sargentini E, Zara L F, Rosa A H, Dos Santos A, Burba P, 2003. Reduction of mercury(II) by tropical river humic substances (Rio Negro)- Part II. Influence of structural features (molecular size, aromaticity, phenolic groups, organically bound sulfur). Talanta, 61: 699-707.

Rugh C L, Senecoff J F, Meagher R B, Merkle S A, 1998. Development of transgenic yellow poplar for mercury phytoremediation. Nature Biotechnology, 16: 925-928.

Rugh C L, Wilde H D, Stack N M, Thompson D M, Summers A O, Meagher R B, 1996. Mercuric ion reduction and resistance in transgenic *Arabidopsis thaliana* plants expressing a modified bacterial merA gene. Proceedings of the National Academy of Sciences of the United States of America, 93: 3182-3187.

Rutter A P, Shakya K M, Lehr R, Schauer J J, Griffin R J, 2012. Oxidation of gaseous elemental mercury in the presence of secondary organic aerosols. Atmospheric Environment, 59: 86-92.

Ryaboshapko A, Bullock R, Ebinghaus R, Ilyin I, Lohman K, Munthe J, Petersen G, Seigneur C, Wangberg I, 2002. Comparison of mercury chemistry models. Atmospheric Environment, 36: 3881-3898.

Schluter K, 2000. Review: Evaporation of mercury from soils. An integration and synthesis of current knowledge. Environmental Geology, 39: 249-271.

Schroeder W H, Anlauf K G, Barrie L A, Lu J Y, Steffen A, Schneeberger D R, Berg T, 1998. Arctic springtime depletion of mercury. Nature, 394: 331-332.

Schroeder W H, Yarwood G, Niki H, 1991. Transformation processes involving mercury species in the atmosphere: Results from a literature survey. Water Air & Soil Pollution, 56: 653-666.

Seigneur C, Vijayaraghavan K, Lohman K, 2006. Atmospheric mercury chemistry: Sensitivity of global model simulations to chemical reactions. Journal of Geophysical Research-Atmospheres, 111: D22306.

Selin N E, Jacob D J, Park R J, Yantosca R M, Strode S, Jaegle L, Jaffe D, 2007. Chemical cycling and deposition of atmospheric mercury: Global constraints from observations. Journal of Geophysical Research-Atmospheres, 112.

Serudo R L, De Oliveira L C, Rocha J C, Paterlini W C, Rosa A H, Da Silva H C, Botero W G, 2007. Reduction capability of soil humic substances from the Rio Negro basin, Brazil, towards Hg(II) studied by a multimethod approach and principal component analysis (PCA). Geoderma, 138: 229-236.

Sherman L S, Blum J D, Douglas T A, Steffen A, 2012. Frost flowers growing in the Arctic ocean-atmosphere-sea ice-snow interface: 2. Mercury exchange between the atmosphere, snow, and frost flowers. Journal of Geophysical Research-Atmospheres, 117: D00R10.

Sherman L S, Blum J D, Johnson K P, Keeler G J, Barres J A, Douglas T A, 2010. Mass-independent fractionation of mercury isotopes in Arctic snow driven by sunlight. Nature Geoscience, 3: 173-177.

Sheu G R, Mason R P, 2004. An examination of the oxidation of elemental mercury in the presence of halide surfaces. Journal of Atmospheric Chemistry, 48: 107-130.

Shia R L, Seigneur C, Pai P, Ko M, Sze N D, 1999. Global simulation of atmospheric mercury concentrations and deposition fluxes. Journal of Geophysical Research-Atmospheres, 104: 23747-23760.

Si L, Ariya P A, 2011. Aqueous photoreduction of oxidized mercury species in presence of selected alkanethiols. Chemosphere, 84: 1079-1084.

Si L, Ariya P A, 2015. Photochemical reactions of divalent mercury with thioglycolic acid: Formation of mercuric sulfide particles. Chemosphere, 119: 467-472.
Si L, Ariya P A, 2008. Reduction of oxidized mercury species by dicarboxylic acids ($C_2 \sim C_4$): Kinetic and product studies. Environmental Science & Technology, 42: 5150-5155.
Sichak S P, Mavis R D, Finkelstein J N, Clarkson T W, 1986. An examination of the oxidation of mercury vapor by rat brain homogenate. Journal of Biochemical Toxicology, 1: 53-68.
Siciliano S D, O'driscoll N J, Lean D R S, 2002. Microbial reduction and oxidation of mercury in freshwater lakes. Environmental Science & Technology, 36: 3064-3068.
Skov H, Christensen J H, Goodsite M E, Heidam N Z, Jensen B, Wahlin P, Geernaert G, 2004. Fate of elemental mercury in the arctic during atmospheric mercury depletion episodes and the load of atmospheric mercury to the arctic. Environmental Science & Technology, 38: 2373-2382.
Smith D H, 1967. A factors mediate resistance to mercury nickel and cobalt. Science, 156: 1114-1116.
Smith T, Pitts K, Mcgarvey J A, Summers A O, 1998. Bacterial oxidation of mercury metal vapor, Hg^0. Applied and Environmental Microbiology, 64: 1328-1332.
Snider G, Raofie F, Ariya P A, 2008. Effects of relative humidity and CO(g) on the O_3-initiated oxidation reaction of Hg^0_g: Kinetic & product studies. Physical Chemistry Chemical Physics, 10: 5616-5623.
Soares L C, Egreja F B, Linhares L A, Windmoller C C, Yoshida M I, 2015. Accumulation and oxidation of elemental mercury in tropical soils. Chemosphere, 134: 181-191.
Soerensen A L, Sunderland E M, Holmes C D, Jacob D J, Yantosca R M, Skov H, Christensen J H, Strode S A, Mason R P, 2010. An improved global model for air-sea exchange of mercury: High concentrations over the North Atlantic. Environmental Science & Technology, 44: 8574-8580.
Solis K L B, Nam G U, Hong Y, 2017. Mercury(II) reduction and sulfite oxidation in aqueous systems: Kinetics study and speciation modeling. Environmental Chemistry, 14: 151-159.
Sommar J, Gardfeldt K, Stromberg D, Feng X B, 2001. A kinetic study of the gas-phase reaction between the hydroxyl radical and atomic mercury. Atmospheric Environment, 35: 3049-3054.
Sommar J, Hallquist M, Ljungstrom E, Lindqvist O, 1997. On the gas phase reactions between volatile biogenic mercury species and the nitrate radical. Journal of Atmospheric Chemistry, 27: 233-247.
Spolaor A, Angot H, Roman M, Dommergue A, Scarchilli C, Varde M, Del Guasta M, Pedeli X, Varin C, Sprovieri F, Magand O, Legrand M, Barbante C, Cairns W R L, 2018. Feedback mechanisms between snow and atmospheric mercury: Results and observations from field campaigns on the Antarctic plateau. Chemosphere, 197: 306-317.
Steffen A, Schroeder W, Bottenheim J, Narayan J, Fuentes J D, 2002. Atmospheric mercury concentrations: Measurements and profiles near snow and ice surfaces in the Canadian Arctic during Alert 2000. Atmospheric Environment, 36: 2653-2661.
Stephens C R, Shepson P B, Steffen A, Bottenheim J W, Liao J, Huey L G, Apel E, Weinheimer A, Hall S R, Cantrell C, Sive B C, Knapp D J, Montzka D D, Hornbrook R S, 2012. The relative importance of chlorine and bromine radicals in the oxidation of atmospheric mercury at Barrow, Alaska. Journal of Geophysical Research-Atmospheres, 117: D00R11.
Subir M, Ariya P A, Dastoor A P, 2012. A review of the sources of uncertainties in atmospheric mercury modeling II. Mercury surface and heterogeneous chemistry: A missing link. Atmospheric Environment, 46: 1-10.
Sugata Y, Clarkson T W, 1979. Exhalation of mercury further evidence for an oxidation-reduction cycle in mammalian tissues. Biochemical Pharmacology, 28: 3474-3476.

Sugio T, Fujii M, Takeuchi F, Negishi A, Maeda T, Kamimura K, 2003. Volatilization of mercury by an iron oxidation enzyme system in a highly mercury-resistant *Acidithiobacillus ferrooxidans* strain MON-1. Bioscience Biotechnology and Biochemistry, 67: 1537-1544.

Sugio T, Iwahori K, Takeuchi F, Negishi A, Maeda T, Kamimura K, 2001. Cytochrome c oxidase purified from a mercury-resistant strain of *Acidithiobacillus ferrooxidans* volatilizes mercury. Journal of Bioscience and Bioengineering, 92: 44-49.

Summers A O, Sugarman L I, 1974. Cell-free mercury(II)-reducing activity in a plasmid-bearing strain of *Escherichia coli*. Journal of Bacteriology, 119: 242-249.

Sun G, Sommar J, Feng X, Lin C-J, Ge M, Wang W, Yin R, Fu X, Shang L, 2016. Mass-dependent and -independent fractionation of mercury isotope during gas-phase oxidation of elemental mercury vapor by atomic Cl and Br. Environmental Science & Technology, 50: 9232-9241.

Sun L M, Lu B Y, Yuan D X, Xue C, 2015. Effect on the photo-production of dissolved gaseous mercury in post-desulfurized seawater discharged from a coal-fired power plant. Water Air and Soil Pollution, 226: 118.

Sun R G, Wang D Y, Mao W, Zhao S B, Zhang C, 2014. Roles of chloride ion in photo-reduction/oxidation of mercury. Chinese Science Bulletin, 59: 3390-3397.

Svensson M, Allard B, Duker A, 2006. Formation of HgS- mixing HgO or elemental Hg with S, FeS or FeS_2. Science of the Total Environment, 368: 418-423.

Tacey S A, Xu L, Mavrikakis M, Schauer J J, 2016. Heterogeneous reduction pathways for Hg(II) species on dry aerosols: A first-principles computational study. Journal of Physical Chemistry A, 120: 2106-2113.

Takenaka N, Ueda A, Daimon T, Bandow H, Dohmaru T, Maeda Y, 1996. Acceleration mechanism of chemical reaction by freezing: The reaction of nitrous acid with dissolved oxygen. Journal of Physical Chemistry, 100: 13874-13884.

Takenaka N, Ueda A, Maeda Y, 1992. Acceleration of the rate of nitrite oxidation by freezing in aqueous solution. Nature, 358: 736-738.

Takeuchi F, Iwahori K, Kamimura K, Negishi A, Maeda T, Sugio T, 2001. Volatilization of mercury under acidic conditions from mercury-polluted soil by a mercury-resistant *Acidithiobacillus ferrooxidans* SUG 2-2. Bioscience Biotechnology and Biochemistry, 65: 1981-1986.

Tas E, Obrist D, Peleg M, Matveev V, Fain X, Asaf D, Luria M, 2012. Measurement-based modelling of bromine-induced oxidation of mercury above the Dead Sea. Atmospheric Chemistry and Physics, 12: 2429-2440.

Tokos J J S, Hall B, Calhoun J A, Prestbo E M, 1998. Homogeneous gas-phase reaction of Hg^0 with H_2O_2, O_3, CH_3I, and $(CH_3)_2S$: Implications for atmospheric Hg cycling. Atmospheric Environment, 32: 823-827.

Tong Y, Eichhorst T, Olson M R, Mcginnis J E, Turner I, Rutter A P, Shafer M M, Wang X, Schauer J J, 2013. Atmospheric photolytic reduction of Hg(II) in dry aerosols. Environmental Science-Processes & Impacts, 15: 1883-1888.

Tong Y, Eichhorst T, Olson M R, Rutter A P, Shafer M M, Wang X, Schauer J J, 2014. Comparison of heterogeneous photolytic reduction of Hg(II) in the coal fly ashes and synthetic aerosols. Atmospheric Research, 138: 324-329.

Toyota K, Dastoor A P, Ryzhkov A, 2014. Air-snowpack exchange of bromine, ozone and mercury in the springtime Arctic simulated by the 1-D model PHANTAS - Part 2: Mercury and its speciation. Atmospheric Chemistry and Physics, 14: 4135-4167.

Tseng C M, Lamborg C, Fitzgerald W F, Engstrom D R, 2004. Cycling of dissolved elemental mercury in Arctic Alaskan lakes. Geochimica Et Cosmochimica Acta, 68: 1173-1184.

Van Otten B, Buitrago P A, Senior C L, Silcox G D, 2011. Gas-phase oxidation of mercury by bromine and chlorine in flue gas. Energy & Fuels, 25: 3530-3536.

Vandal G M, Fitzgerald W F, Rolfhus K R, Lamborg C H, 1995. Modeling the elemental mercury cycle in Pallette Lake, Wisconsin, USA. Water Air and Soil Pollution, 80: 529-538.

Vaughan P P, Blough N V, 1998. Photochemical formation of hydroxyl radical by constituents of natural waters. Environmental Science & Technology, 32: 2947-2953.

Velasquez-Riano M, Benavides-Otaya H D, 2016. Bioremediation techniques applied to aqueous media contaminated with mercury. Critical Reviews in Biotechnology, 36: 1124-1130.

Vijayaraghavan K, Karamchandani P, Seigneur C, Balmori R, Chen S Y, 2008. Plume-in-grid modeling of atmospheric mercury. Journal of Geophysical Research-Atmospheres, 113: D24305.

Von Canstein H, Li Y, Timmis K N, Deckwer W D, Wagner-Dobler I, 1999. Removal of mercury from chloralkali electrolysis wastewater by a mercury-resistant Pseudomonas putida strain. Applied and Environmental Microbiology, 65: 5279-5284.

Vudamala K, Chakraborty P, Sailaja B B V, 2017. An insight into mercury reduction process by humic substances in aqueous medium under dark condition. Environmental Science and Pollution Research, 24: 14499-14507.

Wagner-Dobler I, 2003. Pilot plant for bioremediation of mercury-containing industrial wastewater. Applied Microbiology and Biotechnology, 62: 124-133.

Wagner-Dobler I, Von Canstein H, Li Y, Timmis K N, Deckwer W D, 2000. Removal of mercury from chemical wastewater by microoganisms in technical scale. Environmental Science & Technology, 34: 4628-4634.

Wallschlager D, Kock H H, Schroeder W H, Lindberg S E, Ebinghaus R, Wilken R D, 2000. Mechanism and significance of mercury volatilization from contaminated floodplains of the German river Elbe. Atmospheric Environment, 34: 3745-3755.

Walters N E, Glassford S M, Van Heyst B J, 2016. Mercury flux from naturally enriched bare soils during simulated cold weather cycling. Atmospheric Environment, 129: 134-141.

Wang D Y, He L, Shi X J, Wei S Q, Feng X B, 2006. Release flux of mercury from different environmental surfaces in Chongqing, China. Chemosphere, 64: 1845-1854.

Wang F, Saiz-Lopez A, Mahajan A S, Martin J C G, Armstrong D, Lemes M, Hay T, Prados-Roman C, 2014. Enhanced production of oxidised mercury over the tropical Pacific Ocean: A key missing oxidation pathway. Atmospheric Chemistry and Physics, 14: 1323-1335.

Wang J S, Anthony E J, 2005. An analysis of the reaction rate for mercury vapor and chlorine. Chemical Engineering & Technology, 28: 569-573.

Wang Y, Schaefer J K, Mishra B, Yee N, 2016. Intracellular Hg(0) oxidation in *Desulfovibrio desulfuricans* ND132. Environmental Science & Technology, 50: 11049-11056.

Wang Z, Pehkonen S O, 2004. Oxidation of elemental mercury by aqueous bromine: Atmospheric implications. Atmospheric Environment, 38: 3675-3688.

Wayne R P, Barnes I, Biggs P, Burrows J P, Canosamas C E, Hjorth J, Lebras G, Moortgat G K, Perner D, Poulet G, Restelli G, Sidebottom H, 1991. The nitrate radical - physics, chemistry, and the atmosphere. Atmospheric Environment Part a-General Topics, 25: 1-203.

Whalin L, Kim E H, Mason R, 2007. Factors influencing the oxidation, reduction, methylation and demethylation of mercury species in coastal waters. Marine Chemistry, 107: 278-294.

Whalin L M, Mason R P, 2006. A new method for the investigation of mercury redox chemistry in natural waters utilizing deflatable Teflon(R) bags and additions of isotopically labeled mercury. Analytica Chimica Acta, 558: 211-221.

Wiatrowski H A, Das S, Kukkadapu R, Ilton E, Barkay T, Yee N, 2009. Reduction of Hg(II) to Hg(0) by magnetite. Geochimica Et Cosmochimica Acta, 73: A1436-A1436.

Wiatrowski H A, Ward P M, Barkay T, 2006. Novel reduction of mercury(II) by mercury-sensitive dissimilatory metal reducing bacteria. Environmental Science & Technology, 40: 6690-6696.

Wigfield D C, Perkins S L, 1985. Oxidation of elemental mercury by hydroperoxides in aqueous-solution. Canadian Journal of Chemistry-Revue Canadienne De Chimie, 63: 275-277.

Wigfield D C, Perkins S L, 1983. Oxidation of mercury by catalase and peroxidase in homogeneous solution. Journal of Applied Toxicology: JAT, 3: 185-188.

Wigfield D C, Tse S, 1985. Kinetics and mechanism of the oxidation of mercury by peroxidase. Canadian Journal of Chemistry-Revue Canadienne De Chimie, 63: 2940-2944.

Wigfield D C, Tse S, 1986. The mechanism of biooxidation of mercury. Journal of Applied Toxicology, 6: 73-74.

Windmoller C C, Durao Junior W A, De Oliveira A, Do Valle C M, 2015. The redox processes in Hg-contaminated soils from Descoberto (Minas Gerais, Brazil): Implications for the mercury cycle. Ecotoxicology and Environmental Safety, 112: 201-211.

Xin M, Gustin M, Johnson D, 2007. Laboratory investigation of the potential for re-emission of atmospherically derived Hg from soils. Environmental Science & Technology, 41: 4946-4951.

Xin M, Gustin M S, 2007. Gaseous elemental mercury exchange with low mercury containing soils: Investigation of controlling factors. Applied Geochemistry, 22: 1451-1466.

Yamamoto M, 1996. Stimulation of elemental mercury oxidation in the presence of chloride ion in aquatic environments. Chemosphere, 32: 1217-1224.

Yamamoto M, Hou H, Nakamura K, Yasutake A, Fujisaki T, Nakano A, 1995. Stimulation of elemental mercury oxidation by Sh compounds. Bulletin of Environmental Contamination and Toxicology, 54: 409-413.

Yamamoto S, Kakii K, Nikata T, Kuriyama M, Shirakashi T, 1999. Effects of chemical speciation and light irradiation on the volatilization of mercury by *Chlorella* sp.. Seibutsu-Kogaku Kaishi, 77: 213-218.

Yang Y-K, Zhang C, Shi X-J, Lin T, Wang D-Y, 2007. Effect of organic matter and pH on mercury release from soils. Journal of Environmental Sciences, 19: 1349-1354.

Yee N, Parikh M, Lin C C, Kukkadapu R, Barkay T, 2010. Reduction of Hg(II) to Hg(0) by biogenic magnetite. Geochimica Et Cosmochimica Acta, 74: A1184.

Zhang H, 2006. Photochemical redox reactions of mercury//Atwood D A, (Ed.). Recent Developments in Mercury Science. Berlin, Heidelberg: Springer, 120: 37-79.

Zhang H, Dill C, 2008. Apparent rates of production and loss of dissolved gaseous mercury (DGM) in a southern reservoir lake (Tennessee, USA). Science of the Total Environment, 392: 233-241.

Zhang H, Lindberg S E, 1999. Processes influencing the emission of mercury from soils: A conceptual model. Journal of Geophysical Research-Atmospheres, 104: 21889-21896.

Zhang H, Lindberg S E, 2001. Sunlight and iron(III)-induced photochemical production of dissolved gaseous mercury in freshwater. Environmental Science & Technology, 35: 928-935.

Zhang H, Lindberg S E, Marsik F J, Keeler G J, 2001. Mercury air/surface exchange kinetics of background soils of the Tahquamenon River watershed in the Michigan Upper Peninsula. Water

Air and Soil Pollution, 126: 151-169.

Zhang Y T, Chen X, Yang Y K, Wang D Y, Liu X, 2011. Effect of dissolved organic matter on mercury release from water body. Journal of Environmental Sciences, 23: 912-917.

Zheng J C, 2015. Archives of total mercury reconstructed with ice and snow from Greenland and the Canadian High Arctic. Science of the Total Environment, 509: 133-144.

Zheng W, Hintelmann H, 2010. Isotope fractionation of mercury during its photochemical reduction by low-molecular-weight organic compounds. Journal of Physical Chemistry A, 114: 4246-4253.

Zheng W, Hintelmann H, 2009. Mercury isotope fractionation during photoreduction in natural water is controlled by its Hg/DOC ratio. Geochimica Et Cosmochimica Acta, 73: 6704-6715.

Zheng W, Liang L Y, Gu B H, 2012. Mercury reduction and oxidation by reduced natural organic matter in anoxic environments. Environmental Science & Technology, 46: 292-299.

Zheng W, Lin H, Mann B F, Liang L Y, Gu B H, 2013. Oxidation of dissolved elemental mercury by thiol compounds under anoxic conditions. Environmental Science & Technology, 47: 12827-12834.

Zhou C S, Yang H M, Qi D X, Sun J X, Chen J M, Zhang Z Y, Mao L, Song Z J, Sun L S, 2018. Insights into the heterogeneous Hg^0 oxidation mechanism by H_2O_2 over Fe_3O_4(001) surface using periodic DFT method. Fuel, 216: 513-520.

Zhou J, Wang Z, Zhang X, Sun T, 2017. Investigation of factors affecting mercury emission from subtropical forest soil: A field controlled study in southwestern China. Journal of Geochemical Exploration, 176: 128-135.

Zhu W, Song Y, Adediran G A, Jiang T, Reis A T, Pereira E, Skyllberg U, Bjorn E A, 2018. Mercury transformations in resuspended contaminated sediment controlled by redox conditions, chemical speciation and sources of organic matter. Geochimica Et Cosmochimica Act, 220: 158-179.

第 3 章 汞的甲基化与去甲基化

> **本章导读**
> - 概述甲基汞的物理化学性质、健康效应和环境行为。
> - 系统介绍甲基汞在不同环境介质中的浓度分布与迁移过程。
> - 总结近年来有关生物和化学的汞甲基化与去甲基化反应机制及影响因素。

在各种含汞化合物中甲基汞是对人体和环境健康威胁最大的形态。甲基汞具有发育神经毒性，易被人体吸收并在体内蓄积。它容易穿过血脑屏障，对中枢神经系统造成不可逆的损伤，也可以通过胎盘屏障进入胎儿体内，在胎儿大脑和其他组织中蓄积到超过母体的浓度，损害婴幼儿的神经发育系统。目前，甲基汞的人为排放已经得到有效控制，但无机汞在环境过程中会被转化为甲基汞。甲基汞一旦生成就会随食物链累积放大，其生物浓缩因子可高达 6 个数量级以上。因此，甲基化是汞生物地球化学循环中的重要过程，可使汞的毒性和生物富集能力显著增加。饮食暴露是人体中汞的主要来源，食物中甲基汞的富集和污染程度则取决于甲基汞在环境中的生成和降解。目前，对于环境中汞的甲基化和去甲基化过程及机理还不完全清楚，继续探索自然环境中汞的甲基化和去甲基化过程及其反应机理具有重要意义。

3.1 甲基汞的物理化学性质

甲基汞（methylmercury 或 monomethylmercury）是甲基汞阳离子的简称，通常简写为 MeHg 或 MeHg$^+$，化学式为 CH_3Hg^+，其中汞为+2 价。CH_3Hg^+ 很容易与硫离子（如 S^{2-}、HS^-）、含巯基（—SH）的氨基酸（如半胱氨酸）或蛋白质通过配位共价键结合，也能与氯离子、氢氧根离子等常见阴离子形成配位化合物。

甲基汞化合物在水中的溶解性随阴离子种类的不同而有很大差异。大部分甲基汞化合物溶于水，某些甲基汞化合物（如甲基汞卤化物）能溶于非极性溶剂

（WHO，1990）。甲基汞在室温下有较高的蒸气压，如氯化甲基汞（CH$_3$HgCl）在室温的蒸气压为 1.13 Pa（WHO，1990）。不过，甲基汞的气-水分配系数较小，如 CH$_3$HgCl 在室温下的气-水分配系数为 $(1\sim2)\times10^{-5}$ 量级（Lindqvist et al.，1984）。

3.2 甲基汞的人体暴露与健康风险

除罕见的职业暴露和事故性暴露（经皮肤吸收或呼吸进入）外，甲基汞主要通过饮食暴露途径进入人体。甲基汞进入消化道后，几乎可以完全被吸收进入血液，在血液里容易与半胱氨酸以及含有半胱氨酸的多肽和蛋白质（如血红蛋白）络合，并随血液循环到达全身各器官和组织。特别值得注意的是，甲基汞容易穿过血脑屏障到达脑部，进而影响中枢神经系统的生理状态和功能；也容易通过胎盘，造成出生前胎儿的汞暴露，并影响其中枢神经系统发育（JECFA，2004；WHO，1990）。

汞和甲基汞的人体暴露水平一般通过测量血液（包括脐带血）和头发中总汞和甲基汞的浓度作为生物标志物来进行评价（Ou et al.，2014；Stern and Smith，2003；JECFA，2000）。脐带血甚至脐带组织（Sakamoto et al.，2016；Grandjean et al.，2005）中的汞浓度能更有效地反映胎儿的出生前汞暴露水平（Bjornberg et al.，2003；Grandjean et al.，1999），但这些样品不易获得。因此，研究者通常用母亲血液或头发的汞含量作为替代性指标（Stern and Smith，2003）。其中，头发的采集属于非侵入式取样，简便易行，样品也更易保存和运输。此外，距离头皮不同距离处的发汞含量，还能反映不同时间的汞暴露水平（JECFA，2000；WHO，1990）。当对头发和血液中的汞浓度（总汞或甲基汞）进行换算时，世界卫生组织（WHO）推荐使用 250∶1 的头发/血液比（hair-to-blood ratio）（JECFA，2000；WHO，1990）。虽然这一比值已得到广泛应用，但存在较大的个体差异，在个体水平上进行的汞暴露研究中可能引起较大的不确定性（Liberda et al.，2014；Rothenberg et al.，2013；WHO，1990）。

食用高甲基汞含量的鱼类和海洋哺乳动物等水产品是全球范围内人体甲基汞暴露的主要途径，尤其是食用金枪鱼、剑鱼、鲨鱼等处于较高营养级的掠食性鱼类（WHO，1990）。例如，日本和北极圈附近的国家和地区，居民处于较高的甲基汞暴露水平（Yasutake et al.，2004；Muckle et al.，2001）。研究表明，瑞典育龄妇女血液中甲基汞的浓度随鱼类食用量增加而显著增加（Bjornberg et al.，2005），同时较多的鱼肉摄入也引起了瑞典孕晚期妇女头发中总汞含量及脐带血中甲基汞含量（两者呈显著相关性）的增加，导致胎儿的甲基汞暴露水平上升（Bjornberg et al.，2003）。此外，小型金矿河边的鱼类消费人群以及北极地区海洋哺乳动物的消费人群，其甲基汞暴露水平也较高（Sheehan et al.，2014）。

中国的食用鱼供应总量在世界范围内最高（Driscoll et al., 2013）。在某些人均鱼肉消费量较高的中国沿海/沿河地区，食用水产品是人群甲基汞的主要暴露途径（Tong et al., 2017）。然而，在鱼类消费量较低的中国内地区域（如贵州省），食用稻米则成为人群甲基汞暴露的主要途径（Zhang et al., 2010; Feng et al., 2008）。比如，贵州省铜仁市万山区（中国最主要的汞矿开采及冶炼区）居民头发中的甲基汞含量较高，其主要来源是摄入甲基汞含量较高的稻米（Feng et al., 2008）。2007年，万山区大约有34%的居民，其日均甲基汞摄入量超过美国环境保护署（USEPA）制定的参考值[0.1 μg/(kg 体重·d)]（Zhang et al., 2010）。在万山区及贵州省其他地区（包括未受汞污染的地区），稻米的摄入占居民日均甲基汞摄入量的94%~96%（Zhang et al., 2010）。通过对中国和美国常见的婴儿营养谷粉调查发现，以稻米为原料的谷粉，其甲基汞含量[0.6~13.9 μg/kg；平均（2.3±2.5）μg/kg]显著高于以其他谷物为原料的谷粉（未检出~1.6 μg/kg），稻米谷粉的摄入可能导致婴儿甲基汞暴露的潜在风险（Cui et al., 2017）。

人体的甲基汞浓度主要由饮食暴露水平决定，但同时也受到社会人口学等其他因素影响。例如，西班牙 INMA 出生组研究（2008~2009年）表明，西班牙瓦伦西亚市 580 名四岁学龄前儿童头发中总汞含量不但与脐带血中汞含量（反映出生前暴露水平）以及鱼类的摄入（总量及种类）有关，还与母亲的受孕年龄及是否外出工作有关（Llop et al., 2014）。而美国 NHANES 研究（2011~2012年）则发现，美国人群血液中甲基汞及总汞的浓度水平与人种/族群、性别、年龄、受教育程度等多种人口学因素有关，其中亚裔的血液甲基汞和总汞浓度远高于其他族群，男性血液中的甲基汞浓度高于女性，年龄较大、受教育程度较高的成人血液中甲基汞浓度普遍更高（Mortensen et al., 2014）。

甲基汞对人体健康影响的剂量-效应关系已初步确立。1976 年，WHO 报道，长期日均摄入 3~7 μg/kg 体重的甲基汞（对应的头发中汞浓度约为 50~125 μg/g）不会引起不良健康效应（WHO, 1990）。如果不经常食用受汞污染的水产品，一般人群的甲基汞摄入量[小于 0.1 μg/(kg 体重·d)]远低于这一水平，由甲基汞造成的健康风险极低。然而，怀孕妇女在低于这一水平的甲基汞暴露下，有可能产生健康风险（WHO, 1990）。例如，1971 年伊拉克大规模甲基汞中毒事件中，孕期头发中汞含量超过 10 μg/g 的女性产下的婴儿会表现出神经性运动发育迟缓（WHO, 1990）。研究表明，经常食用鱼类的加拿大人群中，男童出现肌肉紧张度或反射能力异常的频率与其母亲孕期头发中汞含量（均低于 24 μg/g）呈正相关，而女童则没有这一相关性（WHO, 1990）。

在日本水俣病事件和伊拉克大规模甲基汞中毒事件中，研究人员都研究了出生前甲基汞暴露对儿童神经发育的影响。然而，这些研究涉及的是短时间高剂量暴露，与普通人群的长期低剂量暴露情况有所不同。因此，从20世纪80年代起世界范围内陆续开展了多项关于长期低剂量暴露的流行病学研究，包括法罗群岛1986～1987年出生组研究（Oulhote et al.，2017；Debes et al.，2016；Grandjean et al.，1998；Grandjean et al.，1997）、"塞舌尔儿童发育研究"1989年出生组研究（van Wijngaarden et al.，2017；van Wijngaarden et al.，2013；Myers et al.，2009；Myers et al.，2003；Davidson et al.，1998；Myers et al.，1997），以及新西兰（Crump et al.，1998）、美国（Oken et al.，2016；Oken et al.，2005）、英国（Golding et al.，2016）、中国香港（Lam et al.，2013）、中国台湾（Hsi et al.，2014）等开展的相关流行病学调查。然而，这些不同流行病学研究得到的结论不尽相同。例如，法罗群岛出生组研究发现，低剂量出生前甲基汞暴露对7岁龄儿童的神经发育和认知能力（尤其是语言、注意力、记忆力）有负面影响（Grandjean et al.，1998；Grandjean et al.，1997）。而"塞舌尔儿童发育研究"出生组研究则发现，母亲头发中汞含量为0.5～26.7 μg/g（平均5.8 μg/g）的儿童，其学会走路和说话的时间（"发育里程碑"）与出生前甲基汞暴露水平没有关联（Myers et al.，1997）；随后进一步对出生组中5～6岁龄儿童进行的神经心理测试表明，低水平的出生前和出生后甲基汞暴露，都不会对儿童的神经发育和认知能力造成影响（Davidson et al.，1998）。

通过食物源摄入甲基汞的毒性效应往往与食物中共存的其他污染物或营养物质有关（Jacobson et al.，2015）。例如，当受试人群的甲基汞暴露主要源于食用鲸类等海洋哺乳动物时，鲸肉中除了含有高浓度的甲基汞，还含有高浓度的多氯联苯等污染物（Grandjean et al.，1998），研究中观察到的神经发育毒性效应是甲基汞与多氯联苯等共暴露的结果，需要在综合考虑多氯联苯等污染物生物毒性的基础上分析甲基汞的神经发育毒性效应（Rice et al.，2003；Grandjean et al.，2001；Steuerwald et al.，2000；Budtz-Jorgensen et al.，1999）。此外，鱼肉中含有较多硒和 ω-3 多不饱和脂肪酸等营养物质，硒能降低甚至逆转甲基汞的神经发育毒性（Ralston and Raymond，2010），而 ω-3 多不饱和脂肪酸有益于儿童的神经发育（Strain et al.，2008）。当受试人群主要通过食用鱼类摄入甲基汞时，这些营养物质的同时摄入会在一定程度上抵消甲基汞的毒性效应。

3.3　甲基汞的环境行为与归趋

3.3.1　甲基汞的环境分布

环境中甲基汞的浓度主要由汞的甲基化和去甲基化速率以及甲基汞的输入、输出通量大小决定，而这些转化和传输过程的相对幅度则与环境条件密切相关。

例如，硫酸盐还原菌、铁还原菌等厌氧微生物的汞甲基化过程是自然环境中甲基汞的主要来源（Compeau and Bartha，1985），因此湿地（如沼泽、河口、湖泊）中的底泥和缺氧水体是甲基汞生成的"热点"（hot spot）环境，尤其是可分解有机质含量较高、硫酸盐含量适中、无机汞输入量较高、处于厌氧或缺氧条件的湿地环境。下面对不同环境中甲基汞的浓度分布、主要影响因素，以及甲基汞在不同环境介质间的迁移过程分别进行介绍。

1. 河口和海岸带湿地

在河口底泥中，汞的甲基化在较低盐度（0.4%）的厌氧环境中更易发生，而甲基汞的去甲基化则在好氧条件或高盐度的厌氧条件下发生（Compeau and Bartha，1984）。在河口盐沼底泥中，硫酸盐还原菌（*Desulfovibrio desulfuricans*）是汞甲基化的主要微生物，其甲基化活性受硫酸盐含量的影响；若硫酸盐含量过高，硫酸盐还原菌代谢产生的硫化氢（H_2S）会与Hg^{2+}生成硫化汞沉淀，从而降低汞的生物有效性（Compeau and Bartha，1985）。比如，在北美东海岸三处典型河口/海湾生态系统中，底泥的甲基汞浓度主要由汞的甲基化速率（而非去甲基化速率）决定，并且受微生物活性和硫化汞形成的影响（Heyes et al.，2006）。

除硫酸盐浓度外，溶解态和颗粒态天然有机物的浓度和化学组成对甲基汞在沿海湿地的生成和分布也有重要影响。底泥是河口和海岸带湿地中甲基汞的主要来源（Hammerschmidt and Fitzgerald，2004），其中的有机物会影响无机汞在底泥和水相的分配以及生物有效性，从而影响底泥的汞甲基化潜势（Hammerschmidt and Fitzgerald，2006a；2004）。控制水体富营养化的过程可能会使底泥中的有机物含量降低，并使孔隙水中微生物可利用汞的含量升高，从而促进海岸湿地底泥的汞甲基化（Hammerschmidt and Fitzgerald，2004）。比如，长岛海峡底泥的汞甲基化速率常数随有机物含量的升高而逐渐降低。但在美国东北海岸多个河口的观测实验却得到相反的趋势（Schartup et al.，2013），认为河口底泥的甲基化速率并不受有机物含量的影响，而主要由总汞含量决定（Schartup et al.，2013）。此外，水体中甲基汞的浓度受有机物组分的影响，易分解的溶解性有机碳（dissolved organic carbon，DOC）浓度反映了有机物矿化的速率，与甲基汞浓度呈正相关；而腐殖质则能降低无机汞的生物有效性，与甲基汞浓度呈负相关（Soerensen et al.，2017）。有机碳还能通过与甲基汞发生络合作用而影响其在水环境中的分布和迁移（Hall et al.，2008）。特别是其中的芳香性碳，对甲基汞（以及无机汞）的迁移性和生物有效性均有显著影响（Liu et al.，2015）。

美国佛罗里达州南部的 Everglades 湿地是汞甲基化的热点区域，这一地区的汞循环过程和生态影响，尤其是汞的甲基化和去甲基化过程以及甲基汞的营养级传递，受到了广泛关注并得到了系统研究（Tai et al.，2014；Li et al.，2012；Li et al.，2010；Liu et al.，2009；Liu et al.，2008；Rumbold et al.，2008；Duvall and Barron，2000；Cleckner et al.，1998；Gilmour et al.，1998）。Everglades 不同区域的甲基汞浓度有所差异（Liu et al.，2009；Gilmour et al.，1998），在同一地点的不同环境介质（表层水体、土壤、絮状碎屑、周丛生物等）中的分布也不尽相同（Liu et al.，2011；Liu et al.，2009；Liu et al.，2008），而且表现出季节性变化（Liu et al.，2008）。甲基汞的光降解过程对 Everglades 表层水体中的甲基汞浓度有重要影响，表层水中的甲基汞光降解潜势从北向南逐渐提高，与水体中甲基汞的浓度趋势相反，两者呈负相关（Li et al.，2010）。据估算，输入 Everglades 表层水体的甲基汞，其中约 31%可以通过光降解去除，但这一比例远低于其他生态系统，这可能与 Everglades 表层水体中 DOC 浓度较高有关（Li et al.，2010）。对甲基汞源和汇的估算表明，Everglades 各区域土壤均是甲基汞最主要的源，而絮状碎屑和表层水体则是其主要的汇；周丛生物在 Everglades 北部是甲基汞的源，在南部则是汇。

河口和海岸带湿地的甲基汞分布还受到其他因素的影响，如外部输入、植被等。例如，美国加州中部（Black et al.，2009）和南部（Ganguli et al.，2012）海岸的海底地下水流入是海岸湿地海水中甲基汞的重要来源，其甲基汞输入通量与大气沉积或底泥释放的通量在同一数量级。对葡萄牙河口盐沼进行的研究表明，长有植物的盐沼底泥中的微生物活性比无植被的盐沼底泥更高，而前者的甲基汞浓度比后者高 70 倍以上（Canario et al.，2007）；而且植被对盐沼底泥中汞甲基化的促进作用与植物的种类有关（Canario et al.，2017）。根据这项研究推测，植物修复可能促进河口底泥的汞甲基化（Canario et al.，2007）。向无植被的潮汐盐沼湿地中加铁可以显著降低湿地出水的甲基汞输出；但在有植被的湿地中，加铁的效果并不明显（Ulrich and Sedlak，2010）。

内陆咸水湖及其附属沼泽湿地的水化学性质与河口/海岸带湿地有相似之处，如高盐度、偏碱性的 pH（7.5～9）、较高的硫酸盐含量（Johnson et al.，2015），但也存在不同之处，如不同的氧化还原条件和相对静态的水力学特征等。缺氧深层卤水层中的甲基汞生成潜势往往较低，但甲基汞浓度很高，主要是由于该微环境相对封闭，且深水层的（紫外）光照较弱，甲基汞的降解较慢，有利于甲基汞的积累（Johnson et al.，2015）。而在高海拔咸水湖的表层湖水中，甲基汞的光解对甲基汞浓度有重要影响，池水的甲基汞浓度白天降低，夜间升高，而且水体中甲基汞占总汞的比例也随日照时长的增加而降低（Naftz et al.，2011）。这些观测结果表明，光降解对白天甲基汞浓度的变化有决定性作用；而夜间甲基汞浓度的升高，则是由于夜间池水的热对流将池底的甲基汞携带到上层池水（Naftz et al.，2011）。

2. 淡水环境

河口和海岸带湿地的硫酸盐含量（mmol/L 量级）适于硫酸盐还原菌的生长和汞甲基化过程。而在淡水湿地中，硫酸盐含量通常较低，尽管硫酸盐还原菌在底泥中不是优势菌种，但仍能发生汞的甲基化作用，甲基汞浓度（以及硫酸盐还原速率）在底泥和水的界面处往往较高（Gilmour et al.，1992）。硫酸盐浓度是淡水湿地环境甲基汞生成和积累的重要影响因素。比如，在北美大草原的池塘湿地中，底泥和表层水中的甲基汞浓度随硫酸盐浓度的升高而增加（Hoggarth et al.，2015）。向淡水湿地添加硫酸盐可以显著促进甲基汞的生成（Harmon et al.，2004；Gilmour et al.，1992）。在泥炭地（peatlands）沼泽中，孔隙水中甲基汞占总汞的比例在泥炭地与高地集水区的交界处最高，而在泥炭地沼泽中心则较低，这些甲基汞生成"热点"区域的形成主要是由于来自高地径流的无机汞、硫酸盐、DOC 等的输入（Mitchell et al.，2008）。此外，硫酸型酸雨通过增加淡水湿地中硫酸盐的浓度而促进底泥中硫酸盐还原菌的生长代谢和甲基汞的生成。因此，受酸沉降影响，湖泊中鱼类汞含量往往较高（Gilmour and Henry，1991）。由于硝酸盐还原比硫酸盐还原在热力学上更易发生，当湖水中硝酸盐浓度上升时，硫酸盐还原代谢活性以及甲基汞浓度相应下降（Todorova et al.，2009）。另一方面，硫酸原还原过程中产生的硫化物浓度通过改变无机汞的形态以及生物有效性影响淡水湿地中的甲基汞生成（Creswell et al.，2017）。

河流中的甲基汞通常来自流域内沼泽湿地的输入，也可能来自地下水的输入或河流底泥中的汞甲基化过程（Bradley et al.，2012）。河流中溶解态甲基汞（以及总汞）的浓度与集水区盆地的湿地密度及河水中的 DOC 浓度等指标呈显著正相关（Brigham et al.，2009）。流域内的伐木活动能为湿地的汞甲基化微生物提供更多的电子供体（有机物），并且提高湿地中甲基汞（以及无机汞）向溪流的迁移性，导致溪水中甲基汞浓度较高（Skyllberg et al.，2009）。非汛期河流中甲基汞的主要来源为沿岸浅层地下水的输入，其贡献往往高于非沿岸湿地表层径流输入或河流底泥的原位汞甲基化（Bradley et al.，2012）。

湖水中甲基汞主要来自湖泊自身底泥或下层滞水带中微生物的汞甲基化，周围沼泽湿地及河流的甲基汞输入处于次要地位。例如，美国威斯康星州北部湖泊夏季的甲基汞累积量比周围湿地甲基汞输入量高一个数量级，湖水中的甲基汞主要来自下层滞水带中的微生物汞甲基化（Watras et al.，2005）。湖水中硝酸盐浓度升高会抑制硫酸盐还原代谢活性以及甲基汞生成（Todorova et al.，2009），也会引起或加剧湖泊的富营养化，对湖泊中甲基汞生成过程的影响较为复杂。比如，绿藻如刚毛藻（*Cladophora glomerata*）爆发性生长后，脱落和腐败的藻细胞能促进湖底甲基汞的生成并提高湖水中甲基汞的浓度（Lepak et al., 2015）。除营养元素外，甲基汞生成量还受湖泊与主河道的连通状况影响，与河流短暂连通的湖泊，其甲基汞生成量高于长期与河流连通的湖泊（Lazaro et al., 2016）。

与其他淡水环境相比，水库中的鱼类汞含量通常更高，这与水库水位的周期性涨落造成的底泥中汞甲基化增强有关（Eckley et al., 2017）。水位的周期性涨落会使底泥周期性地暴露于空气中，促进底泥中有机物的分解，一方面改变了无机汞在固相与液相（孔隙水）间的分配和生物有效性，另一方面使孔隙水中的DOC含量增加，利于提高汞甲基化微生物的活性（Eckley et al., 2017）。水位的周期性涨落还会促进底泥中还原性硫的氧化过程，导致硫酸盐含量增加，促进甲基汞的生成（Eckley et al., 2017）。此外，富营养的水库与贫营养的水库相比，甲基汞占总汞的比例更高，这主要是因为藻类生长可以创造缺氧环境，并且藻类分解后释放硫酸盐，促进汞甲基化过程（Noh et al., 2016）。水库及其下游河流中的甲基汞浓度往往存在明显的季节变化——旱季浓度高而雨季浓度低（Kasper et al., 2014）。在旱季，水库中的水体会发生强烈的热分层，由此形成的缺氧滞水带中有大量的甲基汞生成；而在雨季，水体沿深度方向充分混合，水中的甲基汞浓度相应降低（Kasper et al., 2014）。

3. 北方森林生态系统

北方森林（Boreal Forest）又称北方针叶林或泰加林（Taiga），分布于欧亚大陆及北美洲北部介于北极苔原和温带森林之间的广阔地带，是地球上仅次于海洋的第二大生物群系。北方森林的生态健全状况对全球生态系统有重要影响。湿地是北方森林生态系统中甲基汞的重要源，而高地则发生甲基汞的滞留和去甲基化，是甲基汞的汇（St. Louis et al., 1996；St. Louis et al., 1994）。虽然湿地的总汞浓度低于高地，但其甲基汞产率是后者的26～79倍（St. Louis et al., 1994）。湿地的甲基汞源强与其内部的水力学特征有关：水量丰沛的时期源强大，水量低时源强小（St. Louis et al., 1996）。人为引水将湿地淹没会导致其甲基汞源强在一年内提高39倍（Kelly et al., 1997），且其影响会持续多年（Rolfhus et al., 2015；St. Louis et al., 2004）。同时，甲基汞产率在碳氮比（C/N）和pH适中的湿地中最高，更高的C/N比和富营养化程度对去甲基化的促进作用更大，反而降低甲基汞的净产率（图3-1）（Tjerngren et al., 2012）。

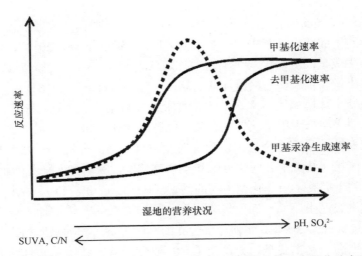

图 3-1 北方森林湿地的营养状态（如 pH、C/N 比）对汞甲基化、去甲基化速率以及甲基汞净生成速率影响的概念模型。电子供体［以天然有机物的比特征紫外吸光值（SUVA）度量］和电子受体（硫酸盐）的浓度变化也可能影响甲基汞净生成速率。图片修改自（Tjerngren et al.，2012）

与其他淡水湿地（Brigham et al.，2009；Skyllberg et al.，2009；Harmon et al.，2004；Gilmour et al.，1992）类似，北方森林湿地的甲基汞浓度也受硫酸盐浓度、硫化物浓度以及有机物含量和种类等因素的影响。例如，添加硫酸盐可以提高北方森林泥炭地沼泽中甲基汞的生成速率和浓度，而停止硫酸盐的添加可以有效降低甲基汞的生成（Wasik et al.，2012）。在北方森林湿地土壤中，甲基汞在固相/吸附相与液相之间的分配主要受天然有机物中巯基组分影响，而二价无机汞的相分配则受天然有机物中巯基组分和无机硫化物的双重影响（Liem-Nguyen et al.，2017）。北方森林湖泊中的甲基汞生成速率和浓度受到湖水中有机物来源和分子组成的影响：湖中浮游植物产生的有机物能提高底泥中微生物的活性，从而促进底泥中甲基汞生成；而陆源有机物的输入降低了湖泊底泥的汞甲基化速率（Bravo et al.，2017）。

北方森林湿地与其他淡水湿地相比最大的特点是湿地的上方或周围长期存在生物量巨大的针叶林树木和大量凋落物（litterfall，如落叶、剥落的树皮等），树冠和凋落物对北方森林湿地的甲基汞分布有重要影响。大气中的甲基汞和无机汞会沉降到树冠，被降水冲刷或者随凋落物到达地面，成为北方森林湿地甲基汞和无机汞的源。树冠下方地面的甲基汞主要来自凋落物，降水冲刷的贡献较小，甲基汞输入通量远高于空旷地面的直接湿沉降输入（Graydon et al.，2008；St. Louis et al.，2001）。树冠增强的干沉降导致的甲基汞输入与树木种类有关。云杉和冷杉下的甲基汞输入较高，而枫树下的甲基汞输入较低（Graydon et al.，2008）。在地表凋落物的分解过程中，会发生汞的甲基化和去甲基化，其相对速率及对甲基汞浓度的影响与土壤含水率（Hall and Louis，2004）、水化学性质以及树木种类（Tsui et al.，2008）有关。例如，在淹没的土壤中，甲基汞的含量会随凋落物的分解而大幅增加；而在未被淹没的土壤中，甲基汞的含量则有所下降（Hall and Louis，2004）。

伐木和土地平整等林业活动对北方森林的甲基汞分布和迁移有重要影响，能增加北方森林地表径流中的甲基汞浓度，并向周边河流、湖泊等地表水环境输出甲基汞（Porvari et al.，2003）。林业活动对甲基汞输出的影响程度与其发生时间（季节）和方式有关（Eklof et al.，2014）。例如，冬季进行的伐木活动没有显著增加地表径流的甲基汞浓度，只是通过增加径流量来提高甲基汞的输出；而夏季进行的伐木与土地平整活动则协同提高了甲基汞浓度（Eklof et al.，2014）。通过比较不同伐木方式（是否挖掘树桩）发现，挖掘树桩形成的积水坑洞以及积水车辙等区域存在丰富的汞甲基化细菌，甲基汞占总汞的比例较高，是甲基汞生成的"热点"位置（Eklof et al.，2018）。森林砍伐后，山坡土壤的凋落物层中甲基汞的浓度及占总汞的比例显著增加（Kronberg et al.，2016）。此外，土壤有机层含水量的增加为汞甲基化微生物提供了大量的电子供体，从而促进了汞甲基化过程以及甲基汞的积累（Kronberg et al.，2016）。

4. 海洋

海洋在全球汞循环中具有重要作用（Driscoll et al.，2013；Mason et al.，2012；Fitzgerald et al.，2007）。食用鱼类、贝类等海产品是人体甲基汞暴露的重要途径。因此，甲基汞在海洋环境中的生成、分布、迁移、转化和生物累积过程是汞环境行为与健康影响研究的重点。目前，关于甲基汞浓度分布的报道已覆盖了全球的主要大洋和海域，包括赤道太平洋（Mason and Fitzgerald，1993）、北太平洋（Hammerschmidt and Bowman，2012）、南大西洋和赤道大西洋（Bratkic et al.，2016；Mason and Sullivan，1999）、北大西洋（Bowman et al.，2015；Mason et al.，1998）、南冰洋（Canario et al.，2017；Mastromonaco et al.，2017；Cossa et al.，2011）等大洋，以及地中海（Cossa et al.，2017；Heimburger et al.，2010；Cossa et al.，1997）、波罗的海（Kuss et al.，2017）等海域。海洋中甲基汞的浓度较低且变化范围较大，一般在 10^{-14} mol/L 到 10^{-12} mol/L 数量级甚至更低（表 3-1）。除甲基汞外，海水中同时还存在二甲基汞（dimethylmercury，DMHg），两者在测定中有时难以区分，因此某些海洋监测研究会测定并报道两者浓度的总和，称为"总甲基化汞"（常表示为 $\sum CH_3Hg$）（Mason et al.，2012）。

表 3-1　海洋环境中甲基汞、总汞浓度

地点	甲基汞（pM）[a]	总汞（pM）	参考文献
赤道太平洋	0.13±0.11	1~2	(Mason and Fitzgerald, 1993)
北太平洋	0.02~0.10	0.2~1.5	(Hammerschmidt and Bowman, 2012)
南大西洋和赤道大西洋	<0.05	1~10 1.7±0.7 [c] 2.9±1.7 [d] 1.7±0.7 [e]	(Mason and Sullivan, 1999)
南大西洋	未检出~0.25	1.5±0.6	(Bratkic et al., 2016)
北大西洋	<0.5（低于检测限）	2.4±1.6 [f]	(Mason et al., 1998)
北大西洋	0.095±0.098 （未检出~0.60）	0.89±0.30 （0.09~1.89）	(Bowman et al., 2015)
南冰洋	0.29±0.21 [b] （0.02~0.86）[b]	0.63~2.76	(Cossa et al., 2011)
南冰洋	1.3±0.6	4.7±3.5	(Canario et al., 2017)
南冰洋	0.07±0.10（春季） 0.06±0.03（春季） 0.14±0.19 [b]（夏季）	2.6±1.3（春季） 2.0±1.0（冬季）	(Mastromonaco et al., 2017)
地中海西部	未检出~0.29 [b]	0.5~6	(Cossa et al., 1997)
地中海西北部	0.30±0.17 [b] （≤0.04~0.82）[b]	未报道	(Heimburger et al., 2010)
地中海西北部	≤0.02~0.48 [b]	0.53~1.45	(Cossa et al., 2017)
波罗的海	0.01~2.3	0.5~11.4	(Kuss et al., 2017)

注：除非特别注明，数据表示整个水层的平均浓度。
a. 摩尔浓度单位，1 pM = 1×10^{-12} mol/L；b. 测定的浓度为"总甲基化汞"，即甲基汞和二甲基汞的总浓度；c. 表层水（混合层）；d. 次表层水（混合层以下，永久温跃层以上）；e. 1500 m 以下；f. 永久温跃层以上。

海洋环境中甲基汞浓度的时间分布由汞的甲基化和去甲基化速率，以及输入、输出的通量大小决定（图 3-2）（Mason et al., 2012）。全球范围内，陆地通过河流向海洋输入的甲基汞通量为 0.1 Mmol/a，另有 0.15 Mmol/a 的甲基汞沉降在海岸带和大陆架的底泥中，海岸带和大陆架的底泥向海洋的甲基汞输入通量约为 0.21 Mmol/a。部分开阔海域（如地中海西北部）会向大陆架海水输入甲基汞和二甲基汞，其通量远高于陆地源和底泥的释放，是大陆架海水中甲基汞的主要来源（Cossa et al., 2017）。大气中的甲基汞浓度占大气总汞浓度的 20%左右（WHO, 1990；Schroeder and Jackson, 1987），与零价汞相比，甲基汞的气-水分配比较低；而在大气沉降的总汞中，甲基汞仅占 0.5%（Mason et al., 2012）。事实上，海洋与大气间的甲基汞传输通量与陆地、底泥等其他传输途径相比低一个数量级，全球大气向海洋的甲基汞沉降通量约为 0.05 Mmol/a，而二甲基汞向大气的逃逸通量约为 0.01 Mmol/a。二甲基汞可以通过去甲基化分解产生甲基汞，这一过程的源强（0.8 Mmol/a）与无机汞甲基化生成甲基汞的源强（0.76 Mmol/a）相当（图 3-2）。

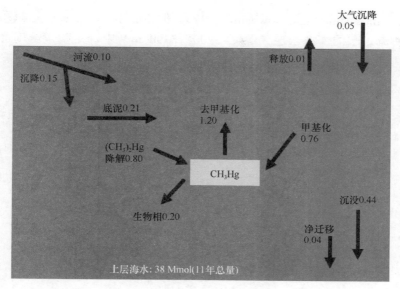

图 3-2　海洋上层（永久温跃层以上，距海面 1000 m 以内）的甲基汞质量平衡。图中通量的单位为 Mmol/a。图片修改自（Mason et al., 2012）

由于上述过程的消长，在全球不同海域以及同一海域的不同深度的甲基汞浓度差异很大，并且随时间变化（表 3-1）。在海洋表层水以及深水区，甲基汞以及二甲基汞的浓度很低；而在中间层，尤其是缺氧水层（如永久温跃层附近），甲基汞的浓度达到最大值（Canario et al., 2017; Cossa et al., 2017; Soerensen et al., 2016; Bowman et al., 2015; Cossa et al., 2011; Heimburger et al., 2010）。表层水体极低的甲基汞浓度主要是由于甲基汞的光解作用以及浮游植物的吸收（Hammerschmidt and Bowman, 2012），而缺氧水层中甲基汞浓度的升高则与有机物的微生物分解以及氧利用度有一定的关联，该水层中可能存在原位汞甲基化反应（Hammerschmidt and Bowman, 2012; Mason et al., 2012; Cossa et al., 2011）。在某些海域，除缺氧层外，甲基汞和二甲基汞浓度在缺氧层上方的次表层水含氧区会出现一个较小的峰值，这种"双峰"分布在地中海西北部（Heimburger et al., 2010）和亚热带北太平洋（Hammerschmidt and Bowman, 2012）。例如，北太平洋的甲基汞以及二甲基汞浓度，在含氧的中间水层（300~700 m）和缺氧区（800~1000 m）分别出现一个峰值（图 3-3），其中缺氧区的甲基汞峰值是由于异养微生物的原位汞甲基化；而含氧中间水层的甲基汞峰值，成因尚不明确，可能与上层颗粒物的沉降、水流传输以及原位生成有关（Diez et al., 2016; Hammerschmidt and Bowman, 2012）。深海的甲基汞浓度一般很低，说明深海底泥中的汞甲基化作用较弱（Hammerschmidt and Bowman，2012）。但在部分海域（如北大西洋），受海底热泉羽流影响的深海水层（深度 3200～3400 m）中，甲基汞浓度可以高达 0.6 pmol/L（Bowman et al.，2015）。

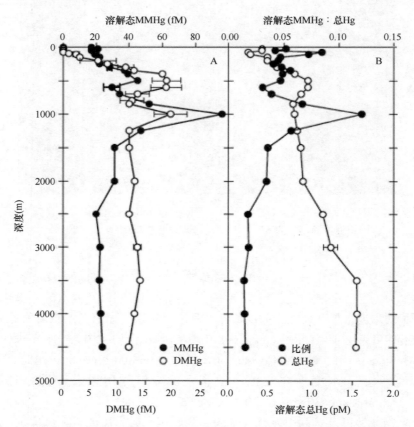

图 3-3　北太平洋某观测站测定的溶解态甲基汞（MMHg）、二甲基汞（DMHg）、总汞以及甲基汞占总汞的比例随深度变化的剖面线。摩尔浓度单位：1 fM = 1×10^{-15} mol/L；1 pM = 1×10^{-12} mol/L。图片修改自（Hammerschmidt and Bowman，2012）

3.3.2　甲基汞的生物积累与放大效应

汞在生物体内的累积及其沿食物链（网）发生的生物放大主要以甲基汞的形态进行（Morel et al.，1998）。由于甲基汞的生成和分布集中于湿地、湖泊、海洋等水环境中，汞生物积累与放大效应相关的研究大多以水生生态系统中的生物为研究对象，如藻类等浮游生物、周丛生物、自游和底栖无脊椎动物、处于不同营养级的鱼类以及以水生生物为食的鸟类和海洋哺乳动物，研究其甲基汞（及无机汞）吸收、排出的途径和机制，以及影响其体内（甲基）汞含量的生物和环境因素。

1. 浮游生物

浮游生物（尤其是浮游植物）通常处于最低的营养级，是水生生态系统食物链和食物网的基石。甲基汞从环境进入食物链（网）最关键的一步就是从水体向浮游植物的富集（Stewart et al.，2008），这一步骤通常具有很高的生物富集因子（Mason et al.，1995）。而且，某些浮游生物体内还会发生汞的甲基化作用（Pucko et al.，2014）。因此，浮游生物的汞吸收和生物累积很早就受到研究人员的关注（Glooschenko，1969），其中以无机汞的研究为主，甲基汞的吸收和生物累积研究相对较少。

甲基汞在浮游植物体内的累积与水体DOC浓度、浮游植物的种类以及细胞死活有关；而上述因素对无机汞的生物积累影响不大（Pickhardt and Fisher，2007）。具体而言，活细胞的甲基汞累积显著高于死细胞，对于活细胞，约60%的甲基汞进入细胞质；而对于死细胞，甲基汞全部存在于细胞表面。这说明活细胞中存在甲基汞的主动运输，可以促进甲基汞进入细胞（Pickhardt and Fisher，2007）。虽然氯化汞（$HgCl_2$）等无机汞和氯化甲基汞都具有亲脂性，但无机汞主要与细胞膜结合，而甲基汞则可在细胞质中累积（Mason et al.，1995）。此外，真核生物的甲基汞累积随DOC的升高而增加，而原核生物（蓝藻）的甲基汞累积受DOC影响较低（Pickhardt and Fisher，2007）。不过，沿海水域中硅藻的甲基汞生物累积与DOC浓度呈负相关（Lee and Fisher，2017），而且DOC的化学组成对硅藻的甲基汞吸收和生物累积有重要影响，其中谷胱甘肽和腐殖酸在环境浓度（分别为10 nmol/L和0.1~0.5 mg-C/L）下就能显著降低硅藻的甲基汞生物富集因子；半胱氨酸和巯基乙酸在较高浓度（高于环境浓度100倍）下才能降低硅藻细胞中甲基汞的积累，甘氨酸和甲硫氨酸则对硅藻的甲基汞吸收没有影响（Lee and Fisher，2017）。

富营养化的水体中通常存在大量的藻类等浮游生物，其甲基汞生物累积受富营养化程度的影响较为复杂。当水生生态系统中的甲基汞主要来自外部输入或底泥生成时，水体富营养化会降低浮游生物的甲基汞含量（Pickhardt et al.，2002）；但在某些水域（如波罗的海），在含氧量正常的低水平富营养化水体中，水层中会发生汞的甲基化并使浮游植物的甲基汞含量升高（Soerensen et al.，2016）。

2. 水生无脊椎动物

底栖的贝类等软体动物体内累积的汞主要为无机汞，甲基汞占的比例通常较低。例如，丹麦西海岸的紫贻贝（*Mytilus edulis*）从洁净水体转移至受汞污染的水体后，体内累积的汞中 75%为无机汞；而同一地区的波罗的海白樱蛤（*Macoma balthica*）体内累积的汞，甲基汞占比低于 6%（Riisgård et al.，1985）。生活于西南太平洋汤加岛弧的海底热泉生态系统中的贝类、螺类等软体动物的总汞含量远高于其他地区，但甲基汞占总汞的比例极低，一般不超过 0.1%（Lee et al.，2015）。而我国香港海域中贝类的甲基汞（0.01~0.02 mg/kg）占总汞（0.03~0.09 mg/kg）的 10%~40%，不同种类间的差异主要是由于甲基汞和无机汞排出生物体速率的不同导致的（Pan and Wang，2011）。

甲基汞在水生无脊椎动物体的生物累积受环境因素的显著影响。富里酸（fulvic acid）能增强无机二价汞和甲基汞在无机颗粒物上的吸附，进而提高紫贻贝对甲基汞的同化效率（assimilation efficiency，AE）（Gagnon and Fisher，1997）。极地苔原湖泊中的 DOC 对汞在片脚类生物体的累积有重要影响。但这一影响是非线性的，当湖水中 DOC 浓度低于 8.5 mg-C/L 时，DOC 能促进汞及甲基汞的生物累积；而当浓度高于 8.5 mg-C/L 时，总汞及甲基汞的生物累积因子反而随 DOC 浓度的升高而降低（French et al.，2014）。这一转变阈值可能与汞的存在形态有关，当 DOC 浓度较低时，汞主要与富里酸结合；而当 DOC 浓度较高时，汞与分子量更大、生物可给性较低的腐殖酸（humic acid）中的强络合位点结合，从而降低了汞的生物有效性（French et al.，2014）。北方森林泥炭地沼泽湿地中蚊子（*Culex* spp.）幼虫体内的甲基汞占总汞含量的 62%±19%，蚊子幼虫体内的汞富集程度随着北方森林硫酸盐沉降量的降低而减少（Wasik et al.，2012）。向河流和湖泊的底泥中投加活性炭可以显著降低底泥孔隙水以及水生寡毛纲环节动物夹杂带丝蚓（*Lumbriculus variegatus*）体内的甲基汞浓度（Gilmour et al.，2013）。

3. 鱼类

鱼体内富集的汞主要以甲基汞形式存在，虽然甲基汞具有亲脂性，但鱼体内的甲基汞主要与蛋白质而非脂肪组织结合（Mason et al.，1995）。在鱼肉中，甲基汞浓度占总汞浓度的比例往往高达 95%以上（Eagles-Smith and Ackerman，2014；Harris et al.，2003；Bloom，1992），但在某些环境中鱼肉中甲基汞占总汞的比例可能较低。例如，在贵州省万山区废弃汞矿区的河流和水库中，鱼类的汞含量较高，肌肉中总汞含量达 0.06~0.68 mg/kg（按湿重计，下同），但其中甲基汞为 0.02~0.10 mg/kg，约占总汞的 28%（Qiu et al.，2009）。造成这一现象的原因可能包括：①鱼类所处的食物网环境与易发生微生物甲基化的厌氧区域隔离；②食物网结构较为简单；③水库富营养化程度较高，食物链底端（如藻类细胞）存在"生物稀释"效应（Liu et al.，2012）。

鱼类对汞的累积主要取决于水体及食物中甲基汞的含量，因而受到环境因素的影响。此外，甲基汞的富集还与鱼的种类、年龄、营养级和进食习惯等生物学因素有关（Arcagni et al.，2018；Kahilainen et al.，2017；Keva et al.，2017；Smylie et al.，2016）。鱼类所处水环境中甲基汞浓度及其生物有效性对鱼类的甲基汞生物累积起决定性作用。当湖泊湿地淹水后，由于微生物甲基化作用的增强，湖水中甲基汞浓度显著升高，掠食性水生昆虫以及鱼类的甲基汞含量也随之增加（Hall et al.，1998）。稻田与其他类型（如作为野生动物栖息地）的人工湿地相比，水相的甲基汞浓度更高，鱼类的甲基汞累积也更显著，是甲基汞生物累积的"热点"区域（Windham-Myers et al.，2014）。历史上受到严重汞污染的水域，即使底泥中汞含量较高，鱼类体内的汞积累并不严重，说明历史污染遗留的汞具有极低的生物有效性，而流域输入和大气沉降可能是生物可利用汞的主要来源（Hodson et al.，2014）。

鱼类所处水环境的化学条件（如 DOC、盐度等）会影响甲基汞的生物有效性，进而对其生物累积效应造成影响（Liu et al.，2009）。例如，佛罗里达州南部的 Everglades 大沼泽地区食蚊鱼的甲基汞生物累积与周丛生物的甲基汞含量呈现相关性，与土壤或表层水体中的甲基汞浓度分布没有关联，而由甲基汞的生物有效性决定。北方森林生态系统中，颜色较深的湖水中鱼类的汞含量较高，这与从湿地中输入的甲基汞和有机质较多有关（St. Louis et al.，1994）。季节性的咸水人工湿地以及高盐度的废弃盐池中鱼类的汞含量远高于其他盐度较低的湿地（Eagles-Smith and Ackerman，2014）。对于洄游性鱼类，其甲基汞生物累积与个体发育不同阶段的栖息地有关。例如，胡瓜鱼（*Osmerus mordax*）的幼鱼和成鱼体内汞同位素组成不同，这主要与其发育过程中栖息地从淡水河流向河口的转换有关；而大西洋鲑（*Salmo salar*）虽然有相当长时间生活于淡水环境中，但其体内的甲基汞主要来自海洋环境（Li et al.，2016a）。

与贫营养的清澈湖泊相比，发生"水华"的富营养化湖泊中鱼类的汞含量更低，这主要是由于大量藻类细胞对甲基汞的"稀释"作用，藻类浓度的增加会显著降低浮游动物的甲基汞含量，从而降低鱼类的甲基汞摄入和累积（Pickhardt et al.，2002）。水库中鱼类的甲基汞含量与富营养化水平呈现负相关关系（Noh et al.，2017）。不过，在某些富营养化的水体中，可能由于微生物汞甲基化的增强以及浮游植物生物量的增加，使经由浮游植物进入食物链（网）的甲基汞总量提升，发生季节性的甲基汞富集（Soerensen et al.，2016）。

4. 鸟类和海洋哺乳动物

鸟类和海洋哺乳动物一般处于较高的营养级，体内的甲基汞含量较高，而且高甲基汞暴露会对其繁殖造成影响（Fuchsman et al., 2017；Evers et al., 2005）。在北极等地区，食用海洋哺乳动物是人群甲基汞暴露的主要途径（Muckle et al., 2001；Andersen et al., 1987）。因此，海洋哺乳动物汞累积的研究主要关注甲基汞摄入对人群健康产生的风险。另一方面，海洋哺乳动物也可以作为环境汞污染的指示生物（Wagemann et al., 1995）。虽然海洋哺乳动物体内的汞主要为甲基汞，但在某些组织和器官，甲基汞占总汞的比例较低。最典型的是肝脏，虽然总汞含量很高，但甲基汞占总汞的比例通常低于15%（Koeman et al., 1973）。例如，港海豹（*Phoca vitulina*）肝脏中总汞浓度高达225～765 mg/kg，但甲基汞仅占总汞的2%～14%（Koeman et al., 1973）。法罗群岛捕获的领航鲸体内汞含量很高；其中肝脏与肌肉、肾脏、鲸脂等器官和组织汞积累严重（Julshamn et al., 1987）。成年领航鲸体内的总汞浓度远高于幼鲸，而且总汞浓度随体型增大而增加，但肝脏中甲基汞的占比随体型增大反而降低（Julshamn et al., 1987）。白鲸、环斑海豹和独角鲸等北极海洋哺乳动物，其肌肉和皮肤中的甲基汞占总汞的百分比分别在95%和90%左右，但在肝脏中只有3%～12%，说明肝脏中可能存在去甲基化作用（Wagemann et al., 1998）。

鸟类体内汞的水平可以反映所处环境的汞污染状况，往往作为汞污染的指示物种（Monteiro and Furness, 1995）。鸟类血液以及卵中的汞浓度反映其短期汞暴露水平，而肝脏和羽毛中的汞浓度则能反映长期汞暴露水平（Evers et al., 2005）。例如，企鹅可以作为南大洋以及南印度洋汞污染的指示生物（Carravieri et al., 2016；Carravieri et al., 2013）。对加拿大东南沿海6种海鸟羽毛及卵中甲基汞和总汞的研究发现，其中的汞几乎全部以甲基汞的形态存在（Bond and Diamond, 2009），加拿大普通潜鸟（*Gavia immer*）卵中的甲基汞浓度占总汞的87%（Scheuhammer et al., 2001）。北美洲水鸟体内的甲基汞含量随年龄而增加，成年鸟体内的甲基汞浓度是雏鸟的5～19倍（Evers et al., 2005）。以淡水湖泊为栖息地的白腹鱼狗（*Ceryle alcyon*）和白头海雕（*Haliaeetus leucocephalus*）体内甲基汞暴露水平很高，以河口和河流为栖息地的物种次之，以海洋为栖息地的物种甲基汞暴露水平最低（Evers et al., 2005）。此外，食鱼性鸟类的甲基汞含量通常较高，以昆虫为食的鸟类体内也测得较高的甲基汞浓度；甲基汞暴露的种间差异主要与其所处的营养级以及栖息地中甲基汞的生物有效性有关（Evers et al., 2005）。

5. 甲基汞的营养级传递及生物放大效应

在水生生态系统中，生物体内的汞（主要是甲基汞）含量一般随其在食物链（网）中所处的营养级而增加，表现出生物放大效应。甲基汞的营养级传递以及生物放大效应广泛存在于湖泊（Arcagni et al., 2018；Fu et al., 2010；Wong et al.,

1997；Cabana et al.，1994)、水库（Stewart et al.，2008)、溪流（Chasar et al.，2009；Mason et al.，2000)、河口（Chen et al.，2014；Kehrig et al.，2009)、海岸带（Kehrig et al.，2010)、海洋（Pucko et al.，2014；Zhu et al.，2013；Cossa et al.，2012)等水生生态系统中。其中，与通过摄食习惯估算的离散式营养级相比，由稳定氮同位素比（$\delta^{15}N$）确定的连续式营养级在预测甲基汞生物放大效应方面更准确，应用也更广泛（Vander Zanden and Rasmussen，1996；Cabana and Rasmussen，1994)。

处于食物链（网）顶端的捕食者（如海洋哺乳动物、鸟类和掠食性鱼类），其体内的甲基汞浓度不但与其所在的营养级水平以及所处环境中甲基汞浓度和生物有效性有关（Chasar et al.，2009)，还与食物网的结构以及具体的营养级传输路径有关。例如，森林火灾会增加湖泊中鱼类的甲基汞含量，一方面是由于火灾后流域集水区向湖泊的汞输入增强，另一方面由于营养物质的输入，湖泊中的食物网结构发生改变，使鱼类处于更高的营养级（Kelly et al.，2006)。需要特别指出的是，食物链（网）底端生物可利用的甲基汞含量对食物链（网）顶端生物的甲基汞浓度有重要影响，其重要性甚至超过生物所处的营养级水平（Chasar et al.，2009；Stewart et al.，2008)。

对于结构复杂的食物网，研究其中甲基汞的来源和营养级传递过程较为困难。近年来，汞同位素技术在精确解析甲基汞来源和营养级传递方面发挥了重要作用。例如，利用汞同位素技术对河流-森林食物网中汞的来源和营养级传递研究发现，河流中的甲基汞来自原位汞甲基化，而森林中的甲基汞主要来自大气沉降；并且河岸蜘蛛体内的甲基汞有大约55%来自河流生态系统，另外45%来自森林生态系统（Tsui et al.，2012)。同时，陆生动物可能被某些溪流中的水生掠食性动物捕食，导致陆源甲基汞进入水生生态系统（Tsui et al.，2014)。对加拿大河口生态系统两条不同食物链的研究表明，不同种类鱼体内的甲基汞分别来自淡水和海洋生态系统，而底泥对河口食物链中甲基汞的贡献很小（Li et al.，2016a)。

3.4 汞的甲基化研究

3.4.1 汞的微生物甲基化

1. 汞微生物甲基化的作用机制

特定的厌氧微生物可以将环境中无机二价汞转化为甲基汞（Li et al.，2016a；Parks et al.，2013；Compeau and Bartha，1985；Jensen and Jernelöv，1969)。早期关于微生物对汞甲基化的认识主要聚焦在硫酸盐还原菌（sulfate-reducing bacteria，SRB)上，包括完全氧化与非完全氧化 SRB（Han et al.，2010)。完全氧化的 SRB

主要通过乙酰辅酶 A 途径实现对 Hg(Ⅱ)甲基化，并由甲基钴胺素蛋白复合物提供甲基（Choi et al.，1994a；Choi et al.，1994b）；而非完全氧化则不通过乙酰辅酶 A 途径（Ekstrom et al.，2003），其机制与乙酰辅酶 A 进行主要的碳代谢途径相关。因此，有些 SRB 菌株没有乙酰辅酶 A 或者在甲基汞产生抑制剂中也能进行汞甲基化（Ekstrom and Morel，2007；Ekstrom et al.，2003）。这些研究说明了 SRB 可能存在不止一种汞甲基化的生物化学途径。之后的研究又发现铁还原菌（iron-reducing bacteria，FeRB）和产甲烷菌（methanogens）也可以对汞进行甲基化（Yu et al.，2013；Yu et al.，2012；Hamelin et al.，2011；Fleming et al.，2006；Kerin et al.，2006）。然而这些不同类群的功能微生物对汞的甲基化分子机制并不清楚。直到 2013 年 Science 杂志上首次报道汞甲基化的基因 *hgcA* 和 *hgcB*（Parks et al.，2013），揭开了汞的微生物甲基化遗传机制（Poulain and Barkay，2013）。研究发现 Hg(Ⅱ)在 *hgcA* 基因编码的咕啉蛋白作用下产生甲基汞，而类咕啉辅因子在 *hgcB* 基因编码的铁氧化还原蛋白作用下还原（图 3-4）（Parks et al.，2013），这与之前不完全异化氧化菌如 *Desulfovibrio desulfuricans* LS 对汞的甲基化途径较为一致（Choi et al.，1994b）。

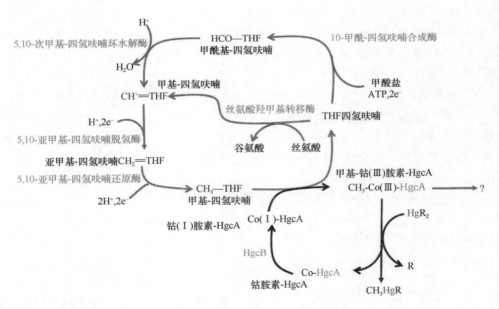

图 3-4　微生物体内汞甲基化的理论途径［图片修改自（Parks et al.，2013）］

当通过敲除 *Desulfovibrio desulfuricans* ND132 和 *Geobacter sulfurreducens* PCA 菌株中的 *hgcA* 或 *hgcB* 基因时，这些微生物丧失汞甲基化能力（Parks et al.，2013）。而重新插入这些基因片段到这两个突变体菌株时，微生物汞甲基化能力得以恢复。目前报道的能甲基化汞的微生物都含有 *hgcAB* 基因，*hgcAB* 基因也成为汞微生物甲基汞的必要条件（Gilmour et al.，2013），我们可以根据微生物是否含有 *hgcAB* 基因来判别其是否具有汞甲基化潜力。

2. 甲基化微生物

甲基化微生物分布在多样的环境中，如淡水、河流、湖泊、沉积物、湿地甚至动物肠道等厌氧生态系统（Podar et al.，2015；Gilmour et al.，2013）。根据 HgcAB 氨基酸序列的进化分析将甲基化微生物分为 5 个进化枝，包括 δ-变形菌的三类（硫酸盐还原菌、铁还原菌、产氢型互营杆菌目）、厚壁菌和部分古菌（产甲烷菌）（Gilmour et al.，2013；Parks et al.，2013），其中 SRB、FeRB 和产甲烷菌是最主要的甲基化微生物类群。目前实验室证明约 50 种甲基化微生物具有汞甲基化能力，但不同种类的微生物对汞的甲基化率差别很大（0.6%~75%）。因此，认识自然环境中汞的甲基化微生物群落组成对于理解甲基汞的产生、食物链传递及健康风险具有重要的意义。例如，研究汞污染稻田土壤 *hgcA* 基因主要分布在 δ-变形菌、厚壁菌、广古菌门和两种未分类的种类（Liu et al.，2014），且 *hgcA* 丰度与土壤甲基汞含量呈极显著相关关系。目前预测的汞甲基化微生物（含有 *hgcAB* 基因）约有 140 种（表 3-2）（Podar et al.，2015；Gilmour et al.，2013；Parks et al.，2013）。下面对主要的甲基化微生物 SRB、FeRB 和产甲烷菌分别进行阐述。

表 3-2　甲基化微生物主要菌属［列表修改自（毕丽等，2018）］

属	门	属	门
Desulfovibrio	Proteobacteria	*Clostridium*	Firmicutes
Desulfocurvus	Proteobacteria	*Alkaliphilus*	Firmicutes
Desulfomicrobium	Proteobacteria	*Clostridium*	Firmicutes
Desulfobulbus	Proteobacteria	*Ruminococcaceae* sp.	Firmicutes
Desulfopila	Proteobacteria	*Syntrophobotulus*	Firmicutes
Desulfofustis	Proteobacteria	*Dehalobacter*	Firmicutes
Desulfococcus	Proteobacteria	*Desulfitobacterium*	Firmicutes
Desulfospira	Proteobacteria	*Desulfosporosinus*	Firmicutes
Desulfotignum	Proteobacteria	*Desulfobacillus*	Firmicutes
Desulfobacterium	Proteobacteria	*Ethanoligenens*	Firmicutes
Desulfobacula	Proteobacteria	*Acetivibrio*	Firmicutes

续表

属	门	属	门
Desulfoluna	Proteobacteria	*Bacteroides*	Firmicutes
Desulfobacterales	Proteobacteria	*Dethiobacter*	Firmicutes
Desulfonatronospira	Proteobacteria	*Bacteroides*	Firmicutes
Desulfonatronovibrio	Proteobacteria	*Natronincola*	Firmicutes
Desulfonatronum	Proteobacteria	*Acetonema*	Firmicutes
Pelobacter	Proteobacteria	*Methanoregula*	Euryarchaeota
Desulfuromonas	Proteobacteria	*Methanosphaerula*	Euryarchaeota
Geobacter	Proteobacteria	*Methanofollis*	Euryarchaeota
Geopsychrobacter	Proteobacteria	*Methanospirillum*	Euryarchaeota
Geoalkalibacter	Proteobacteria	*Methanocorpusculum*	Euryarchaeota
Syntrophorhabdus	Proteobacteria	*Methanomethylovorans*	Euryarchaeota
Desulfomonile	Proteobacteria	*Methanolobus*	Euryarchaeota
Syntrophus	Proteobacteria	*Methanocella*	Euryarchaeota
Smithella	Proteobacteria	*Methanomassiliicoccus*	Euryarchaeota
Desulfacinum	Proteobacteria	*Dehalococcoides*	Chloroflexi
Desulfocarbo	Proteobacteria	*Chloroflexi*	Chloroflexi
Desulfarculus	Proteobacteria	*Anaerolinea*	Chloroflexi
Deltaproteobacteria	Proteobacteria	*Leptolinea*	Chloroflexi
Deltaproteobacterium	Proteobacteria	*Spirochaeta*	Spirochaetes
Deferrisoma	Proteobacteria	*Nitrospira*	Nitrospirae

1）硫酸盐还原菌

SRB 是一类以乳酸或丙酮酸等有机物作为电子供体，在厌氧状态下，把硫酸盐、亚硫酸盐、硫代硫酸盐等还原为硫化氢的细菌总称。SRB 是一类专性厌氧微生物，主要以硫酸盐作为能量产生的电子终端受体。根据 SRB 利用不同底物作为电子供体的差异，将其分为乙酸型（acetate-utilizing SRB）、乳酸型（lactate-utilizing SRB）和丙酮酸型（propionate-utilizing SRB），即分别以乙酸、乳酸和丙酮酸作为电子供体。这三种类型的 SRB 具有不同的汞甲基化能力，乙酸型 SRB 比非乙酸型 SRB 的汞甲基化率高或者相当（Ekstrom et al.，2003；King et al.，2000）。已经证明多种 SRB 具有甲基化能力（Gilmour et al.，2013；Benoit et al.，2001），如 *Desulfovibrio*、*Desulfotomaculum* 和 *Desulfobulbus* 属。同时，一些 SRB 菌株已经成为研究汞微生物甲基化的模式菌株，如 *Desulfovibrio desulfuricans* LS（Choi et al.，1994a）和 *Desulfovibrio desulfuricans* ND132（Gilmour et al.，2011）。最早关于 SRB

对汞甲基化机制的研究主要是以 *Desulfovibrio desulfuricans* LS 为研究对象，可惜该菌株没有在实验室传代下去。后来，从底泥中分离到 *Desulfovibrio desulfuricans* ND132（Gilmour，1986），并对其全基因组进行测序（Brown et al.，2011）。结果表明，*Desulfovibrio desulfuricans* ND132（$5×10^8$ 细胞/mL）在含有 Hg(Ⅱ)（25 nmol/L）的磷酸盐缓冲溶液（PBS）中能够快速产生甲基汞，并在 24 小时内将 28%的无机汞转化为甲基汞（Hu et al.，2013）。但是，并非所有的 SRB 都能进行甲基化，近期研究表明只有含有 *hgcAB* 基因的 SRB 才可能具有汞甲基化能力（Gilmour et al.，2013；Parks et al.，2013）。

SRB 是水体、沉积物和水稻土等环境中重要的甲基化微生物（Liu et al.，2014；Achá et al.，2012；Ekstrom et al.，2003；Compeau and Bartha，1985）。当钼酸盐抑制了硫酸盐还原过程时，汞的甲基化也会完全受到抑制（King et al.，1999；Gilmour et al.，1998）。此外，也有研究发现硫酸盐的还原与汞的甲基化效率之间呈正相关关系（King et al.，2002），说明这两个生物化学过程之间存在着紧密的耦合。硫标记的同位素实验发现红树林沉积物中汞的微生物甲基化过程与硫酸盐还原率相关（Correia and Guimarães，2017）。硫酸盐还原过程受阻导致甲基化减弱，从而证明 SRB 是红树林沉积物中的重要甲基化微生物，但同时还有其他微生物及其复杂的群体直接或间接参与甲基化过程（Correia and Guimarães，2017）。虽然 SRB 在汞甲基化的形成中扮演着重要的作用，但在过去的很长一段时间内我们对 SRB 甲基化汞的功能机制了解十分有限，直到 *hgcAB* 基因的发现（Parks et al.，2013），极大地增强了我们对 SRB 甲基化功能的认识。

2）铁还原菌

在某些沉积物中加入适量 SRB 抑制剂钼酸盐，仍会有甲基汞产生，这一甲基化过程主要由 FeRB 驱动（Fleming et al.，2006）。与 SRB 类似，不是所有的 FeRB 都具有汞甲基化能力，例如，*Shewanella* 不能进行汞甲基化（Kerin et al.，2006）。低浓度的半胱氨酸短时间内会显著增强 *Geobacter sulfurreducens* 的甲基化速率（Schaefer and Morel，2009）。而在没有半胱氨酸的磷酸盐缓冲液中，细胞色素 c 缺失的 *Geobacter sulfurreducens* PCA 突变体产生的甲基汞是野生型 *Geobacter sulfurreducens* PCA 的两倍，且加入的半胱氨酸会抑制 *Geobacter sulfurreducens* PCA 突变体对 Hg(Ⅱ)的甲基化。这些研究说明了野生型 *Geobacter sulfurreducens* PCA 对时间和半胱氨酸浓度的依赖性，同时 Hg(Ⅱ)的形态会影响到微生物的吸收和甲基化（Liu et al.，2015）。最近的研究发现铁还原菌 *Geobacter bemidjiensis* Bem 在厌氧环境下不仅能还原 Hg(Ⅱ)，而且能产生甲基汞和导致甲基汞的去甲基化（Lu et al.，2016）。因此，FeRB 群落是环境中汞形态转化的重要驱动者，并影响水体环境中净甲基汞的含量与环境风险。

3）产甲烷菌

部分产甲烷菌也能进行汞甲基化。最初的研究认为产甲烷菌是主要的甲基化微生物（Wood et al., 1968），因为它产生的甲基钴胺素被认为是 Hg(Ⅱ)唯一的甲基供体。但由于之后对 SRB 和 FeRB 甲基化作用的发现与研究，产甲烷菌的甲基化作用却被忽视，认为 SRB 和 FeRB 是主要的汞甲基化微生物（Kerin et al., 2006），而产甲烷菌在汞甲基化中相对贡献较小（Ullrich et al., 2001）。同位素标记原位点实验发现（Hamelin et al., 2011），藻类生物膜上发生的汞甲基化过程可能与产甲烷微生物有关，而不是 SRB 或 FeRB。虽然他们并不能分离其产甲烷菌以进行甲基化能力测试，但是在固着的活跃微生物群落中检测到了产甲烷微生物的 16S rRNA 序列，其中包括 *Methanococcales*、*Methanobacteria* 和 *Methanosarcinales*，这也表明产甲烷微生物可能是某些特定环境中的主要甲基化微生物菌群。此外，纯培养实验也证明了产甲烷菌 *Methanomethylovorans hollandica*、*Methanolobus tindarius*、*Methanospirillum hungatei* JF-1 的甲基化能力（Gilmour et al., 2013；Yu et al., 2013）。

4）汞甲基化微生物的潜在环境

汞甲基化微生物广泛存在于厌氧环境中。以前的研究主要通过测定微生物活性和甲基化速率（K_m）等方法来预测环境中微生物对汞甲基化的潜能。近年来，随着对汞微生物甲基化分子机制研究的深入，可以根据环境中 *hgcAB* 基因的含量与分布预测甲基汞的产生及其环境风险（Christensen et al., 2016；Podar et al., 2015）。甲基化微生物存在于多样的环境中，如产甲烷环境（如水稻土或者动物消化系统）或极端 pH、盐分等环境。通过对 3500 多个宏基因组数据库分析，发现无脊椎动物的消化道、融化的永久冻土、沉积物和极端环境都是潜在的汞甲基化微生物生存环境（Podar et al., 2015）。此外，在南极海洋冰川环境的宏基因组数据中发现一种海洋兼性厌氧微生物——硝化刺菌属（*Nitrospina*）可能是一种潜在的汞甲基化微生物，而没有检测到常见的厌氧汞甲基化微生物（Gionfriddo et al., 2016）。这些研究在一定程度上说明了有氧环境也可能存在汞甲基化微生物，从而拓宽了对汞微生物甲基化环境多样性的认识。

3. 生物膜中的汞甲基化

生物膜是指通过胞外多聚物（extracellular polymeric substance，EPS）相互附着或附着于某表面（如植物根系、岩石等）的微生物聚合体（Vert et al., 2012）。生物膜系统中通常包含了生产者（如藻类等）与分解者（如细菌、真菌等），是水生态系统中重要的组成部分。生物膜中可发生显著的汞甲基化、去甲基化与汞还原过程，在外部大环境好氧的条件下，形成一个有利于汞甲基化的厌氧微环境（Lin et al.，2013）。生物膜中汞的净甲基化速率往往比沉积物中更高（Hamelin et al.，2015）。不同区域生物膜中甲基汞的积累主要取决于去甲基化速率的差异，而同一地区生物膜中甲基汞的浓度则主要取决于甲基化速率的差异（Olsen et al.，2016）。这一差异受水深、光照、温度等环境因子，以及生物膜自身群落结构的显著影响（Hamelin et al.，2015）。

光合作用抑制剂、产甲烷菌抑制剂、硫酸盐还原抑制剂均可显著降低生物膜中的汞甲基化。在湖滨岩石带生物膜中，硫酸盐还原菌贡献了约 60%的汞甲基化（Desrosiers et al.，2006）。而在湖泊沉水植物根表的生物膜中，产甲烷菌对汞甲基化过程起关键作用（Hamelin et al.，2011）。生物膜中不同微生物间可能通过协同作用促进汞甲基化进程。比如，硫酸盐还原菌及硫氧化菌往往共存于生物膜中，前者是汞微生物甲基化的重要细菌，而后者可以通过将硫氧化，生成硫酸盐，为硫酸盐还原菌提供电子受体，从而促进硫酸盐还原菌主导的汞甲基化（Lin et al.，2013）。

硫酸盐还原菌（*Desulfovibrio desulfuricans*）以生物膜形式聚集存在时，相比于其以浮游形式（planktonic）存在时对汞的甲基化能力更强（Lin et al.，2013；Lin and Jay，2007）。负二价硫浓度的变化对于生物膜形式及浮游形式微生物汞甲基化的影响相同，说明两者甲基化能力的不同主要源自于生物膜中微生物活性的差异，受汞的化学形态影响不大（Lin et al.，2013）。生物膜和浮游形式下汞的甲基化途径有所不同：乙酰钴胺（acetyl-CoA）途径可能是生物膜中汞甲基化较强的原因，生物膜为甲基化酶提供甲基，而浮游形式下则不存在该途径（Lin et al.，2013）。然而，不同自然环境中生物膜汞甲基化的相对重要性可能存在差异。在某些环境，如水库厌氧水层中，浮游态微生物对汞甲基化的贡献可能比生物膜更强（Huguet et al.，2010）。

由于藻类分泌的作用，生物膜附近的小分子巯基浓度远高于周边水体，显著影响无机汞的甲基化活性（Leclerc et al.，2015）。藻类分泌（如胞外分泌物）的小分子巯基，主要用于抵抗光合作用产生的活性氧自由基以及解毒重金属离子。这些胞外多聚物（EPS）主要由多糖、蛋白质、细胞残体等组成，可分为胶囊 EPS（即 capsular EPS，紧附于细胞膜）和胶体 EPS（即 colloidal EPS，与细胞膜的结合更为松散）。生物膜中的总汞约有 3%分布于胞外，且优先与胶囊 EPS 结合。同时，小分子巯基浓度与生物膜中胶体络合态总汞浓度高度相关。因此，藻类分泌物可能通过影响无机汞的活性（如通过与无机汞络合）以及无机汞对于汞甲基化微生物的生物可利用性，控制生物膜汞甲基化。

4. 无机汞的微生物可利用性

微生物对 Hg(Ⅱ)的甲基化反应是一种细胞内反应（Hsu-Kim et al., 2013；Parks et al., 2013；Gilmour et al., 2011；Schaefer et al., 2011）。因此，甲基化微生物细胞对 Hg(Ⅱ)的吸收是微生物实现汞甲基化的一个限速步骤，微生物细胞对 Hg(Ⅱ)的吸收效率直接影响微生物对 Hg(Ⅱ)甲基化潜力。早期研究报道，耐汞微生物主要是通过 *mer* 抗汞操纵子转运系统运输 Hg(Ⅱ)（Barkay et al., 2003）。但是，至今发现的大部分汞甲基化微生物是革兰氏阴性菌，它们都没有 *mer* 抗汞操纵子转运系统。因此，汞甲基化微生物对 Hg(Ⅱ)的吸收可能存在新的机制（Hsu-Kim et al., 2013）。甲基化微生物对 Hg(Ⅱ)的吸收有可能不经过被动扩散而是通过其他的通道，比如借助细胞膜上的转运蛋白进行促进扩散或者主动吸收（Schaefer et al., 2011）。其他证据也表明，微生物可能存在 Hg(Ⅱ)的主动运输机制，例如铁还原菌代谢受阻会导致 Hg(Ⅱ)的吸收和甲基化减少（Schaefer et al., 2011）。此外，锌的添加明显抑制了微生物对汞的吸收与甲基化（Szczuka et al., 2015）。当汞与低分子量的硫醇结合形成络合物时，汞甲基化作用显著增强，由此推论甲基化微生物可能直接吸收汞-巯基络合物（Graham et al., 2012；Zhang et al., 2012；Gilmour et al., 2011）。

区分汞在细胞上的吸附和吸收对于理解汞的生物有效性和甲基化机制尤为重要。谷胱甘肽（glutathione，GSH）可以洗脱 *Geobacter sulfurreducens* PCA 细胞表面吸附的 Hg(Ⅱ)（Schaefer et al., 2011），因此可以用来区分 *Geobacter sulfurreducens* 细胞内外汞的吸收与分配。但是，对于另外一种汞甲基化模式微生物 *Desulfovibrio desulfuricans* ND132 来说，GSH 却促进了 Hg(Ⅱ)的吸收与甲基化，因此，不能作为表面吸附态汞的解吸剂。最近研究建立了以 2,3-二巯基丙磺酸（sodium 2,3-dimercapto-1-propanesulfonate，DMPS）的解吸方法，可以区分许多微生物细胞对汞的吸附与吸收（Lu et al., 2017；Liu et al., 2016b）。例如，当反应体系含有 50 μmol/L DMPS 时，超过 99%的 Hg(Ⅱ)都和 DMPS 共同存在溶液中，而 *Desulfovibrio desulfuricans* ND132 细胞不能获得 Hg(Ⅱ)，从而无法甲基化。

许多研究表明，微生物对汞的甲基化在很短时间内就会达到平衡状态（Hu et al., 2013；Gu et al., 2011；Schaefer et al., 2011），即不再产生甲基汞，这可能与底物（有效汞）的供给与甲基化酶（HgcAB）的活性有关。同位素示踪实验也发现甲基化达到初始平衡后，*Desulfovibrio desulfuricans* ND132 微生物细胞仍然可以吸收并甲基化后添加的 Hg(Ⅱ)，表明该甲基化微生物具有持续的汞甲基化能力（Liu et al., 2016b）。此外，系统中有机配体的浓度也会影响微生物对汞的甲基化效率。例如，*Desulfovibrio desulfuricans* ND132 在达到初始甲基化平衡状态后，半胱氨酸（Cys）、GSH 和青霉胺（PEN）等的添加显著增加了甲基汞的生成（Liu et al.，2016b），很明显微生物获得了更多的有效汞供给。研究发现 90% 以上 Hg(II) 可以进入 *Desulfovibrio desulfuricans* ND132 细胞内，但是进入细胞内的汞仅有约 25% 可以被甲基化。这些不能被甲基化的汞可能与细胞内生物大分子结合，不能进行甲基化过程，但是这些被"固定"的汞在巯基的作用下会重新释放部分汞，从而被微生物继续甲基化。因此，自然环境中巯基的含量与组分都可能影响甲基化微生物对汞的吸收与利用。

5. 环境中汞甲基化的影响因素

环境因子如有机质、硒（亚硒）酸盐、硫酸盐、pH、温度、硫化物等不仅影响汞的生物有效性，也影响甲基化微生物群落结构及其活性（Paranjape and Hall，2017）。其中有机质、硒（亚硒）酸盐、硫酸盐影响较为显著。

近些年，秸秆还田作为一项重要的农业措施，得到了广泛的推广，秸秆分解的有机质可以用来修复重金属污染（Sud et al.，2008）。但有研究发现在汞矿区稻田土壤中添加水稻秸秆后，土壤中甲基汞的浓度显著提高（Liu et al.，2016a；Zhu et al.，2015），主要是秸秆在微生物作用下，逐渐分解为土壤有机质，增加了土壤可溶性碳的含量，从而促进了甲基化微生物的活性。此外，有机质分子组成特征也是环境中甲基汞形成和积累的重要控制因子，通过分析沉积物中有机质分子组成能够在一定程度上预测汞甲基化速率（Bravo et al.，2017）。

硒-汞拮抗作用一直是汞研究的一个热点。已有大量研究证实硒可通过形成稳定的硒-汞络合物等途径降低水生生物对汞的吸收和累积。近年来的研究则进一步指出，硒可显著降低土壤中无机汞的植物吸收及其危害。例如，外源硒（如亚硒酸或者硒酸）可有效抑制无机汞在萝卜（Shanker et al.，1996b）、番茄（Shanker et al.，1996a）、芥菜（Mounicou et al.，2006）、大豆（Yathavakilla and Caruso，2007）、葱（Afton and Caruso，2009）和大蒜（Zhao et al.，2013）中的累积，这很可能是因为硒与无机汞在植物根际环境中形成了大分子聚合物[芥菜，> 70 kDa（Mounicou et al.，2006）；大豆，> 600 kDa（Yathavakilla and Caruso，2007）]或 HgSe 有关。此外，硒-汞在植物体内的作用可能是土壤-植物体系中硒汞拮抗的主要原因。

水稻土施用硒酸钠或者亚硒酸钠可以显著降低籽粒中甲基汞含量（Wang et al.，2016a），且籽粒中甲基汞含量的降低与水稻土甲基汞含量的降低呈现线性相关，说明硒对于土壤中汞甲基化的抑制作用可能是水稻甲基汞降低的关键环节。通过 X 射线吸收近边结构（XANES）光谱进一步分析发现，施硒土壤中汞的形态主要为 HgSe，透射电镜-能量色散 X 射线光谱分析（TEM-EDX）的结果则表明 HgSe 以纳米尺寸存在。基于盆栽实验的若干研究则显示：①施硒后土壤和水稻籽粒中甲基汞浓度降低；②叶面施硒增加了籽粒硒浓度但没有降低籽粒甲基汞浓度；③甲基汞和硒在水稻体内分布截然不同；④施硒土壤中检测到 HgSe 纳米颗粒。因此，推测土壤中硒和汞的相互作用可能是水稻甲基汞累积降低的主要原因。水稻土中硒对甲基汞含量的抑制作用甚至高于某些汞环境行为的关键控制因子（例如硫酸盐）对甲基汞含量的影响（Wang et al., 2016b; Wang et al., 2016c）。

硫酸盐可从多方面影响汞的甲基化过程。一方面，硫酸盐作为一种电子受体，能够显著促进土壤、沉积物中汞甲基化的主要微生物——硫酸盐还原菌的活性。研究发现，硫酸盐加入可促进土壤中汞的甲基化，而单独添加有机质（葡萄糖、乙酸、乳酸等）则没有作用；但同时添加硫酸盐与有机质可更大程度地促进汞甲基化（Martín-Doimeadios et al., 2017），说明硫酸盐（作为电子受体）与有机质（作为电子供体）共同介导了微生物的汞甲基化。此外，大气硫酸盐沉降可显著改变土壤微生物群落组成，从而促进土壤中汞的甲基化，但停止加入硫酸盐后微生物群落会迅速恢复（Strickman et al., 2016）。在不同硫酸盐浓度的池塘中，虽然汞甲基化速率较为接近，但高硫酸盐池塘沉积物中甲基汞占总汞的比例以及水体甲基汞的浓度均较高（Hoggarth et al., 2015）。

另一方面，厌氧条件下硫酸盐还原产生的还原性硫化物（如硫化亚铁、硫离子等）可通过与无机汞（汞甲基化的底物）的强烈结合（如形成惰性的 HgS），降低无机汞对甲基化微生物的生物有效性，从而抑制土壤中汞的甲基化。理论计算表明，硫酸盐还原的主要产物硫离子，可通过与汞形成多硫化物，导致汞活性与生物有效性的提高，从而促进汞的微生物甲基化（Paquette and Helz, 1995）。对贵州汞矿区稻田中汞甲基化研究发现，一些点位的汞甲基化较强（Zhao et al., 2016），其原因可能是硫酸盐刺激了硫酸盐还原菌介导的汞甲基化，以及高浓度的硫离子促进了汞的溶出，导致无机汞的活性提高，有利于甲基化。另一项研究则发现，在硫酸盐还原过程中形成 Hg-S 纳米颗粒可能导致汞的活化（汞进入到金属硫化物的纳米颗粒，如 CuS），导致间隙水中的胶体汞浓度显著升高（Hofacker et al., 2013）；同步辐射的结果显示，主要是由于汞替代了 CuS 纳米颗粒中的铜。此外，在硫酸盐还原开始之前，悬浮细菌也可通过吸附汞-铜合金纳米颗粒（Hg-Cu amalgam nanoparticles），从而活化铜、汞。

3.4.2 汞的非生物甲基化过程

虽然微生物的甲基化作用是环境中甲基汞的主要来源，汞的非生物甲基化过程也不容忽视。1852 年，Frankland 发现零价汞可以与碘甲烷在光照下发生反应，生成碘化甲基汞（Frankland, 1852）。此后，大量研究表明，无机二价汞、一价汞和零价汞在气相或水相中都可能发生化学转化，生成甲基汞或乙基汞等有机汞化合物（阴永光 et al.，2011）。例如，在紫外线或太阳光照射下，甲烷、甲醇、乙酸等有机小分子可以作为甲基供体，与零价汞蒸气或氯化汞发生气相反应，生成甲基汞（阴永光 et al.，2011）。理论计算也表明，零价汞与碘甲烷的气相反应在热力学上可行，但所需的活化能较高，因此需要光照（如可见光）来加速反应（Castro et al.，2011）。此外，Hg^+也能在气相中与碘甲烷、甲苯、乙酸、乙酸乙酯等挥发性有机物反应生成甲基汞，而且该系列反应可在无光条件下发生（He et al.，2015）。

在水相中，无机汞可以在紫外光照下与乙酸（Akagi et al.，1975；Akagi and Takabatake，1973）等有机酸（Falter，1999b）、α-氨基酸（Hayashi et al.，1977）、醛酮类羰基化合物（Yin et al.，2012）等有机小分子反应，生成甲基汞。乙酸盐和无机汞发生的甲基化反应可能是大气湿沉降中甲基汞的重要来源（Hammerschmidt et al.，2007）。此外，很多其他水环境相关的物质（包括多种生物分子、腐殖质和环境污染物）也可作为甲基供体，引发无机汞的甲基化反应，如甲钴胺（甲基维生素B_{12}）、甲基锡、叔丁基化合物（如叔丁醇、叔丁基过氧化氢）、碘甲烷等（Chen et al.，2015；Celo et al.，2006；Falter，1999a）。其中，作为熏蒸剂被广泛使用的碘甲烷在太阳光照射下，可将天然水体中的二价无机汞还原为一价或零价汞，再进一步甲基化生成甲基汞（Yin et al.，2014a）。水相中非生物汞甲基化反应的速率除了与甲基供体的种类和浓度有关，还受水化学条件影响。例如，无机硫化物（Akagi et al.，1975）或巯基化合物（左跃钢和庞叔薇，1985）能促进汞的光化学甲基化过程；乙酸水溶液中二价无机汞的非生物甲基化速率受pH的影响（Gardfeldt et al.，2003）；而水溶液中甲钴胺引起的汞甲基化则受盐度和pH的影响（Chen et al.，2007）。

除了上述实验室研究，在自然水体中也观察到汞的非生物甲基化过程（Celo et al.，2006）。通常认为，底泥中的微生物汞甲基化是水环境中甲基汞的主要来源，而浅层水中的甲基汞光解作用则会降低甲基汞浓度。由此可以预期，浅层水的甲基汞浓度呈现白天浓度降低、夜间浓度升高的趋势（Naftz et al.，2011）。然而，由于光照引起的非生物汞甲基化，加拿大东南部的湖泊在中午时段甲基汞的浓度明显升高（Siciliano et al.，2005）。而且，汞甲基化速率与溶解有机质（DOM）的浓度及分子质量组成有关，分子质量低于5 kDa或在30～300 Da之间的DOM在阳光下能产生甲基汞，而分子质量更高的DOM则不能促进甲基化过程（Siciliano et al.，2005）。

3.5 甲基汞的去甲基化

3.5.1 甲基汞的微生物去甲基化

1. 甲基汞微生物去甲基化的作用机制

环境中甲基汞的产生与降解是同时发生而又对立的两个过程，这两个反应过程的平衡最终决定环境中甲基汞的净含量及其随后的健康风险。在自然环境中，甲基汞降解主要通过生物降解和非生物光化学降解两种途径（Hsu-Kim et al., 2013；BarkayandWagner-Dobler, 2005）。甲基汞的微生物降解过程（去甲基化）最早发现于20世纪70年代（Spangler et al., 1973a；Spangler et al., 1973b）。通过对沉积物培养和纯菌培养实验发现，甲基汞降解产物为 CH_4 和 $Hg(0)$，这个过程被称为沉积物的自我净化（Spangler et al., 1973a）。随后研究揭示甲基汞降解与汞抗性微生物的富集有关（Billen et al., 1974），并发现了微生物的有机汞裂解酶（MerB）（Schottel, 1978；Tezuka and Tonomura, 1978）。甲基汞的微生物降解可作为甲基汞积累的潜在指标（Schottel, 1978；Tezuka and Tonomura, 1978），用于研究不同环境因子对甲基汞积累的影响机制（Steffan et al., 1988）。

微生物去甲基化过程存在还原性去甲基化和氧化性去甲基化两种机制（Marvin-DiPasquale et al., 2000）。还原性去甲基化是甲基汞在微生物还原作用下降解为金属态 $Hg(0)$ 和 CH_4（Schaefer et al., 2004；Oremland et al., 1991）。该过程主要是在耐汞微生物 mer 抗汞操纵子系统驱动下对汞的解毒过程。微生物编码的有机汞裂解酶（MerB）先将 C—Hg 键断裂，产生 CH_4 和无机 $Hg(II)$，随后汞还原酶（MerA）将 $Hg(II)$ 还原成 $Hg(0)$（Schaefer et al., 2004；Barkay et al., 2003）。另一种甲基汞生物降解途径是氧化性去甲基化：微生物将甲基汞氧化成无机 $Hg(II)$、CO_2 和少量 CH_4。然而，调控氧化性去甲基化的关键酶和功能基因尚不明确，对氧化去甲基化微生物类群的认识也非常有限。在不同环境中，两种去甲基化过程的重要性受微生物种类和环境条件的影响。例如，向沉积物中加入硫酸盐和氮素，可以提高 CO_2/CH_4 的比例，表明 SRB 和反硝化菌大量存在时可通过氧化性机制降解甲基汞（Marvin-DiPasquale et al., 2000；Oreml and et al., 1995）。在溶解氧和汞浓度较高的环境中，还原去甲基化是甲基汞生物降解的主要途径（Schaefer et al., 2002）；而在厌氧、低汞浓度的环境中，氧化去甲基化更为显著（Barkay and Wagner-Dobler, 2005；Robinson and Tuovinen, 1984）。

2. 去甲基化微生物

绝大部分甲基化微生物同时具有去甲基化的能力，去甲基化速率与微生物种类及其活性有关（Bridou et al.，2011）。例如，*G. bemidjiensis* Bem 是一株汞甲基化能力很强的细菌（Gilmour et al.，2013），全基因组序列分析发现其含有汞甲基化基因 *hgcAB*。*G. bemidjiensis* Bem 不仅具有合成甲基汞的能力，还具有去甲基化的能力（Lu et al.，2016），但是其去甲基化能力随着反应体系中 Hg(II)和甲基汞浓度的变化而变化。此外，在 *G. bemidjiensis* Bem 不同生长时期，去甲基化效率也有所差别。与稳定期细胞相比，对数期细胞降解甲基汞的能力更强，降低了体系中甲基汞的积累。

微生物对汞的甲基化和去甲基化过程都可能是其在受到有毒金属汞胁迫时的一种解毒机制。微生物通过体内的甲基化或去甲基化过程将有毒的汞进行形态转化后排出细胞外，以达到解毒的目的。能通过还原去甲基化的方式降解甲基汞的微生物往往具有 *merA* 基因，而目前对微生物氧化性去甲基化机制的认识有限，尚未归纳出主要的微生物类群。比如，好氧甲烷氧化菌 *M. trichosporium* OB3b 可通过氧化去甲基化途径降解甲基汞，去甲基化作用随着甲基汞浓度升高而增强。但其他类群的甲烷氧化菌，如 *M. capsulatus* Bath 在多种生长条件、不同甲基汞浓度下都无法降解甲基汞（Lu et al.，2017）。此外，甲醇作为竞争单碳底物能够抑制去甲基化过程，表明甲基汞去甲基化可能通过甲醇脱氢酶裂解 C—Hg 键。因此，*M. trichosporium* OB3b 的去甲基化作用可能是由于 *M. trichosporium* OB3b 细胞误将甲基汞作为碳源，从而吸收代谢甲基汞，然后通过氧化性去甲基化途径降解甲基汞为 Hg(II)（鲁霞，2016）。根据这一现象推测，环境中可能存在更多以单碳进行代谢的好氧菌参与甲基汞的降解（Lu et al.，2017）。

3.5.2 甲基汞在其他生物体内的去甲基化过程

1. 甲基汞在植物体内的去甲基化过程

植物体内的无机汞主要来自于外部环境，即大气和土壤，通过叶片和根系的吸收进入植物。植物体内的汞去甲基化过程会导致植物体内无机汞的增加（Li et al.，2016b；Xu et al.，2016）。例如，将水稻苗暴露于含甲基汞的溶液中（最长暴露时间 20 天）的结果显示，甲基汞暴露后的植物组织中有无机汞的信号，且信号随暴露时间的延长而增强，指示水稻体内可能存在甲基汞向无机汞的转化。同步辐射的结果也显示，甲基汞暴露后的水稻中无机汞形态所占比例极高，且随时间延长而增强，推测是由于甲基汞在体内去甲基化造成的（Xu et al.，2016）。加拿大多伦多大学在随后的一项研究中（Strickman and Mitchell，2017），将水稻暴露于标记了无机汞稳定同位素的土壤中，在水稻不同生长期检测组织与土壤中无机汞与甲基汞浓度。结果显示，甲基汞在土壤中形成之后，被水稻根部吸收进入体内；而水稻体内未发现汞甲基化现象。然而，在水稻的扬花期与成熟期组织中发现了甲基汞的显著下降，表明可能存在体内汞去甲基化。然而，水稻等植物体内的汞去甲基化机制尚不确定，尤其是植物体内检测到的无机汞信号，是否可能来自于外部环境，仍需进一步证实。

2. 甲基汞在动物体内的去甲基化过程

目前研究表明，甲基汞可在人、猴体内发生去甲基化过程。在20世纪50年代日本水俣病发生多年后，在甲基汞暴露的受害者脑组织中仍发现高浓度的无机汞，表明人体内可能存在汞的去甲基化过程（Vahter et al., 1995）。1995年，美国华盛顿大学的研究人员在一项长达18个月的汞暴露研究（Vahter et al., 1995）中发现，猴脑中的无机汞主要来源于甲基汞在脑中的去甲基化过程，而非来自于其他器官的无机汞转移过程。2010年，日本学者研究了人体神经母细胞瘤、胶质母细胞瘤以及肝脏细胞中的汞去甲基化（Nagano et al., 2010），结果显示这些人体细胞可进行较为快速（如72小时内）的汞去甲基化反应，含氧自由基可能驱动了这一过程。

与人体一样，鼠体内也存在汞的去甲基化过程。从鼠的脑部获取星形胶质细胞，并将其在甲基汞溶液中进行24小时的体外培养。结果显示，星形胶质细胞可将甲基汞降解为无机汞，氧化应激（oxidative stress）作用促进这一反应的发生（Shapiro and Chan, 2008）。甲基汞进入鼠体后，可在肾脏和肝脏中发生去甲基化反应。这一过程产生的无机汞可在鼠肝脏中诱导甲醛的产生，因而可通过甲醛来指示鼠体内汞去甲基化（Uchikawa et al., 2016）。鼠体内的汞去甲基化过程可能受到多个因素（如年龄、品种等）的影响。比如，幼鼠无法在体内进行汞去甲基化（Nordenhäll et al., 1998）。其反应机制尚待系统研究。

鱼类（如黑鲷）通过食物链发生甲基汞暴露之后，体内无机汞浓度升高且高达两倍，表明可能存在鱼体内的汞去甲基化现象。其中，消化道是鱼体内汞去甲基化的主要场所，而肝脏中的甲基汞降解速率极低（Wang et al., 2017）。此外，硒暴露可显著促进甲基汞的去甲基化，从而降低甲基汞在鱼体内的累积，这一过程主要在鱼的消化道中发生（Wang and Wang, 2017）。

3.5.3 甲基汞的非生物去甲基化过程

非生物作用导致的甲基汞降解（特别是光降解）是表层水体中最主要的甲基汞降解方式（Hammerschmidt and Fitzgerald, 2006b；Sellers et al., 1996）。甲基汞的非生物去甲基化过程主要包括化学降解和光降解。

1. 非生物去甲基化的作用机制

1) 化学降解

甲基汞中 C—Hg 键具有较强的热力学稳定性（Stumm and Morgan，2012）。环境水体中甲基汞的降解被认为主要是通过光降解和微生物降解进行（Fitzgerald et al.，2007）。已经建立起来的自然界中甲基汞化学降解的途径是通过甲基汞（CH_3Hg^+）和硫化氢（H_2S）反应形成硫化二甲基汞［bis(methylmercuric)sulfide，$(CH_3Hg)_2S$］，而 $(CH_3Hg)_2S$ 并不稳定，会逐渐分解为黑色的硫化汞（β-HgS）和二甲基汞［$(CH_3)_2Hg$］（Craig and Bartlett，1978；Rowland et al.，1977）。其反应方程式如下：

$$2CH_3Hg^+ + H_2S \Longleftrightarrow (CH_3Hg)_2S + 2H^+ \qquad (3\text{-}1)$$

$$(CH_3Hg)_2S \longrightarrow (CH_3)_2Hg + HgS_{(S)} \qquad (3\text{-}2)$$

亚硒酸钠与含巯基化合物［如还原型谷胱甘肽（GSH）］反应生成的 H_2Se 能打断甲基汞的 C—Hg 键，导致甲基汞的降解（Iwata et al.，1982）。产物中的无机汞主要是硒化汞（HgSe）。除亚硒酸盐外，含硒氨基酸（selenoamino acids）也可以发生上述反应，通过产生硒化二甲基汞［$(CH_3Hg)_2Se$］，最终形成硒化汞（HgSe）沉淀（KhanandWang，2010）。

理论计算验证了甲基汞与半胱氨酸（或含硒氨基酸）反应发生去甲基化的可能性，发现甲基汞-半胱氨酸（含硒氨基酸）络合物形成硫化二甲基汞（或硒化二甲基汞），随后降解为二甲基汞和硫化汞（或硒化汞）在热力学上是可行的（Asaduzzaman and Schreckenbach，2011）。

$$CH_3Hg^+ + {}^-Se\text{—}R \Longleftrightarrow CH_3Hg\text{—}Se\text{—}R \qquad (3\text{-}3)$$

$$2CH_3Hg\text{—}Se\text{—}R + {}^-Se\text{—}R \Longleftrightarrow (CH_3Hg)_2Se + R\text{—}Se\text{—}Se\text{—}R + R^- \qquad (3\text{-}4)$$

$$(CH_3Hg)_2Se \longrightarrow (CH_3)_2Hg + HgSe_{(S)} \qquad (3\text{-}5)$$

$$(CH_3)_2Hg + H^+ \longrightarrow CH_3Hg^+ + CH_4 \qquad (3\text{-}6)$$

2) 光降解

1996 年，*Nature* 期刊上首次发文证明了湖水表层存在着甲基汞的光降解，甲基汞光降解速率与甲基汞浓度、光照强度等因素有关（Sellers et al.，1996）。并且，湖泊等水环境中的甲基汞主要是通过光降解方式发生去甲基化，每年由于光降解导致的甲基汞的降低占底泥中甲基汞生成总量的 80%左右（Hammerschmidt and Fitzgerald，2006b）。环境水体中甲基汞光降解的主要机制包括：①甲基汞-天然有机物的络合物通过分子内电子传递导致甲基汞降解（Tai et al.，2014）；②光照导致天然有机物产生单线态分子氧（$^1O_2$）进而推动甲基汞降解（Zhang and Hsu-Kim，2010；Suda et al.，1993）；③通过硝酸盐光分解反应或光-芬顿反应产生羟基自由基（•OH）进而推动甲基汞降解（Hammerschmidt and Fitzgerald，2010；Chen et al.，2003）。

理论上，氯化甲基汞（CH_3HgCl）和氢氧甲基汞（CH_3HgOH）等甲基汞形态，直接通过光照辐射引起 C—Hg 键断裂导致甲基汞降解是非常困难的。从基态到激发态所需能量 ΔE_{S-T}（singlet-triplet excitation energies）是 5.1～5.9 eV（Tossell，1998），而由于臭氧层对光照的吸收，实际到达地面的光照能量约为 4.4 eV（Leighton，1961）。已有实验研究发现甲基汞并不能在去离子水中发生光降解（Tai et al.，2014；Hammerschmidt and Fitzgerald，2010）。但是当甲基汞与含硫有机物形成络合物时，其 ΔE_{S-T} 约为 3.6 eV（Yoon et al.，2005），这表明甲基汞-天然有机物的络合物通过分子内电子传递进行降解在理论上是可能的。

在中国青岛崂山水库和石老人海滩水中甲基汞的光降解并非自由基驱动反应（Han et al.，2017）。DOM 对美国佛罗里达 Everglades 湿地表层水甲基汞光降解起重要作用。光照引起 MeHg-DOM 络合物分子内电子转移导致 C—Hg 键断裂可能是 Everglades 湿地表层水甲基汞降解的主要原因（Tai et al.，2014）。可能的反应如下：

$$(\text{MeHg-DOM})^* \longrightarrow \text{MeHg 降解产物} + \text{DOM} \quad (3\text{-}7)$$

$$\text{MeHg-DOM} \xrightarrow{h\nu} (\text{MeHg-DOM})^* \quad (3\text{-}8)$$

单线态分子氧可能是海洋生态系统中甲基汞和乙基汞光降解的主要动力（Suda et al.，1993）。水体中的天然有机物在光照辐射下产生单线态分子氧，驱动甲基汞的去甲基化过程（Zhang and Hsu-Kim，2010）。甲基汞的光降解速率取决于水体中与甲基汞结合的配体种类和浓度，含硫配体（如谷胱甘肽、巯基乙酸等）结合的甲基汞降解速率较快，而海洋水体中普遍存在的氯化甲基汞降解速率较慢（Zhang and Hsu-Kim，2010）。不同络合形态甲基汞的降解速率常数如表 3-3 所示。

表 3-3　甲基汞络合物降解速率常数［列表修改自（Zhang and Hsu-Kim，2010）］

甲基汞络合物	单线态分子氧 $k_{1_{O_2,CH_3HgL}}$ [$\times 10^6$ L/(mol·s)]	羟基自由基 k_{OH,CH_3HgL} [$\times 10^9$ L/(mol·s)]	紫外光（254 nm） k_D ($\times 10^{-4}$ cm^2/mJ)
CH_3HgCl	<0.32	NA	NA
（CH_3HgOH，$CH_3HgHPO_4^-$ 混合物）	1.9（±0.04）	1.9（±0.09）	<0.23
CH_3Hg-GSH	7.4（±0.2）	1.9（±0.04）	2.2（±0.02）
CH_3Hg-MA	4.9（±0.5）	NA	NA

注：CH_3HgL 表示甲基汞与配体结合的络合物（MeHg-ligand complexes），GSH 表示谷胱甘肽，MA 表示巯基乙酸盐。（CH_3HgOH，$CH_3HgHPO_4^-$ 混合物）表示：在 $^1O_2$ 引起的光降解实验（pH=7.3）中，初始甲基汞形态是 49% CH_3HgOH 和 51% $CH_3HgHPO_4^-$；而在 •OH 和直接的紫外线 C（ultraviolet C，UVC）引起的光降解实验（pH=7.4）中，初始甲基汞形态是 90% CH_3HgOH 和 10% $CH_3HgHPO_4^-$。NA 表示无法得到有效值。

羟基自由基（·OH）是一种强氧化剂，它可以通过硝酸盐光分解（Zepp et al.，1987）、芬顿或光-芬顿反应（Zepp et al., 1992；Fenton，1894）等生成。受农业活动影响的水体往往硝酸盐含量较高但溶解态有机碳含量较低（Zepp et al.，1987）。天然水体表层普遍存在铁元素，特别是溶解态的三价铁，如 Fe^{3+}、$Fe(OH)^{2+}$、$Fe_2(OH)_2^{4+}$等（Kim and Zoh，2013），在酸性条件下（pH<6）易于发生光-芬顿反应（Southworth and Voelker，2003），生成羟基自由基（Zhang et al.，2006；Wu and Deng，2000）。上述环境中，羟基自由基是表层水体甲基汞光降解的主要途径（Hammerschmidt and Fitzgerald，2010；Chen et al.，2003），反应产物包括 Hg^{2+}、Hg^0、$CHCl_3$ 和甲醛。

2. 甲基汞非生物去甲基化的环境影响因素

由于甲基汞在天然水环境中的非生物去甲基化过程主要是通过光化学反应发生的，对于甲基汞非生物去甲基化的环境影响因素研究多集中于光降解过程。尽管目前甲基汞光降解的机理仍存在争议，但是越来越多的学者认识到，不同研究中不一致的结论，可能是由不同体系的水化学条件等环境参数不同导致的。甲基汞的光降解过程很有可能是多反应途径协同进行（Black et al.，2012）。

光照强度和波长范围一直以来被认为是影响甲基汞光降解最主要的因素之一（Li et al.，2010；Garcia et al.，2005）。研究表明，日光虽然可以导致苯基汞的直接光解，但是对于甲基汞、二甲基汞、二乙基汞等非芳香基汞，仅紫外光 C（UVC）（200～280 nm）就可导致其直接光降解（Takizawa et al.，1981）。但是 UVC 的穿透能力弱，在到达地面之前就被臭氧层所吸收，所以环境水体中，日光中的紫外光 B（UVB）（280～320 nm）、紫外光 A（UVA）（320～400 nm）和光合有效辐射（photosynthetically active radiation，PAR，400～800 nm）发挥主要作用。然而，在不同水体中测定的 UVB、UVA 和 PAR 在甲基汞光降解中的相对贡献差异显著。比如，加拿大西北部湖水中甲基汞光降解速率分别为 $0.871×10^{-3}$ $E^{-1}·m^2$（UVB）、$3.22×10^{-3}$ $E^{-1}·m^2$（UVA）、$0.313×10^{-3}$ $E^{-1}·m^2$（PAR），相对比值为 3∶10∶1（UVB∶UVA∶PAR）（Lehnherr and Louis，2009）。而美国旧金山湾水体中甲基汞光降解速率分别为 0.99 $E^{-1}·m^2$（UVB）、0.092 $E^{-1}·m^2$（UVA）、0.0025 $E^{-1}·m^2$（PAR），相对比值为 400∶37∶1（UVB∶UVA∶PAR）（Black et al.，2012）。因此，要准确评价三种不同波长光驱动下的甲基汞降解速率，需要综合考察天然有机质对活性氧自由基（reactive oxygen species，ROS）的形成和甲基汞的络合形态的影响，以及各波段光的光子通量（photon flux）和穿越水层时的光衰减作用等因素（Fernández-Gómez et al.，2013）。

关于甲基汞光降解的研究多集中于湖泊和海洋环境。而溪流具有水浅、快速混合等特点,其中汞含量受点源(采矿和工业废水)排放和大气沉降影响较大,易产生甲基汞的富集(Peterson et al.,2007)。而且,溪流生态系统中的甲基汞光降解过程受光照影响显著,增强光照能显著增强甲基汞的光降解作用(Tsui et al.,2013)。

天然有机物(natural organic matter,NOM)在天然水体中普遍存在,其中大部分是以溶解有机质(DOM)形态存在。天然有机物的种类和浓度是控制水体中甲基汞光降解的重要水化学参数(Chandan et al.,2015;Qian et al.,2014)。光照条件下DOM通过三种方式影响甲基汞的降解过程(Klapstein et al.,2017)(图3-5)。第一,溶解有机质经过太阳光照射后,本身会发生光漂白(其吸光度降低)、光矿化(有机碳损失)反应。因此,高DOM浓度的湖水中,DOM(或DOM-MeHg络合物)分子内会发生光子的竞争,从而抑制了甲基汞的光降解作用,溶解有机质浓度与甲基汞光降解速率常数呈现负相关关系(Klapstein et al.,2018)。天然有机物是多分散混合的体系,其分子质量范围从<100 Da到300 kDa(Mostofa et al.,2013)。高分子质量的DOM引起光衰减、抑制甲基汞光降解的效应更显著(Kim et al.,2017;Zhang et al.,2017;Yin et al.,2014b)。第二,DOM吸收光辐射后产生活性中间产物(photochemically produced reactive intermediates,PPRIs),如活性氧自由基(ROS)等,这些活性中间产物与甲基汞反应导致其降解。第三,甲基汞与溶解有机质形成络合物(MeHg-DOM),通过分子内电子转移使得C—Hg键断裂,从而降解甲基汞(Tai et al.,2014)。

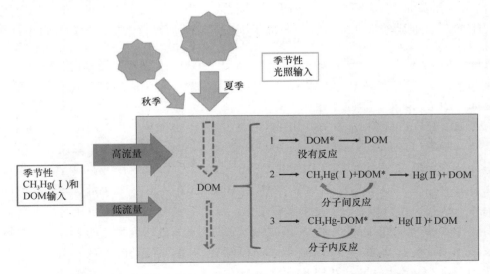

图 3-5　光照下淡水湖泊中可能存在的 DOM 与 MeHg 的反应概念图
[图片修改自(Klapstein et al.,2017)]

无论是甲基汞直接光降解的一级反应（Tai et al.，2014），还是活性氧自由基参与的甲基汞间接光降解的二级反应（Hammerschmidt and Fitzgerald，2010；Zhang and Hsu-Kim，2010），甲基汞的浓度和化学络合形态均是影响其光降解的重要因素。在大多数天然水体中甲基汞浓度较低，易与 DOM 中的还原性硫配体结合。当甲基汞浓度很高或其他结合能力较弱的配体浓度很高时，存在相当部分甲基汞会与 DOM 中弱的配体结合（如 O/N 官能团）或与水体中其他配体结合（如 Cl^-、OH^-）（Dong et al.，2010；Ravichandran，2004）。比如，海洋水中有较高浓度的氯离子（Cl^-），氯化甲基汞（CH_3HgCl）络合物是其主要形态；而大多数淡水含有较低浓度的氯离子和较高浓度的溶解有机质，因此甲基汞与含巯基的天然有机物络合的状态是其主要形态（Amirbahman et al.，2002；Hintelmann et al.，1997）。甲基汞与含巯基天然有机物结合增加了甲基碳原子的电负性，使 C—Hg 键更易断裂（Zhang and Hsu-Kim，2010），促进甲基汞的光降解（Tai et al.，2014）。这可能是导致甲基汞降解速率随着盐度的增加而降低（Black et al.，2012），甲基汞光降解速率在淡水环境中较快（Lehnherr and Louis，2009）、在海洋环境中较慢（Whalin et al.，2007）的主要原因。

除天然有机物外，甲基汞光降解还受到其他水化学参数的影响。比如，水环境的 pH 不仅影响天然有机物（Karlsson and Skyllberg，2003）和甲基汞络合物的化学形态（Chen et al.，2003），也会影响 PPRIs 的生成和分解（Kim et al.，2017；Legrini et al.，1993）。三价铁（Fe^{3+}）、硝酸根（NO_3^-）等离子可以通过提高活性自由基浓度而促进甲基汞光降解（DiMento and Mason，2017）。而碳酸氢根（HCO_3^-）、亚硝酸根（NO_2^-）等阴离子则通过消除活性自由基，如羟基自由基等抑制甲基汞的降解过程（Ma and Graham，2000）。

3.6 展 望

甲基汞在全球范围内广泛分布于水、土壤、大气等各种环境介质，沿食物链和食物网进行营养级传递，产生生物累积和生物放大效应，对处于较高营养级的野生动物以及人体健康产生危害。由微生物和光化学作用驱动的汞甲基化和去甲基化反应共同决定了甲基汞在各环境介质中的浓度，该过程受到汞化学形态的显著影响。为了准确预测甲基汞的环境行为，有效防范甲基汞积累产生的健康风险，亟待加强以下几方面研究：①长时间、低水平甲基汞暴露的健康危害及其作用机制；②导致人体汞暴露的两种主要食物，鱼类和水稻体内汞甲基化和去甲基化的反应机制；③不同环境条件下，无机汞的化学形态和甲基化微生物汞摄入的主控因子；④无光、低汞浓度环境中甲基汞降解的作用机制。

参 考 文 献

毕丽, 贺纪正, 张丽梅, 刘玉荣, 2018. 环境中汞微生物转化研究进展. 环境化学, 37: 2359-2367.

鲁霞, 2016. 环境微生物与铀和汞的相互作用和机理. 兰州: 兰州大学.

阴永光, 李雁宾, 蔡勇, 江桂斌, 2011. 汞的环境光化学. 环境化学, 30: 84-91.

左跃钢, 庞叔薇., 1985. 巯基化合物存在下无机汞的光化学甲基化. 环境科学学报, 5: 239-243.

Achá D, Hintelmann H, Pabón C A, 2012. Sulfate-reducing bacteria and mercury methylation in the water column of the lake 658 of the experimental lake area. Geomicrobiology Journal, 29: 667-674.

Afton S E, Caruso J A, 2009. The effect of Se antagonism on the metabolic fate of Hg in *Allium fistulosum*. Journal of Analytical Atomic Spectrometry, 24: 759-766.

Akagi H, Fujita Y, Takabatake E, 1975. Photochemical methylation of inorganic mercury in the presence of mercuric sulfide. Chemistry Letters, 4: 171-176.

Akagi H, Takabatake E, 1973. Photochemical formation of methylmercuric compounds from mercuric acetate. Chemosphere, 2: 131-133.

Amirbahman A, Reid A L, Haines T A, Kahl J S, Arnold C, 2002. Association of methylmercury with dissolved humic acids. Environmental Science & Technology, 36: 690-695.

Andersen A, Julshamn K, Ringdal O, Mørkøre J, 1987. Trace elements intake in the Faroe Islands II. Intake of mercury and other elements by consumption of pilot whales (*Globicephalus meleanus*). Science of the Total Environment, 65: 63-68.

Arcagni M, Juncos R, Rizzo A, Pavlin M, Fajon V, Arribere M A, Horvat M, Guevara S R, 2018. Species- and habitat-specific bioaccumulation of total mercury and methylmercury in the food web of a deep oligotrophic lake. Science of the Total Environment, 612: 1311-1319.

Asaduzzaman A M, Schreckenbach G, 2011. Degradation mechanism of methyl mercury selenoamino acid complexes: A computational study. Inorganic Chemistry, 50: 2366-2372.

Barkay T, Miller S M, Summers A O, 2003. Bacterial mercury resistance from atoms to ecosystems. FEMS Microbiology Review, 27: 355-384.

Barkay T, Wagner-Dobler I, 2005. Microbial transformations of mercury: Potentials, challenges, and achievements in controlling mercury toxicity in the environment. Advances In Applied Microbiology, 57: 1-52.

Benoit J M, Gilmour C C, Mason R P, 2001. Aspects of bioavailability of mercury for methylation in pure cultures of *Desulfobulbus propionicus* (1pr3). Applied and Environmental Microbiology, 67: 51-58.

Billen G, Joiris C, Wollast R, 1974. A bacterial methylmercury-mineralizing activity in river sediments. Water Research, 8: 219-225.

Bjornberg K A, Vahter M, Petersson-Grawe K, Glynn A, Cnattingius S, Darnerud P O, Atuma S, Aune M, Becker W, Berglund M, 2003. Methyl mercury and inorganic mercury in Swedish pregnant women and in cord blood: Influence of fish consumption. Environmental Health Perspectives, 111: 637-641.

Bjornberg K A, Vahtera M, Grawe K P, Berglund M, 2005. Methyl mercury exposure in Swedish women with high fish consumption. Science of the Total Environment, 341: 45-52.

Black F J, Paytan A, Knee K L, De Sieyes N R, Ganguli P M, Gary E, Flegal A R, 2009. Submarine groundwater discharge of total mercury and monomethylmercury to central california coastal waters. Environmental Science & Technology, 43: 5652-5659.

Black F J, Poulin B A, Flegal A R, 2012. Factors controlling the abiotic photo-degradation of monomethylmercury in surface waters. Geochimica et Cosmochimica Acta, 84: 492-507.

Bloom N S, 1992. On the chemical form of mercury in edible fish and marine invertebrate tissue. Canadian Journal of Fisheries and Aquatic Sciences, 49: 1010-1017.

Bond A L, Diamond A W, 2009. Total and methyl mercury concentrations in seabird feathers and eggs. Archives of Environmental Contamination and Toxicology, 56: 286-291.

Bowman K L, Hammerschmidt C R, Lamborg C H, Swarr G, 2015. Mercury in the North Atlantic Ocean: The US GEOTRACES zonal and meridional sections. Deep Sea Research Part II: Topical Studies In Oceanography, 116: 251-261.

Bradley P M, Journey C A, Lowery M A, Brigham M E, Burns D A, Button D T, Chapelle F H, Lutz M A, Marvin-Dipasquale M C, Riva-Murray K, 2012. Shallow groundwater mercury supply in a coastal plain stream. Environmental Science & Technology, 46: 7503-7511.

Bratkic A, Vahcic M, Kotnik J, Vazner K O, Begu E, Woodward E M S, Horvat M, 2016. Mercury presence and speciation in the South Atlantic Ocean along the 40 degrees S transect. Global Biogeochemical Cycles, 30: 105-119.

Bravo A G, Bouchet S, Tolu J, Bjorn E, Mateos-Rivera A, Bertilsson S, 2017. Molecular composition of organic matter controls methylmercury formation in boreal lakes. Nature Communications, 8: 14255.

Bridou R, Monperrus M, Gonzalez P R, Guyoneaud R, Amouroux D, 2011. Simultaneous determination of mercury methylation and demethylation capacities of various sulfate - reducing bacteria using species - specific isotopic tracers. Environmental Toxicology and Chemistry, 30: 337-344.

Brigham M E, Wentz D A, Aiken G R, Krabbenhoft D P, 2009. Mercury cycling in stream ecosystems. 1. Water column chemistry and transport. Environmental Science & Technology, 43: 2720-2725.

Brown S D, Gilmour C C, Kucken A M, Wall J D, Elias D A, Brandt C C, Podar M, Chertkov O, Held B, Bruce D C, Detter J C, Tapia R, Han C S, Goodwin L A, Cheng J-F, Pitluck S, Woyke T, Mikhailova N, Ivanova N N, Han J, Lucas S, Lapidus A L, Land M L, Hauser L J, Palumbo A V, 2011. Genome sequence of the mercury-methylating strain *Desulfovibrio desulfuricans* ND132. Journal of Bacteriology, 193: 2078-2079.

Budtz-Jorgensen E, Keiding N, Grandjean P, White R F, Wiehe P, 1999. Methylmercury neurotoxicity independent of PCB exposure. Environmental Health Perspectives, 107: A236-A237.

Cabana G, Rasmussen J B, 1994. Modelling food chain structure and contaminant bioaccumulation using stable nitrogen isotopes. Nature, 372: 255-257.

Cabana G, Tremblay A, Kalff J, Rasmussen J B, 1994. Pelagic food chain structure in ontario lakes: A determinant of mercury levels in lake trout (*Salvelinus namaycush*). Canadian Journal of Fisheries and Aquatic Sciences, 51: 381-389.

Canario J, Caetano M, Vale C, Cesario R, 2007. Evidence for elevated production of methylmercury in salt marshes. Environmental Science & Technology, 41: 7376-7382.

Canario J, Santos-Echeandia J, Padeiro A, Amaro E, Strass V, Klaas C, Hoppema M, Ossebaar S, Koch B P, Laglera L M, 2017. Mercury and methylmercury in the Atlantic sector of the Southern Ocean. Deep Sea Research Part II: Topical Studies In Oceanography, 138: 52-62.

Carravieri A, Bustamante P, Churlaud C, Cherel Y, 2013. Penguins as bioindicators of mercury contamination in the Southern Ocean: Birds from the Kerguelen Islands as a case study. Science of the Total Environment, 454: 141-148.

Carravieri A, Cherel Y, Jaeger A, Churlaud C, Bustamante P, 2016. Penguins as bioindicators of mercury contamination in the southern Indian Ocean: geographical and temporal trends. Environmental Pollution, 213: 195-205.

Castro L, Dommergue A, Larose C, Ferrari C, Maron L, 2011. A theoretical study of abiotic methylation reactions of gaseous elemental mercury by halogen-containing molecules. Journal of Physical Chemistry A, 115: 5602-5608.

Celo V, Lean D R S, Scott S L, 2006. Abiotic methylation of mercury in the aquatic environment. Science of the Total Environment, 368: 126-137.

Chandan P, Ghosh S, Bergquist B A, 2015. Mercury isotope fractionation during aqueous photoreduction of monomethylmercury in the presence of dissolved organic matter. Environmental Science & Technology, 49: 259-267.

Chasar L C, Scudder B C, Stewart A R, Bell A H, Aiken G R, 2009. Mercury cycling in stream ecosystems. 3. Trophic dynamics and methylmercury bioaccumulation. Environmental Science & Technology, 43: 2733-2739.

Chen B W, Chen P, He B, Yin Y G, Fang L C, Wang X W, Liu H T, Yang L H, Luan T G, 2015. Identification of mercury methylation product by *tert*-butyl compounds in aqueous solution under light irradiation. Marine Pollution Bulletin, 98: 40-46.

Chen B W, Wang T, Yin Y G, He B, Jiang G B, 2007. Methylation of inorganic mercury by methylcobalamin in aquatic systems. Applied Organometallic Chemistry, 21: 462-467.

Chen C Y, Borsuk M E, Bugge D M, Hollweg T, Balcom P H, Ward D M, Williams J, Mason R P, 2014. Benthic and pelagic pathways of methylmercury bioaccumulation in estuarine food webs of the northeast United States. PLoS One, 9: 1-11.

Chen J, Pehkonen S O, Lin C-J, 2003. Degradation of monomethylmercury chloride by hydroxyl radicals in simulated natural waters. Water Research, 37: 2496-2504.

Choi S C, Chase T, Bartha R, 1994. Metabolic pathways leading to mercury methylation in *Desulfovibrio desulfuricans* LS. Applied and Environmental Microbiology, 60: 4072-4077.

Choi S C, Chase T, Bartha R, 1994. Enzymatic catalysis of mercury methylation by *Desulfovibrio desulfuricans* LS. Applied and Environmental Microbiology, 60: 1342-1346.

Christensen G A, Wymore A M, King A J, Podar M, Hurt R A, Santillan E U, Soren A, Brandt C C, Brown S D, Palumbo A V, Wall J D, Gilmour C C, Elias D A, 2016. Development and validation of broad-range qualitative and clade-specific quantitative molecular probes for assessing mercury methylation in the environment. Applied and Environmental Microbiology, 82: 6068-6078.

Cleckner L B, Garrison P J, Hurley J P, Olson M L, Krabbenhoft D P, 1998. Trophic transfer of methyl mercury in the northern Florida Everglades. Biogeochemistry, 40: 347-361.

Compeau G, Bartha R, 1984. Methylation and demethylation of mercury under controlled redox, pH and salinity conditions. Applied and Environmental Microbiology, 48: 1203-1207.

Compeau G C, Bartha R, 1985. Sulfate-reducing bacteria: Principal methylators of mercury in anoxic estuarine sediment. Applied and Environmental Microbiology, 50: 498-502.

Correia R R S, Guimarães J R D, 2017. Mercury methylation and sulfate reduction rates in mangrove sediments, Rio de Janeiro, Brazil: The role of different microorganism consortia. Chemosphere, 167: 438-443.

Cossa D, De Madron X D, Schafer J, Lanceleur L, Guedron S, Buscail R, Thomas B, Castelle S,

Naudin J J, 2017. The open sea as the main source of methylmercury in the water column of the Gulf of Lions (Northwestern Mediterranean margin). Geochimica et Cosmochimica Acta, 199: 222-237.

Cossa D, Harmelin-Vivien M, Mellon-Duval C, Loizeau V, Averty B, Crochet S, Chou L, Cadiou J F, 2012. Influences of bioavailability, trophic position, and growth on methylmercury in hakes (*Merluccius merluccius*) from northwestern mediterranean and northeastern atlantic. Environmental Science & Technology, 46: 4885-4893.

Cossa D, Heimburger L E, Lannuzel D, Rintoul S R, Butler E C V, Bowie A R, Averty B, Watson R J, Remenyi T, 2011. Mercury in the Southern Ocean. Geochimica et Cosmochimica Acta, 75: 4037-4052.

Cossa D, Martin J M, Takayanagi K, Sanjuan J, 1997. The distribution and cycling of mercury species in the western Mediterranean. Deep Sea Research Part II: Topical Studies In Oceanography, 44: 721-740.

Craig P, Bartlett P, 1978. The role of hydrogen sulphide in environmental transport of mercury. Nature, 275: 635-637.

Creswell J E, Shafer M M, Babiarz C L, Tan S Z, Musinsky A L, Schott T H, Roden E E, Armstrong D E, 2017. Biogeochemical controls on mercury methylation in the Allequash Creek wetland. Environmental Science and Pollution Research, 24: 15325-15339.

Crump K S, Kjellstrom T, Shipp A M, Silvers A, Stewart A, 1998. Influence of prenatal mercury exposure upon scholastic and psychological test performance: Benchmark analysis of a New Zealand cohort. Risk Analysis, 18: 701-713.

Cui W B, Liu G L, Bezerra M, Lagos D A, Li Y B, Cai Y, 2017. Occurrence of methylmercury in rice-based infant cereals and estimation of daily dietary intake of methylmercury for infants. Journal of Agricultural and Food Chemistry, 65: 9569-9578.

Davidson P W, Myers G J, Cox C, Axtell C, Shamlaye C, Sloane-Reeves J, Cernichiari E, Needham L, Choi A, Wang Y N, Berlin M, Clarkson T W, 1998. Effects of prenatal and postnatal methylmercury exposure from fish consumption on neurodevelopment: Outcomes at 66 months of age in the Seychelles Child Development Study. Jama-Journal of the American Medical Association, 280: 701-707.

Debes F, Weihe P, Grandjean P, 2016. Cognitive deficits at age 22 years associated with prenatal exposure to methylmercury. Cortex, 74: 358-369.

Desrosiers M, Planas D, Mucci A, 2006. Mercury methylation in the epilithon of boreal shield aquatic ecosystems. Environmental Science & Technology, 40: 1540-1546.

Diez E G, Loizeau J L, Cosio C, Bouchet S, Adatte T, Amouroux D, Bravo A G, 2016. Role of settling particles on mercury methylation in the oxic water column of freshwater systems. Environmental Science & Technology, 50: 11672-11679.

Dimento B P, Mason R P, 2017. Factors controlling the photochemical degradation of methylmercury in coastal and oceanic waters. Marine Chemistry, 196: 116-125.

Dong W, Liang L, Brooks S, Southworth G, Gu B, 2010. Roles of dissolved organic matter in the speciation of mercury and methylmercury in a contaminated ecosystem in Oak Ridge, Tennessee. Environmental Chemistry, 7: 94-102.

Driscoll C T, Mason R P, Chan H M, Jacob D J, Pirrone N, 2013. Mercury as a global pollutant: sources, pathways, and effects. Environmental Science & Technology, 47: 4967-4983.

Duvall S E, Barron M G, 2000. A screening level probabilistic risk assessment of mercury in Florida

Everglades food webs. Ecotoxicology and Environment Safety, 47: 298-305.

Eagles-Smith C A, Ackerman J T, 2014. Mercury bioaccumulation in estuarine wetland fishes: Evaluating habitats and risk to coastal wildlife. Environmental Pollution, 193: 147-155.

Eckley C S, Luxton T P, Goetz J, Mckernan J, 2017. Water-level fluctuations influence sediment porewater chemistry and methylmercury production in a flood-control reservoir. Environmental Pollution, 222: 32-41.

Eklof K, Bishop K, Bertilsson S, Bjorn E, Buck M, Skyllberg U, Osman O A, Kronberg R M, Bravo A G, 2018. Formation of mercury methylation hotspots as a consequence of forestry operations. Science of the Total Environment, 613: 1069-1078.

Eklof K, Schelker J, Sorensen R, Meili M, Laudon H, Von Bromssen C, Bishop K, 2014. Impact of forestry on total and methyl-mercury in surface waters: Distinguishing effects of logging and site preparation. Environmental Science & Technology, 48: 4690-4698.

Ekstrom E B, Morel F M, 2007. Cobalt limitation of growth and mercury methylation in sulfate-reducing bacteria. Environmental Science & Technology, 42: 93-99.

Ekstrom E B, Morel F M M, Benoit J M, 2003. Mercury methylation independent of the acetyl-coenzyme a pathway in sulfate-reducing bacteria. Applied and Environmental Microbiology, 69: 5414-5422.

Evers D C, Burgess N M, Champoux L, Hoskins B, Major A, Goodale W M, Taylor R J, Poppenga R, Daigle T, 2005. Patterns and interpretation of mercury exposure in freshwater avian communities in northeastern North America. Ecotoxicology, 14: 193-221.

Falter R, 1999a. Experimental study on the unintentional abiotic methylation of inorganic mercury during analysis: Part 1. Localisation of the compounds effecting the abiotic mercury methylation. Chemosphere, 39: 1051-1073.

Falter R, 1999b. Experimental study on the unintentional abiotic methylation of inorganic mercury during analysis: Part 2. Controlled laboratory experiments to elucidate the mechanism and critical discussion of the species specific isotope addition correction method. Chemosphere, 39: 1075-1091.

Feng X B, Li P, Qiu G L, Wang S, Li G H, Shang L H, Meng B, Jiang H M, Bai W Y, Li Z G, Fu X W, 2008. Human exposure to methylmercury through rice intake in mercury mining areas, Guizhou province, China. Environmental Science & Technology, 42: 326-332.

Fenton H, 1894. LXXIII.—Oxidation of tartaric acid in presence of iron. Journal of the Chemical Society, Transactions, 65: 899-910.

FernáNdez-GóMez C, Drott A, BjöRn E, Díez S, Bayona J M, Tesfalidet S, Lindfors A, Skyllberg U, 2013. Towards universal wavelength-specific photodegradation rate constants for methyl mercury in humic waters, exemplified by a boreal lake-wetland gradient. Environmental Science & Technology, 47: 6279-6287.

Fitzgerald W F, Lamborg C H, Hammerschmidt C R, 2007. Marine biogeochemical cycling of mercury. Chemical Reviews, 107: 641-662.

Fleming E J, Mack E E, Green P G, Nelson D C, 2006. Mercury methylation from unexpected sources: molybdate-inhibited freshwater sediments and an iron-reducing bacterium. Applied and Environmental Microbiology, 72: 457-464.

Frankland E, 1852. XIX. On a new series of organic bodies containing metals. Philosophical Transactions of the Royal Society of London, Series A, 142: 417-444.

French T D, Houben A J, Desforges J P W, Kimpe L E, Kokelj S V, Poulain A J, Smol J P, Wang X W,

Blais J M, 2014. Dissolved organic carbon thresholds affect mercury bioaccumulation in arctic lakes. Environmental Science & Technology, 48: 3162-3168.

Fu J J, Wang Y W, Zhou Q F, Jiang G B, 2010. Trophic transfer of mercury and methylmercury in an aquatic ecosystem impacted by municipal sewage effluents in Beijing, China. Journal of Environmental Sciences, 22: 1189-1194.

Fuchsman P C, Brown L E, Henning M H, Bock M J, Magar V S, 2017. Toxicity reference values for methylmercury effects on avian reproduction: Critical review and analysis. Environmental Toxicology and Chemistry, 36: 294-319.

Gagnon C, Fisher N S, 1997. Bioavailability of sediment-bound methyl and inorganic mercury to a marine bivalve. Environmental Science & Technology, 31: 993-998.

Ganguli P M, Conaway C H, Swarzenski P W, Izbicki J A, Flegal A R, 2012. Mercury speciation and transport via submarine groundwater discharge at a southern California coastal lagoon system. Environmental Science & Technology, 46: 1480-1488.

Garcia E, Amyot M, Ariya P A, 2005. Relationship between DOC photochemistry and mercury redox transformations in temperate lakes and wetlands. Geochimica et Cosmochimica Acta, 69: 1917-1924.

Gardfeldt K, Munthe J, Stromberg D, Lindqvist O, 2003. A kinetic study on the abiotic methylation of divalent mercury in the aqueous phase. Science of the Total Environment, 304: 127-136.

Gilmour C C, 1986. Estuarine methylation of tin and its relationship to the microbial sulfur cycle. Maryland: University of Maryland, College Park.

Gilmour C C, Elias D A, Kucken A M, Brown S D, Palumbo A V, Schadt C W, Wall J D, 2011. Sulfate-reducing bacterium *Desulfovibrio desulfuricans* ND132 as a model for understanding bacterial mercury methylation. Applied and Environmental Microbiology, 77: 3938-3951.

Gilmour C C, Henry E A, 1991. Mercury methylation in aquatic systems affected by acid deposition. Environmental Pollution, 71: 131-169.

Gilmour C C, Henry E A, Mitchell R, 1992. Sulfate stimulation of mercury methylation in freshwater sediments. Environmental Science & Technology, 26: 2281-2287.

Gilmour C C, Podar M, Bullock A L, Graham A M, Brown S D, Somenahally A C, Johs A, Hurt R A, Bailey K L, Elias D A, 2013. Mercury methylation by novel microorganisms from new environments. Environmental Science & Technology, 47: 11810-11820.

Gilmour C C, Riedel G, Ederington M, Bell J, Gill G, Stordal M, 1998. Methylmercury concentrations and production rates across a trophic gradient in the northern Everglades. Biogeochemistry, 40: 327-345.

Gionfriddo C M, Tate M T, Wick R R, Schultz M B, Zemla A, Thelen M P, Schofield R, Krabbenhoft D P, Holt K E, Moreau J W, 2016. Microbial mercury methylation in Antarctic sea ice. Nature Microbiology, 1: 16127.

Glooschenko W A, 1969. Accumulation of ^{203}Hg by the marine diatom *Chaetoceros costatum*. Journal of Phycology, 5: 224-226.

Golding J, Gregory S, Iles-Caven Y, Hibbeln J, Emond A, Taylor C M, 2016. Associations between prenatal mercury exposure and early child development in the ALSPAC study. Neurotoxicology, 53: 215-222.

Graham A M, Bullock A L, Maizel A C, Elias D A, Gilmour C C, 2012. Detailed assessment of the kinetics of Hg-cell association, Hg methylation, and methylmercury degradation in several *Desulfovibrio* species. Applied and Environmental Microbiology, 78: 7337-7346.

Grandjean P, Budtz-Jorgensen E, Jorgensen P J, Weihe P, 2005. Umbilical cord mercury concentration as biomarker of prenatal exposure to methylmercury. Environmental Health Perspectives, 113: 905-908.

Grandjean P, Budtz-Jorgensen E, White R F, Jorgensen P J, Weihe P, Debes F, Keiding N, 1999. Methylmercury exposure biomarkers as indicators of neurotoxicity in children aged 7 years. American Journal of Epidemiology, 150: 301-305.

Grandjean P, Weihe P, Burse V W, Needham L L, Storr-Hansen E, Heinzow B, Debes F, Murata K, Simonsen H, Ellefsen P, Budtz-Jorgensen E, Keiding N, White R F, 2001. Neurobehavioral deficits associated with PCB in 7-year-old children prenatally exposed to seafood neurotoxicants. Neurotoxicology and Teratology, 23: 305-317.

Grandjean P, Weihe P, White R F, Debes F, 1998. Cognitive performance of children prenatally exposed to "safe" levels of methylmercury. Environmental Research, 77: 165-172.

Grandjean P, Weihe P, White R F, Debes F, Araki S, Yokoyama K, Murata K, Sorensen N, Dahl R, Jorgensen P J, 1997. Cognitive deficit in 7-year-old children with prenatal exposure to methylmercury. Neurotoxicology and Teratology, 19: 417-428.

Graydon J A, Louis V L S, Hintelmann H, Lindberg S E, Sandilands K A, Rudd J W M, Kelly C A, Hall B D, Mowat L D, 2008. Long-term wet and dry deposition of total and methyl mercury in the remote boreal ecoregion of Canada. Environmental Science & Technology, 42: 8345-8351.

Gu B, Bian Y, Miller C L, Dong W, Jiang X, Liang L, 2011. Mercury reduction and complexation by natural organic matter in anoxic environments. Proceedings of the National Academy of Sciences of the United States of America, 108: 1479-1483.

Hall B D, Aiken G R, Krabbenhoft D P, Marvin-Dipasquale M, Swarzenski C M, 2008. Wetlands as principal zones of methylmercury production in southern Louisiana and the Gulf of Mexico region. Environmental Pollution, 154: 124-134.

Hall B D, Louis V L S, 2004. Methylmercury and total mercury in plant litter decomposing in upland forests and flooded landscapes. Environmental Science & Technology, 38: 5010-5021.

Hall B D, Rosenberg D M, Wiens A P, 1998. Methyl mercury in aquatic insects from an experimental reservoir. Canadian Journal of Fisheries and Aquatic Sciences, 55: 2036-2047.

Hamelin S, Amyot M, Barkay T, Wang Y, Planas D, 2011. Methanogens: Principal methylators of mercury in lake periphyton. Environmental Science & Technology, 45: 7693-7700.

Hamelin S, Planas D, Amyot M, 2015. Mercury methylation and demethylation by periphyton biofilms and their host in a fluvial wetland of the St. Lawrence River (QC, Canada). Science of the Total Environment, 512: 464-471.

Hammerschmidt C R, Bowman K L, 2012. Vertical methylmercury distribution in the subtropical North Pacific Ocean. Marine Chemistry, 132: 77-82.

Hammerschmidt C R, Fitzgerald W F, 2004. Geochemical controls on the production and distribution of methylmercury in near-shore marine sediments. Environmental Science & Technology, 38: 1487-1495.

Hammerschmidt C R, Fitzgerald W F, 2010. Iron-mediated photochemical decomposition of methylmercury in an arctic alaskan lake. Environmental Science & Technology, 44: 6138-6143.

Hammerschmidt C R, Fitzgerald W F, 2006a. Methylmercury cycling in sediments on the continental shelf of southern New England. Geochimica et Cosmochimica Acta, 70: 918-930.

Hammerschmidt C R, Fitzgerald W F, 2006b. Photodecomposition of methylmercury in an arctic Alaskan lake. Environmental Science & Technology, 40: 1212-1216.

Hammerschmidt C R, Lamborg C H, Fitzgerald W F, 2007. Aqueous phase methylation as a potential source of methylmercury in wet deposition. Atmospheric Environment, 41: 1663-1668.

Han S, Narasingarao P, Obraztsova A, Gieskes J, Hartmann A C, Tebo B M, Allen E E, Deheyn D D, 2010. Mercury speciation in marine sediments under sulfate-limited conditions. Environmental Science & Technology, 44: 3752-3757.

Han X, Li Y, Li D, Liu C, 2017. Role of free radicals/reactive oxygen species in MeHg photodegradation: Importance of utilizing appropriate scavengers. Environmental Science & Technology, 51: 3784-3793.

Harmon S M, King J K, Gladden J B, Chandler G T, Newman L A, 2004. Methylmercury formation in a wetland mesocosm amended with sulfate. Environmental Science & Technology, 38: 650-656.

Harris H H, Pickering I J, George G N, 2003. The chemical form of mercury in fish. Science, 301: 1203.

Hayashi K, Kawai S, Ohno T, Maki Y, 1977. Photomethylation of inorganic mercury by aliphatic α-amino-acids. Journal of the Chemical Society, Chemical Communications, 5: 158-159.

He Q, Xing Z, Zhang S C, Zhang X R, 2015. ICP-MS/MS as a tool to study abiotic methylation of inorganic mercury reacting with VOCs. Journal of Analytical Atomic Spectrometry, 30: 1997-2002.

Heimburger L E, Cossa D, Marty J C, Migon C, Averty B, Dufour A, Ras J, 2010. Methyl mercury distributions in relation to the presence of nano- and picophytoplankton in an oceanic water column (Ligurian Sea, North-western Mediterranean). Geochimica et Cosmochimica Acta, 74: 5549-5559.

Heyes A, Mason R P, Kim E-H, Sunderland E, 2006. Mercury methylation in estuaries: Insights from using measuring rates using stable mercury isotopes. Marine Chemistry, 102: 134-147.

Hintelmann H, Welbourn P M, Evans R D, 1997. Measurement of complexation of methylmercury(II) compounds by freshwater humic substances using equilibrium dialysis. Environmental Science & Technology, 31: 489-495.

Hodson P V, Norris K, Berquist M, Campbell L M, Ridal J J, 2014. Mercury concentrations in amphipods and fish of the Saint Lawrence River (Canada) are unrelated to concentrations of legacy mercury in sediments. Science of the Total Environment, 494: 218-228.

Hofacker A F, Voegelin A, Kaegi R, Kretzschmar R, 2013. Mercury mobilization in a flooded soil by incorporation into metallic copper and metal sulfide nanoparticles. Environmental Science & Technology, 47: 7739-7746.

Hoggarth C G J, Hall B D, Mitchell C P J, 2015. Mercury methylation in high and low-sulphate impacted wetland ponds within the prairie pothole region of North America. Environmental Pollution, 205: 269-277.

Hsi H C, Jiang C B, Yang T H, Chien L C, 2014. The neurological effects of prenatal and postnatal mercury/methylmercury exposure on three-year-old children in Taiwan. Chemosphere, 100: 71-76.

Hsu-Kim H, Kucharzyk K H, Zhang T, Deshusses M A, 2013. Mechanisms regulating mercury bioavailability for methylating microorganisms in the aquatic environment: A critical review. Environmental Science & Technology, 47: 2441-2456.

Hu H, Lin H, Zheng W, Tomanicek S J, Johs A, Feng X, Elias D A, Liang L, Gu B, 2013. Oxidation and methylation of dissolved elemental mercury by anaerobic bacteria. Nature Geoscience, 6: 751-754.

Huguet L, Castelle S, Schäfer J, Blanc G, Maury-Brachet R, Reynouard C, Jorand F, 2010. Mercury methylation rates of biofilm and plankton microorganisms from a hydroelectric reservoir in French Guiana. Science of the Total Environment, 408: 1338-1348.

Iwata H, Masukawa T, Kito H, Hayashi M, 1982. Degradation of methylmercury by selenium. Life Sciences, 31: 859-866.

Jacobson J L, Muckle G, Ayotte P, Dewailly E, Jacobson S W, 2015. Relation of prenatal methylmercury exposure from environmental sources to childhood IQ. Environmental Health Perspectives, 123: 827-833.

JECFA, 2000. Methylmercury. *In*: WHO food additives series: 44-Safety evaluation of certain food additives and contaminants. Geneva: World Health Organization (WHO).

JECFA, 2004. Methylmercury (addendum). *In*: WHO food additives series: 52-Safety evaluation of certain food additives and contaminants. Geneva: World Health Organization (WHO).

Jensen S, Jernelöv A, 1969. Biological methylation of mercury in aquatic organisms. Nature, 223: 753-754.

Johnson W P, Swanson N, Black B, Rudd A, Carling G, Fernandez D P, Luft J, Van Leeuwen J, Marvin-Dipasquale M, 2015. Total- and methyl-mercury concentrations and methylation rates across the freshwater to hypersaline continuum of the Great Salt Lake, Utah, USA. Science of the Total Environment, 511: 489-500.

Julshamn K, Andersen A, Ringdal O, Mørkøre J, 1987. Trace elements intake in the Faroe Islands I. Element levels in edible parts of pilot whales (*Globicephalus meleanus*). Science of the Total Environment, 65: 53-62.

Kahilainen K K, Thomas S M, Nystedt E K M, Keva O, Malinen T, Hayden B, 2017. Ecomorphological divergence drives differential mercury bioaccumulation in polymorphic European whitefish (*Coregonus lavaretus*) populations of subarctic lakes. Science of the Total Environment, 599: 1768-1778.

Karlsson T, Skyllberg U, 2003. Bonding of ppb levels of methyl mercury to reduced sulfur groups in soil organic matter. Environmental Science & Technology, 37: 4912-4918.

Kasper D, Forsberg B R, Amaral J H F, Leitao R P, Py-Daniel S S, Bastos W R, Malm O, 2014. Reservoir stratification affects methylmercury levels in river water, plankton, and fish downstream from Balbina Hydroelectric Dam, Amazonas, Brazil. Environmental Science & Technology, 48: 1032-1040.

Kehrig H A, Palermo E F A, Seixas T G, Branco C W C, Moreira I, Malm O, 2009. Trophic transfer of methylmercury and trace elements by tropical estuarine seston and plankton. Estuarine, Coastal and Shelf Science, 85: 36-44.

Kehrig H A, Seixas T G, Baeta A P, Malm O, Moreira I, 2010. Inorganic and methylmercury: Do they transfer along a tropical coastal food web? Marine Pollution Bulletin, 60: 2350-2356.

Kelly C A, Rudd J W M, Bodaly R A, Roulet N P, St.Louis V L, Heyes A, Moore T R, Schiff S, Aravena R, Scott K J, Dyck B, Harris R, Warner B, Edwards G, 1997. Increases in fluxes of greenhouse gases and methyl mercury following flooding of an experimental reservoir. Environmental Science & Technology, 31: 1334-1344.

Kelly E N, Schindler D W, St Louis V L, Donald D B, Vlaclicka K E, 2006. Forest fire increases mercury accumulation by fishes via food web restructuring and increased mercury inputs. Proceedings of the National academy of Sciences of the United States of America, 103: 19380-19385.

Kerin E J, Gilmour C, Roden E, Suzuki M, Coates J, Mason R, 2006. Mercury methylation by dissimilatory iron-reducing bacteria. Applied and Environmental Microbiology, 72: 7919-7921.

Keva O, Hayden B, Harrod C, Kahilainen K K, 2017. Total mercury concentrations in liver and muscle of European whitefish (*Coregonus lavaretus* (L.)) in a subarctic lake - Assessing the factors driving year-round variation. Environmental Pollution, 231: 1518-1528.

Khan M a K, Wang F, 2010. Chemical demethylation of methylmercury by selenoamino acids. Chemical Research In Toxicology, 23: 1202-1206.

Kim M K, Zoh K D, 2013. Effects of natural water constituents on the photo-decomposition of methylmercury and the role of hydroxyl radical. Science of the Total Environment, 449: 95-101.

Kim M K, Won A Y, Zoh K D, 2017. Effects of molecular size fraction of DOM on photodegradation of aqueous methylmercury. Chemosphere, 174: 739-746.

King J K, Harmon S M, Fu T T, Gladden J B, 2002. Mercury removal, methylmercury formation, and sulfate-reducing bacteria profiles in wetland mesocosms. Chemosphere, 46: 859-870.

King J K, Kostka J E, Frischer M E, Saunders F M, 2000. Sulfate-reducing bacteria methylate mercury at variable rates in pure culture and in marine sediments. Applied and Environmental Microbiology, 66: 2430-2437.

King J K, Saunders F M, Lee R F, Jahnke R A, 1999. Coupling mercury methylation rates to sulfate reduction rates in marine sediments. Environmental Toxicology and Chemistry, 18: 1362-1369.

Klapstein S J, Ziegler S E, O'driscoll N J, 2018. Methylmercury photodemethylation is inhibited in lakes with high dissolved organic matter. Environmental Pollution, 232: 392-401.

Klapstein S J, Ziegler S E, Risk D A, O'driscoll N J, 2017. Quantifying the effects of photoreactive dissolved organic matter on methylmercury photodemethylation rates in freshwaters. Environmental Toxicology and Chemistry, 36: 1493-1502.

Koeman J H, Peeters W H M, Koudstaal-Hol C H M, 1973. Mercury-selenium correlations in marine mammals. Nature, 245: 385-386.

Kronberg R M, Jiskra M, Wiederhold J G, Bjorn E, Skyllberg U, 2016. Methyl mercury formation in hillslope soils of Boreal Forests: The role of forest harvest and anaerobic microbes. Environmental Science & Technology, 50: 9177-9186.

Kuss J, Cordes F, Mohrholz V, Nausch G, Naumann M, Kruger S, Schulz-Bull D E, 2017. The impact of the major baltic inflow of december 2014 on the mercury species distribution in the Baltic Sea. Environmental Science & Technology, 51: 11692-11700.

Lam H S, Kwok K M, Chan P H Y, So H K, Li A M, Ng P C, Fok T F, 2013. Long term neurocognitive impact of low dose prenatal methylmercury exposure in Hong Kong. Environment International, 54: 59-64.

Lazaro W L, Diez S, Da Silva C J, Ignacio A R A, Guimaraes J R D, 2016. Waterscape determinants of net mercury methylation in a tropical wetland. Environmental Research, 150: 438-445.

Leclerc M, Planas D, Amyot M, 2015. Relationship between extracellular low-molecular-weight thiols and mercury species in natural lake periphytic biofilms. Environmental Science & Technology, 49: 7709-7716.

Lee C S, Fisher N S, 2017. Bioaccumulation of methylmercury in a marine diatom and the influence of dissolved organic matter. Marine Chemistry, 197: 70-79.

Lee S, Kim S J, Ju S J, Pak S J, Son S K, Yang J, Han S, 2015. Mercury accumulation in hydrothermal vent mollusks from the southern Tonga Arc, southwestern Pacific Ocean. Chemosphere, 127: 246-253.

Legrini O, Oliveros E, Braun A, 1993. Photochemical processes for water treatment. Chemical Reviews, 93: 671-698.
Lehnherr I, Louis V L S, 2009. Importance of ultraviolet radiation in the photodemethylation of methylmercury in freshwater ecosystems. Environmental Science & Technology, 43: 5692-5698.
Leighton P A, 1961. Photochemistry of air pollution New York: Academic Press.
Lepak R F, Krabbenhoft D P, Ogorek J M, Tate M T, Bootsma H A, Hurley J P, 2015. Influence of cladophora-quagga mussel assemblages on nearshore methylmercury production in lake michigan. Environmental Science & Technology, 49: 7606-7613.
Li M L, Schartup A T, Valberg A P, Ewald J D, Krabbenhoft D P, Yin R S, Bolcom P H, Sunderland E M, 2016. Environmental origins of methylmercury accumulated in subarctic estuarine fish indicated by mercury stable isotopes. Environmental Science & Technology, 50: 11559-11568.
Li Y, Mao Y, Liu G, Tachiev G, Roelant D, Feng X, Cai Y, 2010. Degradation of methylmercury and its effects on mercury distribution and cycling in the Florida Everglades. Environmental Science & Technology, 44: 6661-6666.
Li Y, Zhao J, Zhang B, Liu Y, Xu X, Li Y-F, Li B, Gao Y, Chai Z, 2016. The influence of iron plaque on the absorption, translocation and transformation of mercury in rice (*Oryza sativa* L.) seedlings exposed to different mercury species. Plant and Soil, 398: 87-97.
Li Y B, Yin Y G, Liu G L, Tachiev G, Roelant D, Jiang G B, Cai Y, 2012. Estimation of the major source and sink of methylmercury in the Florida Everglades. Environmental Science & Technology, 46: 5885-5893.
Liberda E N, Tsuji L J S, Martin I D, Ayotte P, Nieboer E, 2014. The complexity of hair/blood mercury concentration ratios and its implications. Environmental Research, 134: 286-294.
Liem-Nguyen V, Skyllberg U, Bjorn E, 2017. Thermodynamic modeling of the solubility and chemical speciation of mercury and methylmercury driven by organic thiols and micromolar sulfide concentrations in boreal wetland soils. Environmental Science & Technology, 51: 3678-3686.
Lin C-C, Jay J A, 2007. Mercury methylation by planktonic and biofilm cultures of *Desulfovibrio desulfuricans*. Environmental Science & Technology, 41: 6691-6697.
Lin T Y, Kampalath R A, Lin C-C, Zhang M, Chavarria K, Lacson J, Jay J A, 2013. Investigation of mercury methylation pathways in biofilm versus planktonic cultures of *Desulfovibrio desulfuricans*. Environmental Science & Technology, 47: 5695-5702.
Lindqvist O, Jernelov A, Johansson K, Rodhe H, 1984. Mercury in the Swedish environment: Global and local sources. Solna, Sweden: National Swedish Environment Protection Board.
Liu B, Schaider L A, Mason R P, Shine J P, Rabalais N N, Senn D B, 2015. Controls on methylmercury accumulation in northern Gulf of Mexico sediments. Estuarine, Coastal and Shelf Science, 159: 50-59.
Liu B, Yan H Y, Wang C P, Li Q H, Guedron S, Spangenberg J E, Feng X B, Dominik J, 2012. Insights into low fish mercury bioaccumulation in a mercury-contaminated reservoir, Guizhou, China. Environmental Pollution, 160: 109-117.
Liu G L, Cai Y, Kalla P, Scheidt D, Richards J, Scinto L J, Gaiser E, Appleby C, 2008. Mercury mass budget estimates and cycling seasonality in the Florida Everglades. Environmental Science & Technology, 42: 1954-1960.
Liu G L, Cai Y, Mao Y X, Scheidt D, Kalla P, Richards J, Scinto L J, Tachiev G, Roelant D, Appleby C, 2009. Spatial variability in mercury cycling and relevant biogeochemical controls in the Florida Everglades. Environmental Science & Technology, 43: 4361-4366.

Liu G L, Naja G M, Kalla P, Scheidt D, Gaiser E, Cai Y, 2011. Legacy and fate of mercury and methylmercury in the Florida Everglades. Environmental Science & Technology, 45: 496-501.

Liu Y-R, Dong J-X, Han L-L, Zheng Y-M, He J-Z, 2016. Influence of rice straw amendment on mercury methylation and nitrification in paddy soils. Environmental Pollution, 209: 53-59.

Liu Y-R, Lu X, Zhao L, An J, He J-Z, Pierce E M, Johs A, Gu B, 2016. Effects of cellular sorption on mercury bioavailability and methylmercury production by *Desulfovibrio desulfuricans* ND132. Environmental Science & Technology, 50: 13335-13341.

Liu Y-R, Yu R-Q, Zheng Y-M, He J-Z, 2014. Analysis of the microbial community structure by monitoring an Hg methylation gene (*hgcA*) in paddy soils along an Hg gradient. Applied and Environmental Microbiology, 80: 2874-2879.

Llop S, Murcia M, Aguinagalde X, Vioque J, Rebagliato M, Cases A, Iniguez C, Lopez-Espinosa M J, Amurrio A, Navarrete-Munoz E M, Ballester F, 2014. Exposure to mercury among Spanish preschool children: Trend from birth to age four. Environmental Research, 132: 83-92.

Lu X, Gu W, Zhao L, Haque M F U, Dispirito A A, Semrau J D, Gu B, 2017. Methylmercury uptake and degradation by methanotrophs. Science Advances, 3: 1-6.

Lu X, Liu Y, Johs A, Zhao L, Wang T, Yang Z, Lin H, Elias D A, Pierce E M, Liang L, 2016. Anaerobic mercury methylation and demethylation by *Geobacter bemidjiensis* Bem. Environmental Science & Technology, 50: 4366-4373.

Ma J, Graham N J, 2000. Degradation of atrazine by manganese-catalysed ozonation: Influence of radical scavengers. Water Research, 34: 3822-3828.

Martín-Doimeadios R C R, Mateo R, Jiménez-Moreno M, 2017. Is gastrointestinal microbiota relevant for endogenous mercury methylation in terrestrial animals? Environmental Research, 152: 454-461.

Marvin-Dipasquale M, Agee J, Mcgowan C, Oremland R S, Thomas M, Krabbenhoft D, Gilmour C C, 2000. Methyl-mercury degradation pathways: A comparison among three mercury-impacted ecosystems. Environmental Science & Technology,: 4908-4916.

Mason R P, Choi A L, Fitzgerald W F, Hammerschmidt C R, Lamborg C H, Soerensen A L, Sunderland E M, 2012. Mercury biogeochemical cycling in the ocean and policy implications. Environmental Research, 119: 101-117.

Mason R P, Fitzgerald W F, 1993. The distribution and biogeochemical cycling of mercury in the equatorial Pacific Ocean. Deep Sea Research Part I: Oceanographic Research Papers, 40: 1897-1924.

Mason R P, Laporte J M, Andres S, 2000. Factors controlling the bioaccumulation of mercury, methylmercury, arsenic, selenium, and cadmium by freshwater invertebrates and fish. Archives of Environmental Contamination and Toxicology, 38: 283-297.

Mason R P, Reinfelder J R, Morel F M M, 1995. Bioaccumulation of mercury and methylmercury. Water, Air, and Soil Pollution, 80: 915-921.

Mason R P, Rolfhus K R, Fitzgerald W F, 1998. Mercury in the North Atlantic. Marine Chemistry, 61: 37-53.

Mason R P, Sullivan K A, 1999. The distribution and speciation of mercury in the South and equatorial Atlantic. Deep Sea Research Part II: Topical Studies In Oceanography, 46: 937-956.

Mastromonaco M G N, Gardfeldt K, Assmann K M, Langer S, Delali T, Shlyapnikov Y M, Zivkovic I, Horvat M, 2017. Speciation of mercury in the waters of the Weddell, Amundsen and Ross Seas (Southern Ocean). Marine Chemistry, 193: 20-33.

Mitchell C P J, Branfireun B A, Kolka R K, 2008. Spatial characteristics of net methylmercury production hot spots in peatlands. Environmental Science & Technology, 42: 1010-1016.

Monteiro L R, Furness R W, 1995. Seabirds as monitors of mercury in the marine environment. *In*: Porcella D B, Huckabee J W, Wheatley B, Eds. Mercury as a Global Pollutant. Dordrecht: Springer: 851-870.

Morel F M M, Kraepiel A M L, Amyot M, 1998. The chemical cycle and bioaccumulation of mercury. Annual Review of Ecology and Systematics, 29: 543-566.

Mortensen M E, Caudill S P, Caldwell K L, Ward C D, Jones R L, 2014. Total and methyl mercury in whole blood measured for the first time in the US population: NHANES 2011—2012. Environmental Research, 134: 257-264.

Mostofa K M, Liu C Q, Mottaleb M A, Wan G, Ogawa H, Vione D, Yoshioka T, Wu F, 2013. Dissolved organic matter in natural waters. Photobiogeochemistry of Organic Matter. Heidelberg: Springer, 1-137.

Mounicou S, Shah M, Meija J, Caruso J A, Vonderheide A P, Shann J, 2006. Localization and speciation of selenium and mercury in Brassica juncea-implications for Se-Hg antagonism. Journal of Analytical Atomic Spectrometry, 21: 404-412.

Muckle G, Ayotte P, Dewailly E, Jacobson S W, Jacobson J L, 2001. Determinants of polychlorinated biphenyls and methylmercury exposure in Inuit women of childbearing age. Environmental Health Perspectives, 109: 957-963.

Myers G J, Davidson P W, Cox C, Shamlaye C F, Palumbo D, Cernichiari E, Sloane-Reeves J, Wilding G E, Kost J, Huang L S, Clarkson T W, 2003. Prenatal methylmercury exposure from ocean fish consumption in the Seychelles child development study. Lancet, 361: 1686-1692.

Myers G J, Davidson P W, Shamlaye C F, Axtell C D, Cernichiari E, Choisy O, Choi A, Cox C, Clarkson T W, 1997. Effects of prenatal methylmercury exposure from a high fish diet on developmental milestones in the Seychelles Child Development Study. Neurotoxicology, 18: 819-829.

Myers G J, Thurston S W, Pearson A T, Davidson P W, Cox C, Shamlaye C F, Cernichiari E, Clarkson T W, 2009. Postnatal exposure to methyl mercury from fish consumption: A review and new data from the Seychelles Child Development Study. Neurotoxicology, 30: 338-349.

Naftz D L, Cederberg J R, Krabbenhoft D P, Beisner K R, Whitehead J, Gardberg J, 2011. Diurnal trends in methylmercury concentration in a wetland adjacent to Great Salt Lake, Utah, USA. Chemical Geology, 283: 78-86.

Nagano M, Yasutake A, Miura K, 2010. Demethylation of methylmercury in human neuroblastoma, glioblastoma and liver cells. Journal of Health Science, 56: 326-330.

Noh S, Kim C K, Kim Y, Lee J H, Han S, 2017. Assessing correlations between monomethylmercury accumulation in fish and trophic states of artificial temperate reservoirs. Science of the Total Environment, 580: 912-919.

Noh S, Kim C K, Lee J H, Kim Y, Choi K, Han S, 2016. Physicochemical factors affecting the spatial variance of monomethylmercury in artificial reservoirs. Environmental Pollution, 208: 345-353.

Nordenhäll K, Dock L, Vahter M, 1998. Cross-fostering study of methyl mercury retention, demethylation and excretion in the neonatal hamster. Pharmacology & Toxicology, 82: 132-136.

Oken E, Rifas-Shiman S L, Amarasiriwardena C, Jayawardene I, Bellinger D C, Hibbeln J R, Wright R O, Gillman M W, 2016. Maternal prenatal fish consumption and cognition in mid childhood: Mercury, fatty acids, and selenium. Neurotoxicology and Teratology, 57: 71-78.

Oken E, Wright R O, Kleinman K P, Bellinger D, Amarasiriwardena C J, Hu H, Rich-Edwards J W, Gillman M W, 2005. Maternal fish consumption, hair mercury, and infant cognition in a US cohort. Environmental Health Perspectives, 113: 1376-1380.
Olsen T A, Brandt C C, Brooks S C, 2016. Periphyton biofilms influence net methylmercury production in an industrially contaminated system. Environmental Science & Technology, 50: 10843-10850.
Oremland R S, Culbertson C W, Winfrey M R, 1991. Methylmercury decomposition in sediments and bacterial cultures: Involvement of methanogens and sulfate reducers in oxidative demethylation. Applied and Environmental Microbiology, 57: 130-137.
Oremland R S, Miller L G, Dowdle P, Connell T, Barkay T, 1995. Methylmercury oxidative degradation potentials in contaminated and pristine sediments of the Carson River, Nevada. Applied and Environmental Microbiology, 61: 2745-2753.
Ou L B, Chen L, Chen C, Yang T J, Wang H H, Tong Y D, Hu D, Zhang W, Long W J, Wang X J, 2014. Associations of methylmercury and inorganic mercury between human cord blood and maternal blood: A meta-analysis and its application. Environmental Pollution, 191: 25-30.
Oulhote Y, Debes F, Vestergaard S, Weihe P, Grandjean P, 2017. Aerobic fitness and neurocognitive function scores in young Faroese adults and potential modification by prenatal methylmercury exposure. Environmental Health Perspectives, 125: 677-683.
Pan K, Wang W X, 2011. Mercury accumulation in marine bivalves: Influences of biodynamics and feeding niche. Environmental Pollution, 159: 2500-2506.
Paquette K, Helz G, 1995. Solubility of cinnabar (red HgS) and implications for mercury speciation in sulfidic waters. Water, Air, and Soil Pollution, 80: 1053-1056.
Paranjape A R, Hall B D, 2017. Recent advances in the study of mercury methylation in aquatic systems. Facets, 2: 85-119.
Parks J M, Johs A, Podar M, Bridou R, Hurt R A, Smith S D, Tomanicek S J, Qian Y, Brown S D, Brandt C C, Palumbo A V, Smith J C, Wall J D, Elias D A, Liang L, 2013. The genetic basis for bacterial mercury methylation. Science, 339: 1332-1335.
Peterson S A, Van Sickle J, Herlihy A T, Hughes R M, 2007. Mercury concentration in fish from streams and rivers throughout the western United States. Environmental Science & Technology, 41: 58-65.
Pickhardt P C, Fisher N S, 2007. Accumulation of inorganic and methylmercury by freshwater phytoplankton in two contrasting water bodies. Environmental Science & Technology, 41: 125-131.
Pickhardt P C, Folt C L, Chen C Y, Klaue B, Blum J D, 2002. Algal blooms reduce the uptake of toxic methylmercury in freshwater food webs. Proceedings of the National academy of Sciences of the United States of America, 99: 4419-4423.
Podar M, Gilmour C C, Brandt C C, Soren A, Brown S D, Crable B R, Palumbo A V, Somenahally A C, Elias D A, 2015. Global prevalence and distribution of genes and microorganisms involved in mercury methylation. Science Advances, 1: 1-13.
Porvari P, Verta M, Munthe J, Haapanen M, 2003. Forestry practices increase mercury and methyl mercury output from boreal forest catchments. Environmental Science & Technology, 37: 2389-2393.
Poulain A J, Barkay T, 2013. Cracking the mercury methylation code. Science, 339: 1280-1281.
Pucko M, Burt A, Walkusz W, Wang F, Macdonald R W, Rysgaard S, Barber D G, Tremblay J E, Stern

G A, 2014. Transformation of mercury at the bottom of the arctic food web: An overlooked puzzle in the mercury exposure narrative. Environmental Science & Technology, 48: 7280-7288.

Qian Y, Yin X, Lin H, Rao B, Brooks S C, Liang L, Gu B , 2014. Why dissolved organic matter enhances photodegradation of methylmercury. Environmental Science & Technology Letters, 1: 426-431.

Qiu G L, Feng X B, Wang S F, Fu X W, Shang L H, 2009. Mercury distribution and speciation in water and fish from abandoned Hg mines in Wanshan, Guizhou province, China. Science of The Total Environment, 407: 5162-5168.

Ralston N V C, Raymond L J, 2010. Dietary selenium's protective effects against methylmercury toxicity. Toxicology, 278: 112-123.

Ravichandran M, 2004. Interactions between mercury and dissolved organic matter: A review. Chemosphere, 55: 319-331.

Rice D C, Schoeny R, Mahaffey K, 2003. Methods and rationale for derivation of a reference dose for methylmercury by the US EPA. Risk Analysis, 23: 107-115.

Riisgård H U, Kiørboe T, Møhlenberg F, Drabæk I, Pheiffer Madsen P, 1985. Accumulation, elimination and chemical speciation of mercury in the bivalves *Mytilus edulis* and *Macoma balthica*. Marine Biology, 86: 55-62.

Robinson J B, Tuovinen O H, 1984. Mechanisms of microbial resistance and detoxification of mercury and organomercury compounds: Physiological, biochemical, and genetic analyses. Microbiological Reviews, 48: 95-124.

Rolfhus K R, Hurley J P, Bodaly R A, Perrine G, 2015. Production and retention of methylmercury in Inundated boreal forest soils. Environmental Science & Technology, 49: 3482-3489.

Rothenberg S E, Yu X D, Zhang Y M, 2013. Prenatal methylmercury exposure through maternal rice ingestion: Insights from a feasibility pilot in Guizhou Province, China. Environmental Pollution, 180: 291-298.

Rowland I R, Davies M J, Grasso P, 1977. Volatilisation of methylmercuric chloride by hydrogen sulphide. Nature, 265: 718-719.

Rumbold D G, Lange T R, Axelrad D M, Atkeson T D, 2008. Ecological risk of methylmercury in Everglades National Park, Florida, USA. Ecotoxicology, 17: 632-641.

Sakamoto M, Murata K, Domingo J L, Yamamoto M, Oliveira R B, Kawakami S, Nakamura M, 2016. Implications of mercury concentrations in umbilical cord tissue in relation to maternal hair segments as biomarkers for prenatal exposure to methylmercury. Environmental Research, 149: 282-287.

Schaefer J K, Letowski J, Barkay T, 2002. mer-Mediated resistance and volatilization of Hg (II) under anaerobic conditions. Geomicrobiology Journal, 19: 87-102.

Schaefer J K, Morel F M, 2009. High methylation rates of mercury bound to cysteine by *Geobacter sulfurreducens*. Nature Geoscience, 2: 123-126.

Schaefer J K, Rocks S S, Zheng W, Liang L, Gu B, Morel F M M, 2011. Active transport, substrate specificity, and methylation of Hg(II) in anaerobic bacteria. Proceedings of the National academy of Sciences of the United States of America, 108: 8714-8719.

Schaefer J K, Yagi J, Reinfelder J R, Cardona T, Ellickson K M, Tel-Or S, Barkay J, 2004. Role of the bacterial organomercury lyase (MerB) in controlling methylmercury accumulation in mercury-contaminated natural waters. Environmental Science & Technology, 38: 4304-4311.

Schartup A T, Mason R P, Balcom P H, Hollweg T A, Chen C Y, 2013. Methylmercury production in estuarine sediments: Role of organic matter. Environmental Science & Technology, 47: 695-700.

Scheuhammer A M, Perrault J A, Bond D E, 2001. Mercury, methylmercury, and selenium concentrations in eggs of common loons (Gavia immer) from Canada. Environmental Monitoring and Assessment, 72: 79-94.

Schottel J, 1978. The mercuric and organomercurial detoxifying enzymes from a plasmid-bearing strain of *Escherichia coli*. Journal of Biological Chemistry, 253: 4341-4349.

Schroeder W H, Jackson R A, 1987. Environmental measurements with an atmospheric mercury monitor having speciation capabilities. Chemosphere, 16: 183-199.

Sellers P, Kelly C A, Rudd J W M, Machutchon A R, 1996. Photodegradation of methylmercury in lakes. Nature, 380: 694-697.

Shanker K, Mishra S, Srivastava S, Srivastava R, Daas S, Prakash S, Srivastava M M, 1996. Effect of selenite and selenate on plant uptake and translocation of mercury by tomato (*Lycopersicum esculentum*). Plant and Soil, 183: 233-238.

Shanker K, Mishra S, Srivastava S, Srivastava R, Dass S, Prakash S, Srivastava M M, 1996. Study of mercury-selenium (HgSe) interactions and their impact on Hg uptake by the radish (*Raphanus sativus*) plant. Food and Chemical Toxicology, 34: 883-886.

Shapiro A M, Chan H M, 2008. Characterization of demethylation of methylmercury in cultured astrocytes. Chemosphere, 74: 112-118.

Sheehan M C, Burke T A, Navas-Acien A, Breysse P N, Mcgready J, Fox M A, 2014. Global methylmercury exposure from seafood consumption and risk of developmental neurotoxicity: A systematic review. Bulletin Of The World Health Organization, 92: 254-269.

Siciliano S D, O'driscoll N J, Tordon R, Hill J, Beauchamp S, Lean D R S, 2005. Abiotic production of methylmercury by solar radiation. Environmental Science & Technology, 39: 1071-1077.

Skyllberg U, Westin M B, Meili M, Bjorn E, 2009. Elevated concentrations of methyl mercury in dtreams after forest clear-cut: A consequence of mobilization from soil or new methylation? Environmental Science & Technology, 43: 8535-8541.

Smylie M S, Mcdonough C J, Reed L A, Shervette V R, 2016. Mercury bioaccumulation in an estuarine predator: Biotic factors, abiotic factors, and assessments of fish health. Environmental Pollution, 214: 169-176.

Soerensen A L, Schartup A T, Gustafsson E, Gustafsson B G, Undeman E, Bjorn E, 2016. Eutrophication increases phytoplankton methylmercury concentrations in a coastal sea-A baltic sea case study. Environmental Science & Technology, 50: 11787-11796.

Soerensen A L, Schartup A T, Skrobonja A, Bjorn E, 2017. Organic matter drives high interannual variability in methylmercury concentrations in a subarctic coastal sea. Environmental Pollution, 229: 531-538.

Southworth B A, Voelker B M, 2003. Hydroxyl radical production *via* the photo-Fenton reaction in the presence of fulvic acid. Environmental Science & Technology, 37: 1130-1136.

Spangler W J, Spigarelli J L, Rose J M, Flippin R S, Miller H H, 1973a. Degradation of methylmercury by bacteria isolated from environmental samples. Applied Microbiology, 25: 488-493.

Spangler W J, Spigarelli J L, Rose J M, Miller H M, 1973b. Methylmercury: Bacterial degradation in lake sediments. Science, 180: 192-193.

St. Louis V L, Rudd J W M, Kelly C A, Beaty K G, Bloom N S, Flett R J, 1994. Importance of wetlands as sources of methylmercury to boreal forest ecosystem. Canadian Journal of Fisheries and Aquatic Sciences, 51: 1065-1076.

St. Louis V L, Rudd J W M, Kelly C A, Beaty K G, Flett R J, Roulet N T, 1996. Production and loss of methylmercury and loss of total mercury from boreal forest catchments containing different types of wetlands. Environmental Science & Technology, 30: 2719-2729.

St. Louis V L, Rudd J W M, Kelly C A, Bodaly R A, Paterson M J, Beaty K G, Hesslein R H, Heyes A, Majewski A R, 2004. The rise and fall of mercury methylation in an experimental reservoir. Environmental Science & Technology, 38: 1348-1358.

St. Louis V L, Rudd J W M, Kelly C A, Hall B D, Rolfhus K R, Scott K J, Lindberg S E, Dong W, 2001. Importance of the forest canopy to fluxes of methyl mercury and total mercury to boreal ecosystems. Environmental Science & Technology, 35: 3089-3098.

Steffan R, Korthals E T, Winfrey M, 1988. Effects of acidification on mercury methylation, demethylation, and volatilization in sediments from an acid-susceptible lake. Applied and Environmental Microbiology, 54: 2003-2009.

Stern A H, Smith A E, 2003. An assessment of the cord blood∶maternal blood methylmercury ratio: Implications for risk assessment. Environmental Health Perspectives, 111: 1465-1470.

Steuerwald U, Weihe P, Jorgensen P J, Bjerve K, Brock J, Heinzow B, Budtz-Jorgensen E, Grandjean P, 2000. Maternal seafood diet, methylmercury exposure, and neonatal neurologic function. Journal of Pediatrics, 136: 599-605.

Stewart A R, Saiki M K, Kuwabara J S, Alpers C N, Marvin-Dipasquale M, Krabbenhoft D P, 2008. Influence of plankton mercury dynamics and trophic pathways on mercury concentrations of top predator fish of a mining-impacted reservoir. Canadian Journal of Fisheries and Aquatic Sciences, 65: 2351-2366.

Strain J J, Davidson P W, Bonham M P, Duffy E M, Stokes-Riner A, Thurston S W, Wallace J M W, Robson P J, Shamlaye C F, Georger L A, Sloane-Reeves J, Cernichiari E, Canfield R L, Cox C, Huang L S, Janciuras J, Myers G J, Clarkson T W, 2008. Associations of maternal long-chain polyunsaturated fatty acids, methyl mercury, and infant development in the Seychelles Child Development Nutrition Study. Neurotoxicology, 29: 776-782.

Strickman R J, Mitchell C P J, 2017. Accumulation and translocation of methylmercury and inorganic mercury in *Oryza sativa*: An enriched isotope tracer study. Science of the Total Environment, 574: 1415-1423.

Strickman R J S, Fulthorpe R R, Coleman Wasik J K, Engstrom D R, Mitchell C P J, 2016. Experimental sulfate amendment alters peatland bacterial community structure. Science of the Total Environment, 566: 1289-1296.

Stumm W, Morgan J J, 2012. Aquatic chemistry. New York: John Wiley & Sons.

Sud D, Mahajan G, Kaur M, 2008. Agricultural waste material as potential adsorbent for sequestering heavy metal ions from aqueous solutions: A review. Bioresource Technology, 99: 6017-6027.

Suda I, Suda M, Hirayama K, 1993. Degradation of methyl and ethyl mercury by singlet oxygen generated from sea water exposed to sunlight or ultraviolet light. Archives of Toxicology, 67: 365-368.

Szczuka A, Morel F M, Schaefer J K, 2015. Effect of thiols, zinc, and redox conditions on Hg uptake in *Shewanella oneidensis*. Environmental Science & Technology, 49: 7432-7438.

Tai C, Li Y B, Yin Y G, Scinto L J, Jiang G B, Cai Y, 2014. Methylmercury photodegradation in surface water of the florida everglades: Importance of dissolved organic matter-methylmercury complexation. Environmental Science & Technology, 48: 7333-7340.

Takizawa Y, Minagawa K, Hisamatsu S, 1981. Studies on mercury behaviour in man's environment:(report V) Photodegradation of methylmercury in the atmosphere by ultraviolet rays with sterilization. Japanese Journal of Public Health, 28: 313-320.

Tezuka T, Tonomura K, 1978. Purification and properties of a second enzyme catalyzing the splitting of carbon-mercury linkages from mercury-resistant *Pseudomonas* K-62. Journal of Bacteriology, 135: 138-143.

Tjerngren I, Meili M, Bjorn E, Skyllberg U, 2012. Eight boreal wetlands as sources and sinks for methyl mercury in relation to soil acidity, C/N ratio, and small-scale flooding. Environmental Science & Technology, 46: 8052-8060.

Todorova S G, Driscoll C T, Matthews D A, Effler S W, Hines M E, Henry E A, 2009. Evidence for regulation of monomethyl mercury by nitrate in a seasonally stratified, eutrophic lake. Environmental Science & Technology, 43: 6572-6578.

Tong Y D, Wang M Z, Bu X G, Guo X, Lin Y, Lin H M, Li J, Zhang W, Wang X J, 2017. Mercury concentrations in China's coastal waters and implications for fish consumption by vulnerable populations. Environmental Pollution, 231: 396-405.

Tossell J, 1998. Theoretical study of the photodecomposition of methyl Hg complexes. Journal of Physical Chemistry A, 102: 3587-3591.

Tsui M T K, Blum J D, Finlay J C, Balogh S J, Kwon S Y, Nollet Y H, 2013. Photodegradation of methylmercury in stream ecosystems. Limnology and Oceanography, 58: 13-22.

Tsui M T K, Blum J D, Finlay J C, Balogh S J, Nollet Y H, Palen W J, Power M E, 2014. Variation in terrestrial and aquatic sources of methylmercury in stream predators as revealed by stable mercury isotopes. Environmental Science & Technology, 48: 10128-10135.

Tsui M T K, Blum J D, Kwon S Y, Finlay J C, Balogh S J, Nollet Y H, 2012. Sources and transfers of methylmercury in adjacent river and forest food webs. Environmental Science & Technology, 46: 10957-10964.

Tsui M T K, Finlay J C, Nater E A, 2008. Effects of stream water chemistry and tree species on release and methylation of mercury during litter decomposition. Environmental Science & Technology, 42: 8692-8697.

Uchikawa T, Kanno T, Maruyama I, Mori N, Yasutake A, Ishii Y, Yamada H, 2016. Demethylation of methylmercury and the enhanced production of formaldehyde in mouse liver. Journal of Toxicological Sciences, 41: 479-487.

Ullrich S M, Tanton T W, Abdrashitova S A, 2001. Mercury in the aquatic environment: A review of factors affecting methylation. Critical Reviews in Environmental Science and Technology, 31: 241-293.

Ulrich P D, Sedlak D L, 2010. Impact of iron amendment on net methylmercury export from tidal wetland microcosms. Environmental Science & Technology, 44: 7659-7665.

Vahter M E, Mottet N K, Friberg L T, Lind S B, Charleston J S, Burbacher T M, 1995. Demethylation of methyl mercury in different brain sites of *Macaca fascicularis* monkeys during long-term subclinical methyl mercury exposure. Toxicology and Applied Pharmacology, 134: 273-284.

Van Wijngaarden E, Thurston S W, Myers G J, Harrington D, Cory-Slechta D A, Strain J J, Watson G E, Zareba G, Love T, Henderson J, Shamlaye C F, Davidson P W, 2017. Methyl mercury exposure and neurodevelopmental outcomes in the Seychelles Child Development Study main cohort at age 22 and 24 years. Neurotoxicology and Teratology, 59: 35-42.

Van Wijngaarden E, Thurston S W, Myers G J, Strain J J, Weiss B, Zarcone T, Watson G E, Zareba G,

Mcsorley E M, Mulhern M S, Yeates A J, Henderson J, Gedeon J, Shamlaye C F, Davidson P W, 2013. Prenatal methyl mercury exposure in relation to neurodevelopment and behavior at 19 years of age in the Seychelles Child Development Study. Neurotoxicology and Teratology, 39: 19-25.

Vander Zanden M J, Rasmussen J B A, 1996. A trophic position model of pelagic food webs: Impact on contaminant bioaccumulation in lake trout. Ecological Monographs, 66: 451-477.

Vert M, Doi Y, Hellwich K-H, Hess M, Hodge P, Kubisa P, Rinaudo M, Schué F, 2012. Terminology for biorelated polymers and applications (IUPAC Recommendations 2012). Pure and Applied Chemistry, 84: 377-410.

Wagemann R, Lockhart W L, Welch H, Innes S, 1995. Arctic marine mammals as integrators and indicators of mercury in the Arctic. In: Porcella D B, Huckabee J W, Wheatley B,Eds. Mercury as a Global Pollutant. Dordrecht: Springer: 683-693.

Wagemann R, Trebacz E, Boila G, Lockhart W L, 1998. Methylmercury and total mercury in tissues of arctic marine mammals. Science of the Total Environment, 218: 19-31.

Wang X, Wang W-X, 2017. Selenium induces the demethylation of mercury in marine fish. Environmental Pollution, 231: 1543-1551.

Wang X, Wu F, Wang W-X, 2017. *In vivo* mercury demethylation in a marine fish (*Acanthopagrus schlegeli*). Environmental Science & Technology, 51: 6441-6451.

Wang Y-J, Dang F, Zhao J-T, Zhong H, 2016. Selenium inhibits sulfate-mediated methylmercury production in rice paddy soil. Environmental Pollution, 213: 232-239.

Wang Y, Dang F, Evans R D, Zhong H, Zhao J, Zhou D, 2016a. Mechanistic understanding of MeHg-Se antagonism in soil-rice systems: the key role of antagonism in soil. Scientific Reports, 6: 19477.

Wang Y, Dang F, Zhong H, Wei Z, Li P, 2016b. Effects of sulfate and selenite on mercury methylation in a mercury-contaminated rice paddy soil under anoxic conditions. Environmental Science and Pollution Research, 23: 4602-4608.

Wasik J K C, Mitchell C P J, Engstrom D R, Swain E B, Monson B A, Balogh S J, Jeremiason J D, Branfireun B A, Eggert S L, Kolka R K, Almendinger J E, 2012. Methylmercury declines in a boreal peatland when experimental sulfate deposition decreases. Environmental Science & Technology, 46: 6663-6671.

Watras C J, Morrison K A, Kent A, Price N, Regnell O, Eckley C, Hintelmann H, Hubacher T, 2005. Sources of methylmercury to a wetland-dominated lake in northern Wisconsin. Environmental Science & Technology, 39: 4747-4758.

Whalin L, Kim E-H, Mason R, 2007. Factors influencing the oxidation, reduction, methylation and demethylation of mercury species in coastal waters. Marine Chemistry, 107: 278-294.

WHO, 1990. Environmental health criteria No. 101: Methylmercury. Geneva: World Health Organization (WHO).

Windham-Myers L, Fleck J A, Ackerman J T, Marvin-Dipasquale M, Stricker C A, Heim W A, Bachand P A M, Eagles-Smith C A, Gill G, Stephenson M, Alpers C N, 2014. Mercury cycling in agricultural and managed wetlands: A synthesis of methylmercury production, hydrologic export, and bioaccumulation from an integrated field study. Science of the Total Environment, 484: 221-231.

Wong A H K, Mcqueen D J, Williams D D, Demers E, 1997. Transfer of mercury from benthic invertebrates to fishes in lakes with contrasting fish community structures. Canadian Journal of

Fisheries and Aquatic Sciences, 54: 1320-1330.
Wood J M, Kennedy F S, Rosen C, 1968. Synthesis of methyl-mercury compounds by extracts of a methanogenic bacterium. Nature, 220: 173-174.
Wu F, Deng N, 2000. Photochemistry of hydrolytic iron(III) species and photoinduced degradation of organic compounds: A minireview. Chemosphere, 41: 1137-1147.
Xu X, Zhao J, Li Y, Fan Y, Zhu N, Gao Y, Li B, Liu H, Li Y-F, 2016. Demethylation of methylmercury in growing rice plants: An evidence of self-detoxification. Environmental Pollution, 210: 113-120.
Yasutake A, Matsumoto M, Yamaguchi M, Hachiya N, 2004. Current hair mercury levels in Japanese for estimation of methylmercury exposure. Journal of Health Science, 50: 120-125.
Yathavakilla S K V, Caruso J A, 2007. A study of Se-Hg antagonism in *Glycine max* (soybean) roots by size exclusion and reversed phase HPLC-ICPMS. Analytical and Bioanalytical Chemistry, 389: 715-723.
Yin Y, Li Y, Tai C, Cai Y, Jiang G, 2014. Fumigant methyl iodide can methylate inorganic mercury species in natural waters. Nature Communications, 5: 4633.
Yin Y, Shen M, Zhou X, Yu S, Chao J, Liu J, Jiang G, 2014. Photoreduction and stabilization capability of molecular weight fractionated natural organic matter in transformation of silver ion to metallic nanoparticle. Environmental Science & Technology, 48: 9366-9373.
Yin Y G, Chen B W, Mao Y X, Wang T, Liu J F, Cai Y, Jiang G B, 2012. Possible alkylation of inorganic Hg(II) by photochemical processes in the environment. Chemosphere, 88: 8-16.
Yoon S-J, Diener L M, Bloom P R, Nater E A, Bleam W F, 2005. X-ray absorption studies of CH_3Hg^+-binding sites in humic substances. Geochimica et Cosmochimica Acta, 69: 1111-1121.
Yu R-Q, Flanders J R, Mack E E, Turner R, Mirza M B, Barkay T, 2012. Contribution of coexisting sulfate and iron reducing bacteria to methylmercury production in freshwater river sediments. Environmental Science & Technology, 46: 2684-2691.
Yu R-Q, Reinfelder J R, Hines M E, Barkay T, 2013. Mercury methylation by the methanogen *Methanospirillum hungatei*. Applied and Environmental Microbiology, 79: 6325-6330.
Zepp R G, Braun A M, Hoigne J, Leenheer J A, 1987. Photoproduction of hydrated electrons from natural organic solutes in aquatic environments. Environmental Science & Technology, 21: 485-490.
Zepp R G, Faust B C, Hoigne J, 1992. Hydroxyl radical formation in aqueous reactions (pH 3～8) of iron(II) with hydrogen peroxide: the photo-Fenton reaction. Environmental Science & Technology, 26: 313-319.
Zhang C, Wang L, Wu F, Deng N, 2006. Quantitation of hydroxyl radicals from photolysis of Fe(III)-citrate complexes in aerobic water (5 pp). Environmental Science and Pollution Research, 13: 156-160.
Zhang D, Yin Y, Li Y, Cai Y, Liu J, 2017. Critical role of natural organic matter in photodegradation of methylmercury in water: Molecular weight and interactive effects with other environmental factors. Science of The Total Environment, 578: 535-541.
Zhang H, Feng X B, Larssen T, Qiu G L, Vogt R D, 2010. In inland China, rice, rather than fish, is the major pathway for methylmercury exposure. Environmental Health Perspectives, 118: 1183-1188.
Zhang T, Hsu-Kim H, 2010. Photolytic degradation of methylmercury enhanced by binding to natural organic ligands. Nature Geoscience, 3: 473-476.
Zhang T, Kim B, Levard C M, Reinsch B C, Lowry G V, Deshusses M A, Hsu-Kim H, 2012.

Methylation of mercury by bacteria exposed to dissolved, nanoparticulate, and microparticulate mercuric sulfides. Environmental Science & Technology, 46: 6950-6958.

Zhao J, Gao Y, Li Y-F, Hu Y, Peng X, Dong Y, Li B, Chen C, Chai Z, 2013. Selenium inhibits the phytotoxicity of mercury in garlic (*Allium sativum*). Environmental Research, 125: 75-81.

Zhao L, Qiu G, Anderson C W N, Meng B, Wang D, Shang L, Yan H, Feng X, 2016. Mercury methylation in rice paddies and its possible controlling factors in the Hg mining area, Guizhou province, Southwest China. Environmental Pollution, 215: 1-9.

Zhu A J, Zhang W, Xu Z Z, Huang L M, Wang W X, 2013. Methylmercury in fish from the South China Sea: Geographical distribution and biomagnification. Marine Pollution Bulletin, 77: 437-444.

Zhu H, Zhong H, Evans D, Hintelmann H, 2015. Effects of rice residue incorporation on the speciation, potential bioavailability and risk of mercury in a contaminated paddy soil. Journal of Hazardous Materials, 293: 64-71.

第4章 同位素技术在汞分子转化及区域传输研究中的应用

本章导读

- 介绍汞同位素分馏的基本概念和专业术语,阐述质量分馏、奇数同位素非质量分馏和偶数同位素非质量分馏的主要理论和判断依据。
- 介绍实验室研究所揭示的汞分子在物理、化学和生物转化过程中的同位素分馏规律以及如何利用这些规律研究环境中的汞循环。
- 介绍大气不同形态汞循环过程的同位素示踪,主要包括大气气态汞、大气降水汞和大气颗粒态汞的采样方法、实验室预处理、源区和区域传输示踪。
- 模型研究是从空间大尺度上模拟和预测汞分子转化与区域传输的有效手段,介绍基于汞同位素分馏的汞循环模型在定量全球汞收支方面的未来应用。

4.1 汞同位素分馏的基本概念和理论

汞在自然界有七个稳定同位素:^{196}Hg、^{198}Hg、^{199}Hg、^{200}Hg、^{201}Hg、^{202}Hg 和 ^{204}Hg,其自然丰度分别为 0.15%、9.97%、16.87%、23.10%、13.18%、29.86%和6.87%(Cohen et al.,2008;Meija et al.,2016)(表4-1)。

表4-1 稳定汞同位素基本物理参数(Cohen et al.,2008)

同位素	原子质量 m_a(u)	丰度(%,原子分数)	核自旋量子数(I)	核磁矩(μ/μ_N)
^{196}Hg	195.965833(3)	0.15(1)	0+	0
^{198}Hg	197.9667690(4)	9.97(20)	0+	0
^{199}Hg	198.9682799(4)	16.87(22)	½−	+0.5058855(9)
^{200}Hg	199.9683260(4)	23.10(19)	0+	0
^{201}Hg	200.9703023(6)	13.18(9)	$^3/_2$−	−0.5602257(14)
^{202}Hg	201.9706430(6)	29.86(26)	0+	0
^{204}Hg	203.9734939(5)	6.87(15)	0+	0

汞的同位素比值用符号 δ 来表示：

$$\delta^{xxx}\mathrm{Hg}(‰) = \left(\frac{R^{xxx/198}\mathrm{Hg}_{\mathrm{Sample}}}{R^{xxx/198}\mathrm{Hg}_{\mathrm{NIST\,3133}}} - 1\right) \times 1000 = \left(\frac{(^{xxx}\mathrm{Hg}/^{198}\mathrm{Hg})_{\mathrm{Sample}}}{(^{xxx}\mathrm{Hg}/^{198}\mathrm{Hg})_{\mathrm{NIST\,3133}}} - 1\right) \times 1000 \quad (4\text{-}1)$$

式中，xxx 是除了 198 以外的其他汞同位素的质量数，NIST 3133 是用于校正和标准化样品汞同位素比值的标准汞溶液，$(^{xxx}\mathrm{Hg}/^{198}\mathrm{Hg})_{\mathrm{Sample}}$ 是测得的样品汞同位素比值，$(^{xxx}\mathrm{Hg}/^{198}\mathrm{Hg})_{\mathrm{NIST\,3133}}$ 是测得的标准溶液 NIST 3133 的汞同位素比值。汞同位素不但具有常见的质量分馏（mass-dependent fractionation，MDF），而且还具有奇数同位素的非质量分馏（mass-independent fractionation of odd isotopes，odd-MIF）（Bergquist and Blum，2007；Blum and Bergquist，2007）。此外，近期越来越多的研究表明特殊的大气汞物理化学过程还可能产生非常明显的偶数汞同位素的非质量分馏（mass-independent fractionation of even isotopes，even-MIF）（Chen et al.，2012）。汞同位素的非质量分馏值用符号 Δ 来表示，代表实测的与理论预测的 $\delta'^{xxx}\mathrm{Hg}$ 值之间的差异（Blum and Bergquist，2007）：

$$\Delta^{xxx}\mathrm{Hg}(‰) = \delta'^{xxx}\mathrm{Hg} - {}^{xxx}\beta \times \delta'^{202}\mathrm{Hg} \quad (4\text{-}2)$$

当 δ 小于 10‰，δ' 与 δ 具有近似等价的关系：

$$\delta'^{xxx}\mathrm{Hg} = \ln\left(\frac{R^{xxx/198}\mathrm{Hg}_{\mathrm{Sample}}}{R^{xxx/198}\mathrm{Hg}_{\mathrm{NIST\,3133}}}\right) \times 1000$$

$$\approx \left(\frac{R^{xxx/198}\mathrm{Hg}_{\mathrm{Sample}}}{R^{xxx/198}\mathrm{Hg}_{\mathrm{NIST\,3133}}} - 1\right) \times 1000 = \delta^{xxx}\mathrm{Hg} \quad (4\text{-}3)$$

因此，非质量分馏值也可以表示为

$$\Delta^{xxx}\mathrm{Hg}(‰) \approx \delta^{xxx}\mathrm{Hg} - {}^{xxx}\beta \times \delta^{202}\mathrm{Hg} \quad (4\text{-}4)$$

式中，$^{xxx}\beta$ 是动力学质量分馏比例因子（scaling factor），对于 $^{199}\mathrm{Hg}$、$^{200}\mathrm{Hg}$、$^{201}\mathrm{Hg}$ 和 $^{204}\mathrm{Hg}$ 分别为 0.2520、0.5024、0.7520 和 1.4930。一般而言，$\delta^{202}\mathrm{Hg}$、$\Delta^{199}\mathrm{Hg}$ 和 $\Delta^{200}\mathrm{Hg}$ 分别用来表示质量分馏、奇数同位素的非质量分馏和偶数同位素的非质量分馏。

汞同位素在含汞物质 A 和 B 之间分馏幅度的大小，一般用分馏系数 $^{xxx/198}\alpha_{\mathrm{A\text{-}B}}$ 来表示：

$$^{xxx/198}\alpha_{\mathrm{A\text{-}B}} = \frac{R^{xxx/198}\mathrm{Hg}_{\mathrm{A}}}{R^{xxx/198}\mathrm{Hg}_{\mathrm{B}}} = \frac{(^{xxx}\mathrm{Hg}/^{198}\mathrm{Hg})_{\mathrm{A}}}{(^{xxx}\mathrm{Hg}/^{198}\mathrm{Hg})_{\mathrm{B}}} \quad (4\text{-}5)$$

式中，$^{xxx/198}\alpha_{\mathrm{A\text{-}B}}$ 可进一步细分为 MDF、odd-MIF 以及 even-MIF 三个部分（Sonke，2011）：

$$\alpha^{xxx/198}Hg_{A\text{-}B} = \alpha^{xxx/198}Hg_{A\text{-}B(MDF)} \times \alpha^{xxx/198}Hg_{A\text{-}B(odd\text{-}MIF)} \times \alpha^{xxx/198}Hg_{A\text{-}B(even\text{-}MIF)}$$
(4-6)

对于 $\alpha^{202/198}Hg_{A\text{-}B}$ 而言，$\alpha^{202/198}Hg_{A\text{-}B(odd\text{-}MIF)}$ 和 $\alpha^{202/198}Hg_{A\text{-}B(even\text{-}MIF)}$ 均等于 1；对于 $\alpha^{199/198}Hg_{A\text{-}B}$ 而言，$\alpha^{199/198}Hg_{A\text{-}B(even\text{-}MIF)}$ 等于 1；对于 $\alpha^{200/198}Hg_{A\text{-}B}$ 而言，$\alpha^{200/198}Hg_{A\text{-}B(odd\text{-}MIF)}$ 等于 1。此外，富集系数 ε 和 E 也被用来表示同位素的分馏幅度，其数学表达式如下：

$$\varepsilon^{202}Hg(‰) = 1000 \times (\alpha^{202/198}Hg_{MDF} - 1) \quad (4\text{-}7)$$

$$E^{199}Hg(‰) = 1000 \times (\alpha^{199/198}Hg_{odd\text{-}MIF} - 1) \quad (4\text{-}8)$$

$$E^{200}Hg(‰) = 1000 \times (\alpha^{200/198}Hg_{even\text{-}MIF} - 1) \quad (4\text{-}9)$$

4.1.1 质量分馏

质量分馏（MDF）是最为常见、最为经典的同位素分馏模式。几乎所有的地球化学过程都能够产生汞同位素的质量分馏。质量分馏是由同位素零点转动能（zero-point vibrational energy）的不同所引起的一种量子力学效应，主要受反应物分子的起始态和过渡态的能量所控制。同位素的质量分馏可细分为平衡过程的质量分馏和动力学过程的质量分馏，其理论早在半个多世纪前就已经建立（Bigeleisen，1949；Urey，1947）。

当汞同位素只发生质量分馏时，不同同位素比值所表示的分馏系数具有如下的数学关系：

$$\ln\alpha_{A\text{-}B}^{xxx/198} = {}^{xxx}\beta \times \ln\alpha_{A\text{-}B}^{202/198} \quad (4\text{-}10)$$

比例因子 $^{xxx}\beta$ 的数值大小在同位素平衡［式（4-11）］和动力学［式（4-12）］分馏过程中存在微小的差异：

$$^{xxx}\beta_{equ} = \frac{\left(\dfrac{1}{m^{xxx}Hg} - \dfrac{1}{m^{198}Hg}\right)}{\left(\dfrac{1}{m^{202}Hg} - \dfrac{1}{m^{198}Hg}\right)} \quad (4\text{-}11)$$

$$^{xxx}\beta_{kin} = \frac{\ln\dfrac{m^{xxx}Hg}{m^{198}Hg}}{\ln\dfrac{m^{202}Hg}{m^{198}Hg}} \quad (4\text{-}12)$$

式中，$m^{xxx}Hg$ 为汞同位素的原子量。式（4-10）可以转换为常用的同位素比值 δ'

和 δ 的形式：

$$\delta'^{xxx}\text{Hg} = {}^{xxx}\beta \times \delta'^{202}\text{Hg} \tag{4-13}$$

$$\delta^{xxx}\text{Hg} \approx {}^{xxx}\beta \times \delta^{202}\text{Hg} \tag{4-14}$$

式（4-13）、式（4-14）可用于构建三同位素图解（three-isotope plot），用来判断汞同位素是否发生非质量分馏（Sonke，2011；Young et al.，2002）。对于质量分馏而言，数据点将会落在以 $^{xxx}\beta$ 为斜率的直线上；对于非质量分馏而言，数据点将会偏离以 $^{xxx}\beta$ 为斜率的直线。图 4-1 展示的为我国煤炭汞不同同位素比值的三同位素图解，可以判断出奇数同位素 ^{199}Hg 和 ^{201}Hg 发生了较为明显的非质量分馏。

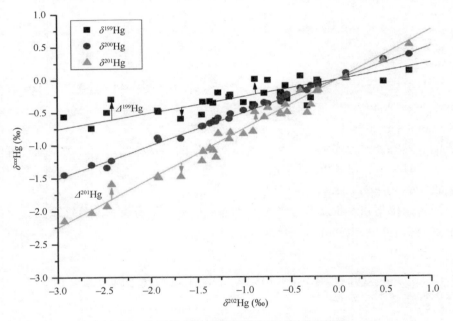

图 4-1　我国煤炭汞同位素组成的三同位素图解

所有样品的 δ^{200}Hg 都沿着质量分馏线分布，说明 ^{200}Hg 只发生了质量分馏；部分样品的 δ^{199}Hg 和 δ^{201}Hg 偏离了质量分馏线，说明 ^{199}Hg 和 ^{201}Hg 发生了非质量分馏（非质量分馏幅度的大小用 Δ^{199}Hg 和 Δ^{201}Hg 表示）

4.1.2　非质量分馏

除了质量分馏外，汞同位素还可能发生非质量分馏（MIF）。奇数汞同位素的非质量分馏效应在自然界广泛存在，而偶数汞同位素的非质量分馏效应主要在大气样品中观测到（Blum et al.，2014）。目前，磁同位素效应（magnetic isotope effect，MIE）（Buchachenko，2001）和核体积效应（nuclear volume effect，NVE；一些文献称 nuclear field shift，NFS）是解释奇数汞同位素非质量分馏的两种主流理论。

1. 磁同位素效应

磁同位素效应是基于化学反应对磁性核素的选择性：奇数汞同位素具有非零的核自旋数和磁矩（^{199}Hg：1/2−，+0.5058855；^{201}Hg：3/2−，−0.5602257，参见表 4-1），而偶数汞同位素的核自旋数和磁矩为零。磁同位素效应是纯粹的动力学过程，主要用于解释光参与下的二价无机汞的还原和甲基汞的降解过程所发生的奇数同位素的非质量分馏（Chandan et al.，2015；Rose et al.，2015；Zheng and Hintelmann，2009；Bergquist and Blum，2007）。根据反应物顺磁性核素中间体初始自旋态的不同，磁同位素效应可以在反应物中富集奇数汞同位素 [通常称为(+)MIE] 或者在产物中富集奇数汞同位素 [通常称为(−)MIE]（Zheng and Hintelmann，2010a）。室内控制实验表明与含 O/N 基团结合 Hg(Ⅱ)的光还原主要受(+)MIE 所控制，而与含—SH 基团结合 Hg(Ⅱ)的光还原主要受(−)MIE 效应所控制（Kritee et al.，2018；Zheng and Hintelmann，2010a）。磁同位素效应能够产生特征的 Δ^{199}Hg/Δ^{201}Hg 比值，一般介于 1.0～1.3 之间（Zheng and Hintelmann，2009；Bergquist and Blum，2007）。

2. 核体积效应

由于重元素同位素之间的相对质量差较小，其理论上计算的同位素质量分馏效应较小。然而，自然界中观测到的部分重元素同位素的分馏幅度却很大。核体积效应常被用来解释重元素（如 U）同位素比值异常的变化（Schauble，2013，2007；Bigeleisen，1996a，1996b）。核体积效应引起的同位素分馏幅度的大小与同位素核电荷半径均方 $\langle r^2 \rangle$ 的差异成正比：

$$\frac{\ln \alpha^{xxx/198}\text{Hg}}{\ln \alpha^{202/198}\text{Hg}} = {}^{xxx}\beta_{\text{NVE}} = \frac{\langle r^{xxx}\text{Hg}^2 \rangle - \langle r^{198}\text{Hg}^2 \rangle}{\langle r^{202}\text{Hg}^2 \rangle - \langle r^{198}\text{Hg}^2 \rangle} \qquad (4\text{-}15)$$

由于汞同位素的 $\langle r^2 \rangle$ 与质量 m 呈非线性的关系，$^{xxx}\beta_{\text{NVE}}$ 与 $^{xxx}\beta_{\text{equ}}$ [式（4-11）] 或 $^{xxx}\beta_{\text{kin}}$ [式（4-12）] 的值之间存在一定的差异，而这种差异在奇数同位素 ^{199}Hg 和 ^{201}Hg 尤为明显（表 4-2）。核体积效应分馏与质量分馏间比例因子的差异幅度，即为核体积效应所造成的非质量分馏幅度的大小。

核体积效应可以是平衡过程或者动力学过程，其同位素分馏方向与质量分馏一致，所造成的汞同位素分馏效应可以从理论上进行计算（Schauble，2013；2007）。核体积效应主要用来解释金属汞挥发（Ghosh et al.，2013；Estrade et al.，2009）、非光致 Hg(Ⅱ)还原（Zheng and Hintelmann，2010b；2009）以及溶解 Hg(Ⅱ)与 SH-Hg(Ⅱ)平衡交换（Wiederhold et al.，2010）过程中的非质量分馏现象。根据实测的汞同位素 $\langle r^2 \rangle$ 值（Angeli，2004；Nadjakov et al.，1994；Ulm et al.，1987；Hahn

表 4-2　各同位素比值相对于 $^{202/198}$Hg 的比例因子

同位素比值	$^{xxx}\beta_{equ}^{a}$	$^{xxx}\beta_{kin}^{b}$	$^{xxx}\beta_{NVE1}^{c}$	$^{xxx}\beta_{NVE2}^{c}$
$^{196/198}$Hg	−0.5151	−0.5074	−0.4660	
$^{199/198}$Hg	0.2539	0.2520	0.1076	0.0804
$^{200/198}$Hg	0.5049	0.5024	0.4966	0.4712
$^{201/198}$Hg	0.7539	0.7520	0.7003	0.6838
$^{202/198}$Hg	1	1	1	1
$^{204/198}$Hg	1.4855	1.4928	1.6543	1.4994

a. 平衡同位素质量分馏的比例因子；b. 动力学同位素质量分馏的比例因子；c. 不同 $\langle r^2 \rangle$ 计算出的核体积效应的比例因子：$^{xxx}\beta_{NVE1}$ 的 $\langle r^2 \rangle$ 引自（Angeli, 2004）；$^{xxx}\beta_{NVE2}$ 的 $\langle r^2 \rangle$ 引自（Hahn et al., 1979）。

et al., 1979），理论计算的 Δ^{199}Hg/Δ^{201}Hg 值应介于 1.65～2.79 之间（Sonke, 2011）；而实验观测到的 Δ^{199}Hg/Δ^{201}Hg 接近 1.6（Schauble, 2013; 2007; Zheng and Hintelmann, 2010b）。因此，Δ^{199}Hg/Δ^{201}Hg 值的大小可用于判断磁同位素效应与核体积效应所造成的非质量分馏。

从表 4-2 可知，偶数汞同位素（^{200}Hg、^{204}Hg）的比例因子在核体积效应及质量分馏效应中的大小比较接近。因此，核体积效应所预测的偶数同位素非质量分馏的幅度较小，难以解释在大气湿沉降中观测到的非常明显甚至超过 1‰的 Δ^{200}Hg 值（Enrico et al., 2017; 2016; Demers et al., 2013; Chen et al., 2012; Gratz et al., 2010）。Chen 等（2012）认为偶数汞同位素的非质量分馏可能与大气对流层顶 Hg(0) 的非均质光氧化有关。但是，实验室研究发现 Hg(0) 的光氧化产生的偶数同位素非质量分馏幅度极其有限（Sun et al., 2016a）。此外，有学者在使用的荧光灯中发现所有汞同位素均存在异常的非质量分馏现象，并认为这种分馏可能与同位素的光化学自吸收效应（按照同位素的丰度进行分馏）有关（Mead et al., 2013）。

4.2　汞分子转化过程中同位素分馏规律及应用

汞同位素在物理、化学和生物转化过程中能够发生特征的质量分馏和非质量分馏现象（Blum et al., 2014）。表 4-3 列出了实验室控制实验得出的各个过程（还原、氧化、甲基化、去甲基化、吸附、挥发、扩散）质量分馏的富集系数（ε^{202}Hg）和奇数同位素非质量分馏的富集系数（E^{199}Hg）。但是，一些关键过程（如可能导致偶数汞同位素非质量分馏的大气/水体零价汞的氧化、大气二价汞和颗粒态汞之间的转化）的汞同位素分馏机制仍不清楚。需要指出的是，实验室控制实验得出的同位素富集系数比较理想化，具有很大的局限性，可能并不完全适用于解释自然界所观测到的同位素分馏效应。

第4章 同位素技术在汞分子转化及区域传输研究中的应用

表4-3 汞在物理、化学和生物转化过程的同位素质量分馏和奇数同位素的非质量分馏的富集系数

反应物	产物	$\varepsilon^{202}Hg$	$E^{199}Hg$	参考文献	实验条件
Hg(0)光致氧化					
Hg(0)(g)	Hg(II)(g)	0.74	-0.25^a	(Sun et al., 2016a)	Hg(0)+Br•, UVB
Hg(0)(g)	Hg(II)(g)	-0.59	-0.37^b	(Sun et al., 2016a)	Hg(0)+Cl•, UVB
Hg(II)光致还原					
Hg(II)(aq)	Hg(0)(aq)	-0.60	NA	(Bergquist and Blum, 2007)	富里酸, Hg/DOC=102800 ng/mg, 自然光
Hg(II)(aq)	Hg(0)(aq)	-0.60	-0.99	(Bergquist and Blum, 2007)	富里酸, Hg/DOC=103000 ng/mg, 自然光
Hg(II)(aq)	Hg(0)(aq)	-1.09	-2.01	(Zheng and Hintelmann, 2009)	湖水DOM, Hg/DOC=8330 ng/mg, UVA/B
Hg(II)(aq)	Hg(0)(aq)	-1.06	-2.18	(Zheng and Hintelmann, 2009)	湖水DOM, Hg/DOC=8330 ng/mg, UVA/B
Hg(II)(aq)	Hg(0)(aq)	-0.99	-5.71	(Zheng and Hintelmann, 2009)	湖水DOM, Hg/DOC=833 ng/mg, 自然光
Hg(II)(aq)	Hg(0)(aq)	-1.26	-6.61	(Zheng and Hintelmann, 2009)	湖水DOM, Hg/DOC=833 ng/mg, UVA/B
Hg(II)(aq)	Hg(0)(aq)	-0.72	-3.56	(Zheng and Hintelmann, 2009)	湖水DOM, Hg/DOC=167 ng/mg, 自然光
Hg(II)(aq)	Hg(0)(aq)	-0.77	-2.73	(Zheng and Hintelmann, 2009)	湖水DOM, Hg/DOC=35 ng/mg, UVA/B
Hg(II)(aq)	Hg(0)(aq)	-0.55	NS	(Yang and Sturgeon, 2009)	甲酸, Hg/DOC=102800 ng/mg, UVC
Hg(II)(aq)	Hg(0)(aq)	-1.20	-0.70	(Rose et al., 2015)	富里酸, Hg/DOC=15000 ng/mg, 滤除UVB后自然光
Hg(II)(aq)	Hg(0)(aq)	-0.80	-1.40	(Rose et al., 2015)	富里酸, Hg/DOC=15000 ng/mg, 自然光
Hg(II)(aq)	Hg(0)(aq)	-1.80	-0.50	(Rose et al., 2015)	富里酸, Hg/DOC=15000 ng/mg, 滤除UVB后自然光
Hg(II)(aq)	Hg(0)(aq)	-1.50	-0.70	(Rose et al., 2015)	富里酸, Hg/DOC=15000 ng/mg, 自然光
Hg(II)(aq)	Hg(0)(aq)	-1.32	1.02	(Zheng and Hintelmann, 2010a)	半胱氨酸, 300~800 nm 光
Hg(II)(aq)	Hg(0)(aq)	-1.71	NA	(Zheng and Hintelmann, 2010a)	丝氨酸, 300~800 nm 光
Hg(II)非光致还原					
Hg(II)(aq)	Hg(0)(aq)	-2.00	NS	(Bergquist and Blum, 2007)	富里酸, 无光
Hg(II)(aq)	Hg(0)(aq)	-1.30	NS	(Bergquist and Blum, 2007)	富里酸, 无光
Hg(II)(aq)	Hg(0)(aq)	-1.86	NS	(Bergquist and Blum, 2007)	湖水DOM, 无光

续表

反应物	产物	ε^{202}Hg	E^{199}Hg	参考文献	实验条件
Hg(II)非光致还原					
Hg(II)(aq)	Hg(0)(aq)	−1.52	0.18	(Zheng and Hintelmann, 2010b)	DOM, 无光
Hg(II)(aq)	Hg(0)(aq)	−1.77	0.26	(Zheng and Hintelmann, 2010b)	SnCl$_2$溶剂, 无光
Hg(II)(aq)	Hg(0)(aq)	−1.56	0.17	(Zheng and Hintelmann, 2010b)	SnCl$_2$溶剂, 无光
Hg(II)(aq)	Hg(0)(aq)	−0.43	NS	(Yang and Sturgeon, 2009)	SnCl$_2$溶剂, 无光
Hg(II)(aq)	Hg(0)(aq)	−0.49	NS	(Yang and Sturgeon, 2009)	NaBH$_4$溶剂, 无光
Hg(II)非光致、生物还原					
Hg(II)(aq)	Hg(0)(aq)	−1.50	NS	(Kritee et al., 2007)	抗Hg(II)厌氧菌 *Escherichia coli* JM109/pPB117 (37℃)
Hg(II)(aq)	Hg(0)(aq)	−1.70	NS	(Kritee et al., 2007)	抗Hg(II)厌氧菌 *Escherichia coli* JM109/pPB117 (30℃)
Hg(II)(aq)	Hg(0)(aq)	−1.90	NS	(Kritee et al., 2007)	抗Hg(II)厌氧菌 *Escherichia coli* JM109/pPB117 (22℃)
Hg(II)(aq)	Hg(0)(aq)	−1.30	NS	(Kritee et al., 2007)	自然培育的抗Hg(II)细菌 (30℃)
Hg(II)(aq)	Hg(0)(aq)	−0.40	NS	(Kritee et al., 2007)	未自然培育的抗Hg(II)细菌 (30℃)
Hg(II)(aq)	Hg(0)(aq)	−1.20	NS	(Kritee et al., 2007)	抗Hg(II)厌氧菌 *Bacillus cereus* 5 (30℃)
Hg(II)(aq)	Hg(0)(aq)	−1.40	NS	(Kritee et al., 2008)	抗Hg(II)厌氧菌 *Anoxybacillus* sp. FB9 (60℃)
Hg(II)(aq)	Hg(0)(aq)	−1.80	NS	(Kritee et al., 2008)	抗Hg(II)厌氧菌 *Shewanella oneidensis* MR-1 (30℃)
其他无机(挥发、扩散、吸附)过程					
Hg(0)(aq)	Hg(0)(g)	−0.45	NS	(Zheng et al., 2007)	溶液溶解态Hg(0)挥发
Hg(0)(g)	Hg(0)(g)	−1.28	NS	(Koster van Groos et al., 2013)	气态Hg(0)扩散
Hg(0)(liq)	Hg(0)(g)	−0.86	NS	(Estrade et al., 2009)	金属态Hg(0)气化
Hg(0)(liq)	Hg(0)(g)	−1.21	NS	(Ghosh et al., 2013)	金属态Hg(0)气化
Hg(II)(aq)	Hg(II)−SH(s)	−0.58	NS	(Wiederhold et al., 2010)	溶解态Hg(II)吸附至有机物
Hg(II)(aq)	Hg(II)−FeOOH(s)	−0.36	NS	(Jiskra et al., 2012)	溶解态Hg(II)吸附至矿物

续表

反应物	产物	$\varepsilon^{202}Hg$	$E^{199}Hg$	参考文献	实验条件
甲基汞（MMHg）降解					
MMHg(aq)	Hg(0)(aq)	-1.30	-3.32	(Bergquist and Blum, 2007)	富里酸，MMHg/DOC=6200 ng/mg，自然光
MMHg(aq)	Hg(0)(aq)	-1.70	-7.93	(Bergquist and Blum, 2007)	富里酸，MMHg/DOC=9350 ng/mg，自然光
MMHg(aq)	Hg(0)(aq)	-1.10	-2.30	(Rose et al., 2015)	富里酸，MMHg/DOC=12500 ng/mg，自然光
MMHg(aq)	Hg(0)(aq)	-3.00	-3.60	(Rose et al., 2015)	富里酸，MMHg/DOC=12500 ng/mg，自然光
MMHg(aq)	Hg(0)(aq)	-1.80	-14.90	(Chandan et al., 2015)	低 MMHg/Sred-DOM 实验（n=4），模拟光（UVA & UVB）
MMHg(aq)	Hg(II)(aq)	-0.25	NA	(Malinovsky et al., 2010)	5组不同溶液组分实验的均值，254 nm UV
MMHg(aq)	Hg(0)(aq)	-0.40	NS	(Kritee et al., 2009)	抗Hg(II)厌氧菌 Escherichia coli JM109/pPB117 (37℃)
Hg(II)烷基化					
Hg(II)(aq)	MMHg(aq)	-3.98	NS	(Jiménez-Moreno et al., 2013)	非生物，黑暗，MeCo作为甲基供体（MeCo/Hg(II)摩尔比=10，20℃，pH=4，0.1 mol/L 醋酸缓冲液，0.5 mol/L 氯离子）
Hg(II)(aq)	MMHg(aq)	-8.92	NS	(Jiménez-Moreno et al., 2013)	非生物，黑暗，可见光，MeCo作为甲基供体（MeCo/Hg(II)摩尔比=10，20℃，pH=4，0.1 mol/L 醋酸缓冲液，0.5 mol/L 氯离子）
Hg(II)(aq)	MMHg(aq)	-0.70	NS	(Malinovsky and Vanhaecke, 2011)	非生物，黑暗，MeCo作为甲基供体（无羟基自由基）
Hg(II)(aq)	MMHg(aq)	-0.75	NS	(Malinovsky and Vanhaecke, 2011)	非生物，黑暗，甲基锡作为甲基供体（无羟基自由基）
Hg(II)(aq)	MMHg(aq)	-1.25	NS	(Malinovsky and Vanhaecke, 2011)	非生物，黑暗，醋酸作为甲基供体（羟基自由基）
Hg(II)(aq)	MMHg(aq)	-0.90	NS	(Malinovsky and Vanhaecke, 2011)	非生物，黑暗，二甲基亚砜作为甲基供体（羟基自由基）
Hg(II)(aq)	MMHg(aq)	-2.59	NS	(Rodríguez-González et al., 2009)	厌氧菌 Desulfobulbus propionicus UMD10，黑暗/发酵环境（丙酮酸盐作为碳源和电子受体，30℃，pH=7.2）
Hg(II)(aq)	MMHg(aq)	-3.09	NS	(Perrot et al., 2015)	硫酸盐还原厌氧菌 Desulfovibrio dechloracetivorans，硫酸盐还原和无硫酸盐环境
Hg(II)(aq)	MMHg(aq)	-0.90	NS	(Janssen et al., 2016)	厌氧菌 Geobacter sulfurreducens PCA，生长停滞期
Hg(II)(aq)	MMHg(aq)	-1.10	NS	(Janssen et al., 2016)	厌氧菌 Desulfovibrio desulfuricans ND132，生长停滞期
Hg(II)(aq)	CH₃CH₂Hg(aq)	-1.20	NS	(Yang and Sturgeon, 2009)	由 NaBEt4 乙基化

注：NS，不显著；NA，未提供。$\varepsilon^{202}Hg$ 正/负表示产物集聚或缺失重的汞同位素；$E^{199}Hg$ 正/负表示产物集聚或缺失奇数汞同位素。

a. 根据该反应中 $\delta^{202}Hg$ 和 $\Delta^{199}Hg$ 的线性相关系数计算，$\Delta^{199}Hg/\delta^{202}Hg=-0.32$；b. 根据该反应中 $\delta^{202}Hg$ 和 $\Delta^{199}Hg$ 的线性相关系数计算，$\Delta^{199}Hg/\delta^{202}Hg=0.63$。

4.2.1 汞在无机转化过程中的同位素分馏规律

1. 离子态无机汞的还原

溶液中与溶解有机质（dissolved organic matter, DOM）结合 Hg(II)的还原过程包括生物和非生物（光照、非光照）途径，其产生的同位素分馏效应研究最为广泛。Hg(II)不同的还原途径，产生的质量分馏方向是一致的，都使得反应物 Hg(II)富集重的汞同位素而产物 Hg(0)富集轻的汞同位素。然而，Hg(II)不同的还原途径产生的奇数同位素非质量分馏的机理却显著不同。

依据微生物种类和反应物 Hg(II)浓度的不同，Hg(II)生物还原产生的质量分馏富集系数（$\varepsilon^{202}Hg$）的变化范围为-0.4‰~-2.0‰；Hg(II)生物还原并不产生非质量分馏（Janssen et al., 2016; Kritee et al., 2013; 2008; 2007）。大部分实验发现，与 DOM 结合的 Hg(II)光致还原的非质量分馏受(+)MIE 的控制，在反应物 Hg(II)中富集奇数汞同位素（Rose et al., 2015; Yang and Sturgeon, 2009; Zheng and Hintelmann, 2009; Bergquist and Blum, 2007）。这很好地解释了湿沉降（降水、降雪）（Enrico et al., 2016; Wang et al., 2015; Yuan et al., 2015; Sherman et al., 2015; 2012b; Demers et al., 2013; Donovan et al., 2013; Chen et al., 2012; Gratz et al., 2010）和海水（Štrok et al., 2015; 2014）中偏正的 $\Delta^{199}Hg$ 值。但是，Zheng 等（2010a）发现与 Hg(II)结合的 DOM 官能团的不同可以造成非质量分馏方向的不同：含 O/N 官能团造成反应物 Hg(II)中富集奇数汞同位素[(+)MIE]；含—SH 官能团造成反应物 Hg(II)中缺失奇数汞同位素[(−)MIE]。土壤和叶片中观测到的偏负 $\Delta^{199}Hg$ 值可能与土壤或者植物叶片中与—SH 结合 Hg(II)的光还原有关（Demers et al., 2013; Sonke, 2011）。溶液中 Hg(II)的非光致还原（DOM 或者 $SnCl_2$ 作为还原物）所造成的非质量分馏的幅度比较小，但却被发现奇数汞同位素在反应物 Hg(II)中相对缺失，可能与汞同位素的核体积效应（NVE）有关（Zheng and Hintelmann, 2010b）。

汞同位素非质量分馏的方向和幅度，可以用于判断 Hg(II)的不同还原途径。具体来说，根据非质量分馏的有无，可以判断 Hg(II)的生物和非生物还原途径；根据非质量分馏的 $\Delta^{199}Hg/\Delta^{201}Hg$ 比值，可以判断 Hg(II)的光致还原（$\Delta^{199}Hg/\Delta^{201}Hg=1$~1.3）和非光致还原（$\Delta^{199}Hg/\Delta^{201}Hg=1.5$~1.6）途径；根据 $\Delta^{199}Hg$ 和 $\Delta^{201}Hg$ 在反应物和产物的相对大小，可以判断与 Hg(II)结合的官能团的种类。此外，Hg(II)的不同还原途径产生的 $\Delta^{199}Hg/\delta^{202}Hg$ 比值也具有显著区别（Blum et al., 2014）。

2. 气态零价汞的氧化

由于 Hg(0)的氧化难以控制反应速度且生成的产物难以定量分离，Hg(0)氧化过程中同位素分馏的研究较少。Sun 等（2016a）发现标准状态下（25℃，1 个大气压）卤素自由基（Br•、Cl•）参与的 Hg(0)光氧化不但能够产生质量分馏，还能够产生奇数同位素的非质量分馏。虽然该实验也观测到偶数同位素的非质量分馏，但是分馏幅度不是非常明显。Sun 等（2016a）的研究表明，Cl•和 Br•参与的 Hg(0)光氧化所造成的同位素分馏机理存在显著的不同。在 Cl•参与的 Hg(0)光氧化情况下，反应物 Hg(0)富集重的、奇数的汞同位素；而在 Br•参与的 Hg(0)光氧化情况下，反应物 Hg(0)富集轻的、奇数的汞同位素。在 Br•参与的 Hg(0)氧化过程中，反应物 Hg(0)富集轻的汞同位素不是很正常，因为 Hg(0)光氧化一般来说是动力学反应，会使得反应物 Hg(0)富集重的汞同位素。另外，Cl•参与的 Hg(0)光氧化的 $\Delta^{199}Hg/\Delta^{201}Hg$ 值约为 1.89±0.18，明显不同于 Br•参与的 Hg(0)光氧化的 $\Delta^{199}Hg/\Delta^{201}Hg$ 值（1.64±0.30）以及 NVE 造成的 $\Delta^{199}Hg/\Delta^{201}Hg$ 值（约 1.6），且显著高于 MIE 造成的 $\Delta^{199}Hg/\Delta^{201}Hg$ 值（1.0～1.3）。因此，Hg(0)光氧化过程中同位素分馏的机理还需要进一步的实验研究。

3. 溶液-有机物/矿物表面汞的转化

在溶液中，离子态无机 Hg(Ⅱ)容易吸附在有机物和矿物颗粒上转化为颗粒态 Hg(P)。实验室控制实验发现溶解态 Hg(Ⅱ)在吸附到铁氧化物（针铁矿）(Jiskra et al.，2012) 和含—SH 基团的有机物（Wiederhold et al.，2010）的过程中，能够造成较为明显的质量分馏。但是，该过程造成的非质量同位素分馏极其有限（$\Delta^{199}Hg<0.1‰$）。相对于溶解态 Hg(Ⅱ)，吸附态 Hg(Ⅱ)［或颗粒态 Hg(P)］相对富集轻的汞同位素，其质量分馏富集系数 $\varepsilon^{202}Hg$ 介于–0.6‰～–0.3‰之间。

4. 气态零价汞的动力学扩散

气态零价 Hg(0)可以在溶液和空气介质中从高浓度向低浓度进行扩散，该扩散过程主要伴随着质量分馏。Zheng 等（2007）发现溶液中溶解 Hg(0)在向大气缓慢扩散的过程中能够使挥发的 Hg(0)富集轻的汞同位素，其 $\varepsilon^{202}Hg$ 值大小介于 –0.5‰～–0.4‰之间。Koster van Groos 等（2013）发现大气中的 Hg(0)从高浓度向低浓度扩散过程中能够使扩散的 Hg(0)富集较轻的汞同位素，其 $\varepsilon^{202}Hg$ 值约为 –1.3‰。

5. 大气-陆地界面汞的交换

大气-陆地界面间的汞交换主要包括大气汞的干沉降（叶片通过气孔吸收）、湿沉降（降雨、降雪）以及陆地土壤/植被汞的再释放，是控制大气汞和土壤汞含量、分布的主要因素。大气汞通过植物叶片气孔吸收的干沉降过程能够产生非常大的质量分馏，该过程被认为是控制全球汞同位素分异的关键因素。植物叶片吸收 Hg(0) 后，会在叶片细胞内将其氧化成 Hg(Ⅱ)。随着叶片的凋零，叶片中的汞会转移至上层有机土壤中，从而使得有机土壤继承了叶片的汞同位素组成（Zheng et al., 2016；Jiskra et al., 2015）。诸多研究发现，植物叶片的 δ^{202}Hg 值要比同区域大气汞（主要是 Hg(0)）的 δ^{202}Hg 值低 2‰~3‰（Obrist et al., 2017；Enrico et al., 2016；Yu et al., 2016；Demers et al., 2013）。根据大气 Hg(0) 含量和 δ^{202}Hg 值在沼泽地和邻近山区间的差异，Enrico 等（2016）估算叶片 Hg(0) 吸收过程中的质量分馏富集系数 ε^{202}Hg 约为-2.6‰。此外，研究还发现叶片要比同区域大气 Hg(0) 的 Δ^{199}Hg 值低达 0.3‰（Yu et al., 2016；Demers et al., 2013）；这种系统的差异可能与叶片组织—SH 官能团结合 Hg(Ⅱ) 的光致还原（即汞的再释放）有关。最近的一个实验室控制实验也表明微藻细胞中的 Hg(Ⅱ) 的光致还原能够产生较为显著的质量分馏（ε^{202}Hg=−0.75‰）和(−)MIE 控制的奇数同位素的非质量分馏（Kritee et al., 2018）。另外，汞主要与土壤中有机质的—SH 官能团结合，其 Hg(Ⅱ) 的光致还原、Hg(0) 再释放过程也可能产生(−)MIE 控制的奇数同位素的非质量分馏（Zheng and Hintelmann, 2010a）。

4.2.2 汞在无机-有机转化过程中的同位素分馏规律

汞的甲基化能够使毒性较小的无机汞转化为剧毒的甲基汞，从而在生物体内进行累积和传递；去甲基化能够使甲基汞转化为无机汞，从而缓解甲基汞的生物毒性。汞的甲基化与去甲基化过程同位素分馏规律的研究对于认清生物体中甲基汞的来源、汞的甲基化/去甲基化动力学以及甲基汞的生物暴露途径具有重要意义。研究汞的甲基化与去甲基化过程中的同位素分馏存在的一个主要技术障碍就是实验过程中无机 Hg(Ⅱ) 和 MMHg 之间的相互转化以及其他同时进行的平行反应、竞争反应［如甲基化过程中，Hg(Ⅱ) 连续转化为 MMHg 和二甲基汞；去甲基化过程中，同时存在以 Hg(Ⅱ) 和 Hg(0) 为主的产物］，难以有效地隔离研究甲基化和去甲基化过程，从而分别获得甲基化和去甲基化过程的同位素分馏系数。另外一个技术障碍是同位素分馏的抑制效应：当细胞浓度或者反应液中某种物质的浓度达到或低于一定阈值时，反应体系就会限制或者改变同位素分馏的幅度甚至方向。汞的甲基化与去甲基化过程能够产生不同程度的质量分馏，但只有光参与下的去甲基化过程才能产生明显的奇数同位素的非质量分馏（Perrot et al., 2015；Jiménez-Moreno et al., 2013；Rodríguez-González et al., 2009；Kritee et al., 2009；2008；2007）。虽然前期部分研究认为生物体中汞的无机-有机转化过程也可能产生奇数同位素的非质量分馏（Jackson, 2015；Das et al., 2009；Ghosh et al., 2008；Jackson et al., 2008；Buchachenko et al., 2004），但最近的实验室控制实验发现生物体在汞的甲基化、去甲基化、积累和传递过程中并不产生非质量分馏（Perrot et al., 2015; Kwon et al., 2012; Kritee et al., 2009; Rodríguez-González et al., 2009）。因此，奇数同位素的非质量分馏能够很好地区分生物和非生物过程。

1. 汞的甲基化

汞的甲基化包括生物甲基化和非生物甲基化途径。这两种甲基化途径都能够产生非常明显的质量分馏而无非质量分馏，使得反应物 Hg(Ⅱ)富集重的汞同位素而生成物 MMHg 富集轻的汞同位素（Janssen et al., 2016; Perrot et al., 2015, 2013; Jiménez-Moreno et al., 2013; Rodríguez-González et al., 2009）。

1) 生物途径

Rodríguez-González 等（2009）发现在厌氧的条件下，硫酸盐还原细菌（SRB）*Desulfobulbus propionicus* UMD10 的甲基化过程能够产生较大的质量分馏，ε^{202}Hg 约为–2.6‰。同样，Perrot 等（2015）发现另外一种硫酸盐还原细菌（SRB）*Desulfovibrio dechloracetivorans* BerOc1 无论是在硫酸盐还原环境中还是在非硫酸盐环境中的甲基化过程都能够产生相同幅度的质量分馏，ε^{202}Hg 约为–3‰。但是，Janssen 等（2016）却在厌氧条件下发现静止生长的铁还原细菌（FeRB）*Geobacter sulfurreducens* PCA 和硫酸盐还原细菌（SRB）*Desulfovibrio desulfuricans* ND132 在甲基化过程中发生了明显减弱的质量分馏，ε^{202}Hg 介于–1.1‰～–0.9‰之间。因此，生物甲基化过程中质量分馏幅度的大小不但取决于菌种的类别，还受培养基质和培养条件的影响。

2) 非生物途径

Jiménez-Moreno 等（2013）以甲基钴胺素（MeCo）作为甲基基团供体化合物，发现富氯离子水溶液中的 Hg(Ⅱ)在甲基化（包括光照和非光照条件）的过程中能够产生非常大的质量分馏。与之类似的是，Perrot 等（2013）发现在无氯离子的水溶液中，MeCo 能够快速、连续地使 Hg(Ⅱ)发生甲基化和二甲基化（包括光照和非光照条件），也产生较大的质量分馏。但是，由于可逆反应和平行反应的同时进行，难以准确计算出汞甲基化过程中的同位素分馏系数。根据数值模拟结果，汞甲基化过程（MeCo 作为供体、0.5 mol/L Cl⁻）的质量分馏富集系数 ε^{202}Hg 值在非光照条件下约为–4‰，而在光照条件下高达–9‰（Jiménez-Moreno et al., 2013）。

2. 甲基汞（MMHg）的去甲基化

MMHg 的去甲基化途径包括生物的新陈代谢以及非生物的光降解和化学反应，其初始产物为 Hg(Ⅱ)，在还原剂存在的条件下可以进一步还原成 Hg(0)。值得注意的是，产物 Hg(Ⅱ)可能进行逆向甲基化，干扰去甲基化过程；而产物 Hg(0)

能够被较快地移出实验体系，减少对去甲基化过程的干扰。去甲基化过程能够产生非常明显的质量分馏，使得反应物 MMHg 富集重的汞同位素而生成物 Hg(Ⅱ)/Hg(0)富集轻的汞同位素。另外，MMHg 的光降解还能够产生特征的奇数汞同位素的非质量分馏现象。

1）生物途径

生物过程中甲基汞降解的同位素分馏效应研究较少。根据细菌 *Escherichia coli* JM109 的培养实验，Kritee 等（2009）发现细菌多拷贝质粒 pPB117 所携带的抗汞操纵子（*mer*）转运系统在厌氧的条件下能够降解 MMHg，并产生 Hg（0）和甲烷；该过程能够产生约 $\varepsilon^{202}Hg= -0.4‰$ 的质量分馏。另外，根据非质量分馏的机理和反应条件，Kritee 等（2009）推测非 *mer* 细菌介导的 MMHg 降解也只能够产生质量分馏。

2）非生物途径

非生物途径中光照条件下 MMHg 降解所产生同位素分馏的研究最为充分。Bergquist 和 Blum（2007）率先通过实验室控制实验发现与 DOM 结合的 MMHg 的光降解能够产生非常明显的质量分馏和奇数同位素的非质量分馏，使得剩余的 MMHg 富集重的、奇数的汞同位素。继而，Malinovsky 等（2010）、Rose 等（2015）和 Chandan 等（2015）通过进一步细化 MMHg 降解过程中实验参数（如反应溶液基质、光源的波长、MMHg 与 DOM 中还原态 S 的比值即 MMHg/S_{red}-DOM），探究了不同实验条件对同位素分馏幅度的影响。总体来说，①MMHg 光降解产生的奇数同位素的分馏幅度要大于质量分馏的幅度；②$\Delta^{199}Hg/\Delta^{201}Hg$ 值约为 1.3，明显大于无机汞 Hg(Ⅱ)的光还原过程。迄今为止，并没有专门针对 MMHg 避光条件下的化学去甲基化实验，但是 Jiménez-Moreno 等（2013）和 Perrot 等（2013）在利用 MeCo 作为供体进行 Hg(Ⅱ)甲基化实验的过程发现同时进行的化学去甲基化过程，且该过程伴随着非常明显的质量分馏。根据 Jiménez-Moreno 等（2013）对甲基化（MeCo 作为供体、0.5 mol/L Cl⁻）过程中同时进行的去甲基化过程模拟结果，化学去甲基化的质量分馏富集系数 $\varepsilon^{202}Hg$ 值在非光照条件下约为–4‰，而在光照条件下高达–7‰。

4.2.3 汞分子转化过程中同位素分馏效应的应用

认清汞同位素在汞分子物理、化学、生物转化过程中的同位素分馏规律对于利用汞同位素进行准确溯源和定量研究各过程对汞循环的影响具有重要的意义。其中，以下两个关于非质量分馏的重要发现使得汞同位素的应用得到了极大的拓展，改变了以前依靠汞浓度和形态所理解的汞循环过程。

（1）同位素的非质量分馏信号在生物营养级传递与新陈代谢过程中不会发生

改变，生物体内的非质量分馏信号主要反映的是非生物过程。

研究发现溶液中无机 Hg(Ⅱ) 的光还原和 MMHg 的光降解是奇数汞同位素非质量分馏的主要过程，而生物新陈代谢过程（吸收、传递、富集）所造成的汞分子转化并不产生明显的非质量分馏，这为利用非质量分馏来定量示踪生物体汞的暴露来源与途径提供了最坚实的理论依据（Kwon et al.，2012）。其中，研究最为广泛的是鱼体与人体组织汞暴露来源的定量研究。

鱼体中的汞主要以 MMHg 形式存在，MMHg 的比例一般随着鱼类营养级的升高而增加。鱼体中汞同位素的非质量分馏值（$\varDelta^{xxx}Hg$）主要继承于水体的 MMHg；利用控制实验中 MMHg 光降解过程中的同位素分馏系数，可以大致估算出水体中 MMHg 的光降解程度。例如，根据鱼体中的 $\varDelta^{201}Hg$ 值，Bergquist 和 Blum（2007）计算出北美密西根湖和新英格兰湖湖水中的 MMHg 分别光降解了大约 70%和 25%～55%。另外，鱼体中的 MMHg 还可能来源于其他储库（如沉积物、深层海水），而这部分储库原位产生的 MMHg 可能并没有或者很少发生光降解，其非质量分馏值与基底无机 Hg(Ⅱ) 相似，这时就不可以简单地利用鱼体中的非质量分馏值来估算附近水体中 MMHg 的光降解幅度。一个比较典型的例子是北太平洋不同深度鱼体中的 $\varDelta^{199}Hg$ 和 $\varDelta^{201}Hg$ 值随着海水深度的增加而降低（Blum et al.，2013）。如果还利用上述计算方法，就会得到表层水域（约 10 m）产生的 MMHg 约 46%～80%被光降解，而深部海洋（约 600 m）产生的 MMHg 约 9%～21%被光降解。事实上，深部鱼体的非质量分馏值主要继承于表层海水已发生光降解的 MMHg，而表层海水中的 MMHg 在向下层海洋传递的过程中，其非质量分馏值被深层海洋产生的未发生光降解的 MMHg（非质量分馏值接近于零）所改变，从而使得深部鱼体的非质量分馏值变小。

人类头发和尿液中的汞同位素组成可用于对人体的汞暴露评价（Du et al.，2018；Li et al.，2017；Rothenberg et al.，2017；Bonsignore et al.，2015；Li et al.，2014；Sherman et al.，2013；Laffont et al.，2011；2009）。头发中的汞能够很好地反映 MMHg 的暴露，而尿液中的汞更多地反映无机汞的暴露。前人的研究得出的一个比较明确的规律是：头发相对于食物中汞的质量分馏值（$\delta^{202}Hg$）大约增加了 2‰，而非质量分馏值并没有发生明显的变化。Laffont 等（2009）首先发现玻利维亚当地人头发的质量分馏值（$\delta^{202}Hg$）相对于当地主食的淡水鱼类富集了大约 2‰，而非质量分馏值却非常类似，从而证实了非质量分馏信号在各类型营养级传递过程中的保守性（Laffont et al.，2009）。继而，Laffont 等（2011）研究了不同来源汞暴露（金矿开采挥发的金属汞蒸气、淡水鱼中的 MMHg、海鱼中的 MMHg）人群中头发的汞同位素特征，发现：①以海鱼为食的欧洲居民的头发表现出与以淡水鱼为食的玻利维亚当地的人头发相似的汞同位素分馏特征（$\delta^{202}Hg$ 从鱼体到头发发生了约 2‰分馏；非质量分馏基本上不变）；②以金属汞蒸气为主要暴露源的金矿工人头发中的 δ^{202}Hg 相对于汞蒸气并没有发生 2‰分馏，但是其非质量分馏值要远远高于汞蒸气源。利用上述三个汞暴露源的头发样品为同位素限定端元，Laffont 等（2011）计算了金属 Hg(0)、淡水鱼 MMHg、海鱼 MMHg 在不同人群头发中的贡献比例。由于中国部分地区主要以稻米为主食，中国人群的汞暴露具有明显不同于欧美国家人群的特点，其头发的汞同位素组成继承了稻米的汞同位素特征。稻米中的汞主要来源于耕种土壤和周围大气，其同位素质量分馏值和非质量分馏值（接近于零）明显低于鱼体（Du et al.，2018；Li et al.，2017；Rothenberg et al.，2017；Feng et al.，2016）。因此，与经常食鱼的人群头发相比，以稻米为主食的人群头发中更富集轻的汞同位素并相对缺失奇数汞同位素（Du et al.，2018；Li et al.，2017；Rothenberg et al.，2017）。由于非质量分馏信号在人体新陈代谢过程中的保守性，根据人群头发以及主食中汞同位素非质量分馏值的大小，就可以估算出各个食物对人体的汞暴露风险。例如：Du 等（2018）发现贵阳市民约 60%的汞暴露来自于非鱼食物（主要是稻米），而对于生活在贵州汞矿区附近以及偏远地区的人们，非鱼食物汞暴露的比例可增加至 90%以上。

（2）陆地生态系统并不产生明显的偶数同位素的非质量分馏，陆地样品观测到偶数同位素的非质量分馏信号主要反映大气源和大气过程。

迄今为止，只有在大气湿沉降（降雨、降雪）样品中发现非常显著的偶数同位素非质量分馏信号（Δ^{200}Hg）（Yuan et al.，2018；Enrico et al.，2016；Wang et al.，2015；Yuan et al.，2015；Sherman et al.，2015；2012b；Demers et al.，2013；Donovan et al.，2013；Chen et al.，2012；Gratz et al.，2010），因而推测偶数同位素的非质量分馏主要是由于上层大气 Hg(0)的非均质光氧化（Cai and Chen，2015；Chen et al.，2012）。虽然偶数汞同位素的分馏机理还未有理论和实验证实，但这并不妨碍其在大气汞示踪上的应用，特别是大气汞的干、湿沉降对生态系统的贡献。植物叶片同时接收大气的干沉降[通过叶片气孔吸收大气 Hg(0)]和湿沉降（通过大气降水）的汞，但是叶片在汞吸收和沉降后发生的一系列物理化学过程改变了大气汞的质量分馏和奇数同位素非质量分馏的混合信号（Enrico et al.，2017；Obrist et al.，2017；Enrico et al.，2016；Yu et al.，2016；Zheng et al.，2016；Jiskra et al.，2015；Demers et al.，2013），因此利用这两种同位素信号进行源示踪时必须对过程分馏的幅度进行校正，从而带来了极大的不确定性。但是，大气汞偶数同位素的非质量分馏信号并不受叶片分馏过程的影响。利用 Δ^{200}Hg 和 Δ^{204}Hg，Enrico 等（2016）得出法国 Pyrenees 山脉地区泥炭藓中的汞分别有大约 80%和 20%来自于大气汞的干、湿沉降，而全世界淡水湖泊沉积物中汞约有 60%来自于通过土壤和植被转移的大气干沉降。该结果对全球汞循环的研究具有重要意义，它表明：①陆地生态系统上大气汞的干沉降要远远大于大气汞的湿沉降；②大气汞干沉降通量的增加可能导致大气汞的生命周期由原来的公认的一年缩减到四个月。此外，基于植被主要记录大气汞干沉降这个认知，Enrico 等（2017）利用泥炭柱的同位素组成重建了全新世以来大气 Hg(0)的浓度；Obrist 等（2017）发现北极的汞污染主要是由于苔原吸收了大量的大气 Hg(0)并转移至周围的冰冻土壤中。

4.3 汞同位素在大气汞转化和跨区域传输方面的应用

汞主要借助大气作为介质进行区域和全球的传输。一般来说，气态 Hg(0)约占大气总汞的 90%以上，而气态 Hg(Ⅱ)和颗粒态 Hg(P)在一些特殊地区和特殊的气象条件下才可达到大气总汞的 10%以上（Fu et al.，2015）。气态 Hg(0)难溶于水，相对不太活泼，在大气中的滞留时间约为 0.5～2 年，但在光照的情况下遇到强氧化物（如卤素自由基、臭氧）能够较快转化为气态 Hg(Ⅱ)和颗粒态 Hg(P)（Selin et al.，2008；Driscoll et al.，2013）。气态 Hg(Ⅱ)和颗粒态 Hg(P)易溶于水，大气滞留时间只有几天到数周，在还原剂（如溶解性有机物）参与下可以还原为零价汞（Horowitz et al.，2017）。大气形态汞之间的相互转化会造成汞同位素的分馏，使得大气中汞同位素信号相对于释放源发生较大的偏移，给源示踪带来极大的不确定性。但是，大气汞同位素的分馏效应也同时为定量研究不同形态汞之间的转化提供了一个全新的化学示踪手段。

4.3.1 大气形态汞同位素样品的采集与预处理技术

大气形态汞同位素分析测试的关键在于如何定量捕获足够量的大气中不同形态汞，并采取适宜的物理化学流程对捕获的形态汞进行预处理，使其转化为易于分析的可溶性离子态汞。

1. 用于同位素分析的大气形态汞采集

大气气态 Hg(0)和 Hg(Ⅱ)（合称大气气态汞）在采样的过程中难以定量分离，一般一起采集。由于 Hg(Ⅱ)所占的比例非常小，所测试的大气气态汞的同位素组成可以代表大气 Hg(0)的同位素组成（Xu et al.，2017；Enrico et al.，2016；Yu et al.，2016；Fu et al.，2016，2014；Demers et al.，2015，2013；Gratz et al.，2010；Sherman et al.，2010）。镀金石英砂/石英珠是大气气态汞最经典的采集手段，但是受限于采样的气流通量（<2 L/min）限制，在低汞背景的大气中难以在短时间内收集足够多的汞（>10 ng）用于高精度的同位素分析。为此，一些同位素研究者把数个镀金石英砂管进行并联，以期在短时间内收集到足够多的大气气态汞

（Demers et al.，2015；2013；Gratz et al.，2010；Sherman et al.，2010）。但是，长时间地将镀金石英砂管暴露于大流量的空气（特别是水汽和酸气比较高的空气）中容易造成汞捕获效率的降低，从而造成汞同位素的分馏。针对以上问题，Fu 等（2014）开发了一种溴化/氯化活性炭管法来富集大气气态汞。通过与镀金石英砂管法以及其他方法进行对比，溴化/氯化活性炭管法在较高的大气气流通量（10～20 L/min）的情况下仍具有出色的大气汞捕获效率且不易受到气体成分的干扰。

由于大气气态 Hg(Ⅱ)的浓度较低且不易与 Hg(0)和 Hg(P)进行分离，目前大气气态 Hg(Ⅱ)的采集和同位素分析测试比较少。Rolison 等（2013）利用高通量颗粒物采样器（1.2 m^3/min）和 KCl 溶液浸透的石英纤维滤膜对墨西哥湾 Grand Bay 区域的大气气态 Hg(Ⅱ)进行富集，并进行了同位素测试。由于含量较低，该研究 24 小时只富集了很微量的大气 Hg(Ⅱ)（1～13 ng），且富集效率约为同时进行大气形态汞采集的 Terkan 2537A/1130/1135 的 1/3。大气颗粒态 Hg(P)一般利用大流量颗粒物采样器配合石英纤维滤膜进行采集，采样方法与一般大气颗粒物的采集类似，采样时间由当地大气汞的浓度决定。

大气降水（雨、雪）含有溶解的大气气态 Hg(Ⅱ)和颗粒态 Hg(P)，其采样方法较为直观，可以直接利用自动的大气降水采集器或者人工改造的降水采集器进行采集。但是由于大气降水中的汞含量一般比较低（几十 ng/L），须采集几升至数十升的降水样品进行汞同位素测试，且须注意采样过程中的汞污染问题。采集后的雨水样品一般直接加入 BrCl 溶液把所有形态的汞转化为可溶性二价汞离子，分析总汞含量及其同位素组成（Enrico et al.，2016；Wang et al.，2015；Yuan et al.，2015；Sherman et al.，2015；2012b；Gratz et al.，2010）。另外，为了更精细地了解雨水中溶解态汞和颗粒态汞的含量和同位素组成，在化学处理和测试之前，可以利用 0.45 μm 的滤膜对溶解态和颗粒态的汞进行分离（Yuan et al.，2018；Chen et al.，2012）。

2. 大气形态汞样品的预富集技术

汞预富集技术的关键是把采集到的微量大气汞通过一系列物理化学方法定量转化为可用于汞同位素测试的溶解离子态汞。对于采集到固体介质上的大气形态汞，一般采用热挥发-强氧化溶液（如 H_2SO_4 酸化的 $KMnO_4$ 溶液、HNO_3 和 HCl 的混合液）富集法。例如：Sherman 等（2010）和 Gratz（2010）把镀金石英砂管采集的大气气态汞通过金属线圈逐步加热至 500℃（20～100℃：10 min；100～300℃：90 min；300～500℃：40 min；500℃：10 min），使挥发的汞缓慢富集至酸化的 $KMnO_4$ 溶液中用于汞同位素测试。Rolison 等（2013）则是利用管式炉把镀金石英砂管采集的大气汞加热至 500℃（20～500℃：10 min；500℃：20 min）进行解吸附，然后把挥发的汞同样富集至酸化的 $KMnO_4$ 溶液中。参考双管式炉-

酸液吸收预富集固体样品汞的方法（Sun et al.，2013），Fu 等（2014）将镀金石英砂管和氯化/溴化活性炭管采集的大气气态汞在燃烧炉分别逐步加热至 525℃和 1000℃，继而使挥发的汞和其他物质在串联的高温裂解炉中纯化，最后采用 40%的 HNO_3：HCl（体积比=2：1）富集最终挥发的汞。石英滤膜上采集的 Hg(P)和 Hg(Ⅱ)既可以利用燃烧炉-强氧化溶液进行预富集，也可以利用传统的酸消解法进行预处理。值得一提的是，相对于湿法酸消解，管式炉-强氧化溶液法能够有效地消除样品基质对汞同位素分析的干扰。

对于大气降水中汞同位素的分析，汞的预富集一般采用以下两种方法。第一种方法是 $SnCl_2$ 还原法：把大气降水与含有 $SnCl_2$ 的化学试剂在气-液分离器中进行匀速混合或者在盛有大气降水的瓶子中缓慢地加入一定量的 $SnCl_2$，同时通入惰性气体（如 Ar）把还原产生的气态 Hg(0)缓慢带入强氧化溶液中（Gratz et al.，2010；Sherman et al.，2010）。第二种方法是树脂分离富集法。Chen 等（2010）通过实验室控制实验发现 AG 1×4 离子交换树脂能够有效地分离和富集汞溶液中的汞，并具有很好的回收率，其大致步骤如下：①0.05% L-半胱氨酸、4 mol/L HNO_3 和去离子水清洗树脂；②0.1 mol/L HCl 调节树脂；③加入 0.1 mol/L HCl+0.5% 0.2 mol/L BrCl 使样品中的汞转化为稳定的二价溶解态，进而进行树脂过滤；④加入 0.1 mol/L HCl 去除树脂上捕获的样品基质；⑤加入 0.05% L-半胱氨酸+0.5 mol/L HNO_3 淋滤树脂吸附的汞。Chen 等（2012）和 Yuan 等（2018）利用离子交换树脂法分别富集了加拿大彼得伯勒（Peterborough）地区和中国贵阳地区降水中的汞并分析了其同位素组成。

4.3.2 实地观测的大气形态汞同位素组成

表 4-4 总结了目前为止全球各个区域报道的大气气态汞[主要是 Hg(0)]（Obrist et al.，2017；Xu et al.，2017；Yamakawa et al.，2017；Enrico et al.，2016；Fu et al.，2016；Yu et al.，2016；Demers et al.，2015；2013；Rolison et al.，2013；Sherman et al.，2010；Gratz et al.，2010）、大气降水汞［包含气态 Hg(Ⅱ)和颗粒态 Hg(P)］（Yuan et al.，2018；Obrist et al.，2017；Enrico et al.，2016；Wang et al.，2015；Yuan et al.，2015；Sherman et al.，2015；2012a；2012b；2010；Demers et al.，2013；Donovan et al.，2013；Chen et al.，2012；Gratz et al.，2010）和大气颗粒态 Hg(P)（Yuan et al.，2018；Xu et al.，2017；Das et al.，2016；Yu et al.，2016；Huang et al.，2016；2015；Rolison et al.，2013）的汞同位素组成，包括质量分馏（$\delta^{202}Hg$）、奇数同位素的非质量分馏值（$\Delta^{199}Hg$、$\Delta^{201}Hg$）和偶数同位素的非质量分馏值（$\Delta^{200}Hg$、$\Delta^{204}Hg$）。由于大气气态 Hg(Ⅱ)的同位素组成报道较少，且采样方法存在较大的争议，暂不做介绍。

表 4-4 文献报道的大气气态汞[主要为 Hg(0)]、大气降水汞[Hg(Ⅱ)+Hg(P)]和大气颗粒态汞[Hg(P)]的同位素组成

大气形态汞	统计值	$\delta^{202}Hg$（‰）	$\Delta^{199}Hg$（‰）	$\Delta^{200}Hg$（‰）	$\Delta^{201}Hg$（‰）	$\Delta^{204}Hg$（‰）
大气气态汞	范围	$-3.88\sim1.60$	$-0.41\sim0.19$	$-0.24\sim0.27$	$-0.32\sim0.12$	$-0.23\sim0.30$
	均值±标准差	-0.10 ± 0.93	-0.12 ± 0.12	-0.04 ± 0.06	-0.11 ± 0.10	0.06 ± 0.09
	中值	-0.07	-0.13	-0.04	-0.12	0.07
	样品数	259	259	259	259	114
大气降水汞	范围	$-4.37\sim0.79$	$-5.08\sim1.57$	$-0.20\sim1.24$	$-4.65\sim1.53$	$-0.73\sim0.27$
	均值±标准差	-0.67 ± 1.03	0.16 ± 0.82	0.15 ± 0.19	0.15 ± 0.77	-0.26 ± 0.21
	中值	-0.39	0.33	0.13	0.30	-0.28
	样品数	214	214	214	214	69
大气颗粒态汞	范围	$-3.48\sim1.61$	$-0.53\sim1.36$	$-0.05\sim0.28$	$-0.53\sim1.20$	$-0.81\sim0.41$
	均值±标准差	-1.03 ± 0.75	0.04 ± 0.24	0.04 ± 0.05	0.01 ± 0.22	-0.20 ± 0.30
	中值	-0.84	0.01	0.03	-0.01	-0.18
	样品数	211	211	211	211	25

结合同位素值的直方图（图 4-2），可以看出：①$\delta^{202}Hg$ 在大气气态汞、大气降水汞和大气颗粒态汞中的变化范围均高达 5‰；$\delta^{202}Hg$ 在大气气态汞（-0.10‰±0.93‰）、大气降水汞（-0.67‰±1.03‰）和大气颗粒态汞（-1.03‰±0.75‰）中的均值呈逐渐降低的趋势，其主要变化区间分别为-1‰~1‰、-1.5‰~0.5‰和-1.5‰~0.0‰。②$\Delta^{199}Hg$ 和 $\Delta^{201}Hg$ 在大气降水汞中的变化范围超过 6‰，在大气颗粒态汞中的变化范围近 2‰，而在大气气态汞中的变化范围仅为 0.5‰；$\Delta^{199}Hg$ 和 $\Delta^{201}Hg$ 在大气降水汞（0.16‰±0.82‰；0.15‰±0.77‰）、大气颗粒态汞（0.04‰±0.24‰；0.01‰±0.22‰）和大气气态汞（-0.12‰±0.12‰；-0.11‰±0.10‰）中的均值呈逐渐降低的趋势，其主要变化区间分别为 0‰~1‰、0‰~0.5‰和-0.5‰~0.0‰。③$\Delta^{200}Hg$ 只有在大气降水汞中的变化范围较大（近 1.5‰），而在大气气态汞和大气颗粒态汞中的变化范围仅为 0.5‰和 0.3‰；$\Delta^{200}Hg$ 在大气降水汞（0.15‰±0.19‰）、大气颗粒态汞（0.04‰±0.05‰）和大气气态汞（-0.04‰±0.06‰）中的均值呈逐渐降低的趋势，其主要变化区间分别为-0.10‰~0.30‰、-0.05‰~0.10‰和-0.10‰~0.05‰。④由于部分多接收器-电感耦合等离子体质谱（MC-ICPMS）法拉第杯的构造不允许同时测量所有的汞同位素且 ^{204}Hg 的丰度较低，只有少许样品测试了 $\Delta^{204}Hg$ 的值。$\Delta^{204}Hg$ 在大气降水汞和大气颗粒态汞中的变化范围约为 1‰，在大气气态汞中的变化范围约为 0.5‰；与 $\Delta^{200}Hg$ 的变化趋势相反，$\Delta^{204}Hg$ 在大气降水汞（-0.26‰±0.21‰）、大气颗粒态汞（-0.20‰±0.30‰）和大气气态汞（0.06‰±0.09‰）中的均值呈逐渐增加的趋势。

为了更加清楚地厘清不同大气形态汞的同位素变化规律，图 4-3 和图 4-4 分别列出了不同地区大气汞的 δ^{202}Hg vs. Δ^{199}Hg，Δ^{199}Hg vs. Δ^{200}Hg 图解。从中可以看出一些显著异常的值，例如：①来自于美国密西西比 Grand Bay 海岸带的大气气态汞和大气颗粒态汞样品分别具有非常明显的负 δ^{202}Hg 值和正 Δ^{199}Hg 值；②受燃煤汞排放影响的佛罗里达州克里斯特尔（Crystal）河周围的大气降水具有明显偏负的

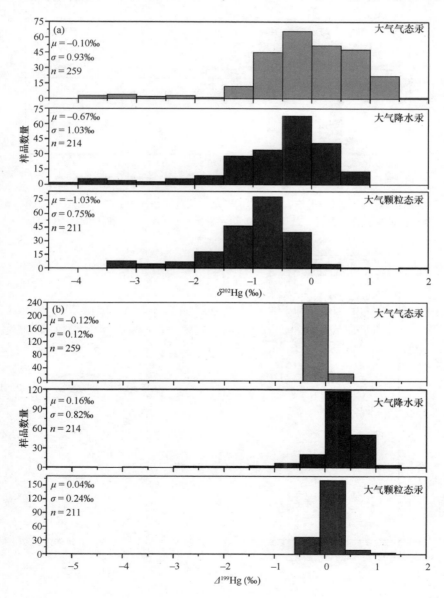

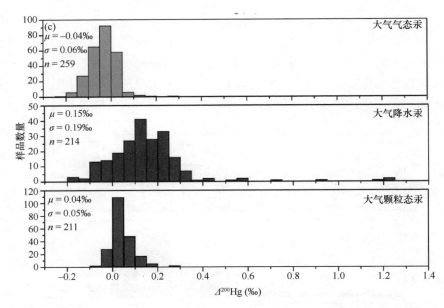

图 4-2 文献报道的大气气态汞、大气降水汞以及大气颗粒态汞的质量分馏值（$\delta^{202}Hg$，a），奇数同位素的非质量分馏值（$\Delta^{199}Hg$，b）以及偶数同位素的非质量分馏值（$\Delta^{200}Hg$，c）

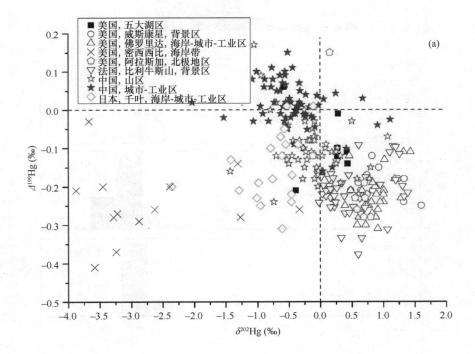

第 4 章 同位素技术在汞分子转化及区域传输研究中的应用

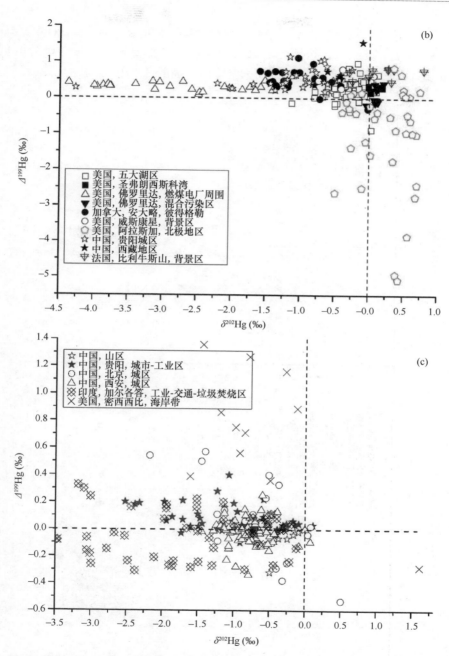

图 4-3 世界不同地区的大气气态汞（a）、大气降水（b）以及大气颗粒态汞（c）的质量分馏值（δ^{202}Hg）与奇数同位素的非质量分馏（Δ^{199}Hg）的散点分布图

δ^{202}Hg 值；③由于大气汞"亏损"事件（atmospheric mercury deposition event，AMDE）及随后的光化学反应，北极地区的大气降水样品（主要是雪）具有极为偏负的 Δ^{199}Hg 值；④加拿大彼得格勒地区的大气降雨样品的 Δ^{200}Hg 值高达 1.24‰，明显

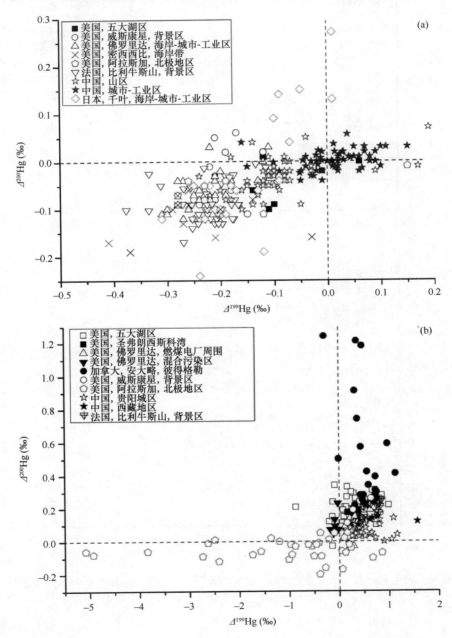

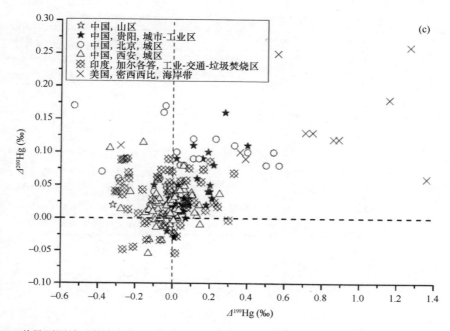

图 4-4 世界不同地区的大气气态汞（a）、大气降水（b）以及大气颗粒态汞（c）奇数同位素的非质量分馏（Δ^{199}Hg）与偶数同位素的非质量分馏（Δ^{200}Hg）的散点分布图

高于其他地区的大气降水样品。另外，通过对比不同区域、不同污染模式下的汞同位素组成，发现：①相对于中国的大气气态汞样品，欧美地区（美国和法国）的大气气态汞具有较高的 δ^{202}Hg 值和较低的 Δ^{199}Hg 值；②相对于受人为影响比较严重的城市大气气态汞，背景区大气气态汞总体上具有较高的 δ^{202}Hg 值和较低的 Δ^{199}Hg 值。

4.3.3 人为/自然源大气汞排放的同位素数据库

1. 自然源大气汞的同位素组成

自然源的汞主要来自于火山的主动和被动去气过程以及岩石的自然风化过程。目前，只有少数研究报道了被动去气火山喷发物的汞同位素组成（Sun et al., 2016b; Zambardi et al., 2009）（见表 4-5）。Zambardi 等（2009）发现意大利火山气孔喷发物总汞的 δ^{202}Hg 均值为 −0.74‰±0.18‰（2SD, $n=4$），与 Sun 等（2016）所报道的印度尼西亚火山气孔喷发物总汞的 δ^{202}Hg 值非常类似（−0.83‰～−0.58‰）。以上两个研究均未发现火山气孔喷发物具有明显的非质量分馏信号。另外，Zambardi 等（2009）发现释放到大气中的火山喷发物的气态汞相对于总汞负偏移了约 1.0‰，而颗粒态汞相对于总汞正偏移了约 0.6‰，表明火山喷发物的汞在附近大气中发生了非常明显的分馏。

表 4-5 文献报道的被动去气火山气孔喷发物的汞同位素组成

火山气孔编号	火山名	$\delta^{202}Hg$	$\Delta^{199}Hg$	$\Delta^{200}Hg$	$\Delta^{201}Hg$	参考文献
C2	Merapi	−0.81	0.08	0.02	0.07	(Sun et al., 2016b)
C4	Merapi	−0.58	0.00	0.13	−0.03	(Sun et al., 2016b)
C5	Merapi	−0.82	0.04	0.04	0.02	(Sun et al., 2016b)
C9	Merapi	−0.82	0.09	0.06	0.09	(Sun et al., 2016b)
C193	Papandayan	−0.83	0.07	−0.01	−0.02	(Sun et al., 2016b)
FA	Vulcano	−0.27	0.01	−0.04	0.04	(Zambardi et al., 2009)
F0	Vulcano	−0.79	0.08	0.05	0.01	(Zambardi et al., 2009)
F5	Vulcano	−1.09	−0.05	−0.03	−0.10	(Zambardi et al., 2009)
F11	Vulcano	−0.79	0.13	0.03	0.02	(Zambardi et al., 2009)
均值±2SD		−0.76±0.22	0.05±0.06	0.03±0.05	0.01±0.06	

2. 人为源大气汞的同位素组成

Sun 等（2016b）根据人为源大气形态汞[Hg(0)、Hg(Ⅱ)、Hg(P)]历史排放清单、源物质的汞同位素组成以及源物质在燃烧和工业利用过程中的形态汞同位素分馏规律（主要发生质量分馏而无非质量分馏值），建立了工业革命以来（1850～2010 年）人为源大气形态汞同位素数据库。需要强调的是该数据库中除了煤炭燃烧外，大部分排放源只考虑了源物质的汞同位素组成，而缺少对汞排放过程中同位素分馏效应的校正。该数据库包括 10 个 "by-product（汞因作为副产物而排放）" 源和 16 个 "intentional Hg use（汞因作为化工和生产原料而排放）" 源。表 4-6 列出了 Sun 等（2016b）模拟的所有人为源（"by-product" + "intentional Hg use"）以及 "by-product" 源的总汞和大气形态汞[Hg(0)、Hg(Ⅱ)、Hg(P)]的质量分馏值（$\delta^{202}Hg$）。相应的人为源大气形态汞排放的奇数同位素的非质量分馏值（$\Delta^{199}Hg$）列于表 4-7。由于人为源物质的偶数同位素非质量分馏值（$\Delta^{200}Hg$）接近于零，因而人为源排放大气汞的 $\Delta^{200}Hg$ 值也接近于零。

由表 4-6 可以看到，"by-product" 源总汞的 $\delta^{202}Hg$ 呈现一个递增的趋势，由 1850 年的−2.1‰增加到 2010 年的−0.9‰。总汞的 $\delta^{202}Hg$ 变化趋势大致上可以分为三个阶段：①1850～1930 年 $\delta^{202}Hg$ 值从−2.1‰增加至−1.3‰，代表主要汞排放源由金属汞的开采和工业利用逐步转向煤炭燃烧和有色金属冶炼；1930～1960 年 $\delta^{202}Hg$ 值稳定在−1.5‰～−1.3‰，代表各个汞污染源的汞排放的比例相对稳定；1960～2010 年 $\delta^{202}Hg$ 值从−1.5‰增加至−0.9‰，代表全球全面转向以燃煤为主的汞排放。大气气态 Hg(0)（−2.1‰～−0.8‰）、气态 Hg(Ⅱ)（−2.1‰～−0.8‰）和颗粒态 Hg(P)（−2.0‰～−1.4‰）的 $\delta^{202}Hg$ 值基本上与总汞的变化趋势一致，但是近

表 4-6 所有人为源以及"by-product"源的总汞和大气形态汞 [气态 Hg(0)、气态 Hg(Ⅱ)、颗粒态 Hg(P)] 的 δ^{202}Hg 值　　　　　　　　　　　　　　（单位：‰）

年份	所有人为源				"by-product"源			
	P10	P90	均值	中值（P50）	P10	P90	均值	中值（P50）
大气总汞								
1850	−1.79	−0.70	−1.20	−1.14	−3.10	−1.37	−2.19	−2.11
1860	−1.66	−0.69	−1.14	−1.09	−2.94	−1.36	−2.10	−2.04
1870	−1.51	−0.50	−1.00	−0.91	−3.14	−1.33	−2.18	−2.10
1880	−1.69	−0.59	−1.10	−1.01	−3.18	−1.31	−2.19	−2.10
1890	−1.45	−0.53	−0.98	−0.91	−2.80	−1.41	−2.07	−2.01
1900	−1.37	−0.55	−0.95	−0.92	−2.58	−1.34	−1.93	−1.88
1910	−1.25	−0.64	−0.93	−0.91	−2.33	−1.24	−1.75	−1.70
1920	−1.04	−0.64	−0.83	−0.82	−2.03	−1.16	−1.56	−1.52
1930	−0.88	−0.60	−0.74	−0.74	−1.63	−1.10	−1.35	−1.33
1940	−0.78	−0.54	−0.66	−0.66	−1.61	−1.12	−1.36	−1.35
1950	−0.89	−0.62	−0.75	−0.75	−1.64	−1.06	−1.35	−1.35
1960	−0.91	−0.64	−0.78	−0.77	−1.76	−1.20	−1.48	−1.47
1970	−0.90	−0.64	−0.77	−0.76	−1.76	−1.18	−1.47	−1.46
1980	−1.05	−0.70	−0.87	−0.87	−1.75	−1.05	−1.39	−1.39
1990	−1.06	−0.69	−0.87	−0.87	−1.64	−0.95	−1.29	−1.29
2000	−0.92	−0.52	−0.72	−0.72	−1.31	−0.60	−0.96	−0.97
2010	−0.94	−0.41	−0.68	−0.68	−1.28	−0.42	−0.86	−0.87
气态 Hg(0)								
1850	−1.76	−0.69	−1.18	−1.12	−3.14	−1.38	−2.21	−2.14
1860	−1.66	−0.68	−1.14	−1.07	−3.00	−1.36	−2.14	−2.09
1870	−1.51	−0.50	−0.76	−0.90	−3.18	−1.33	−2.21	−2.13
1880	−1.69	−0.58	−1.09	−1.00	−3.27	−1.31	−2.22	−2.13
1890	−1.47	−0.52	−0.99	−0.91	−2.86	−1.41	−2.09	−2.03
1900	−1.34	−0.54	−0.93	−0.90	−2.67	−1.33	−1.96	−1.91
1910	−1.22	−0.61	−0.90	−0.88	−2.42	−1.21	−1.78	−1.73
1920	−0.97	−0.56	−0.75	−0.74	−2.19	−1.07	−1.58	−1.53
1930	−0.80	−0.52	−0.66	−0.65	−1.66	−0.99	−1.31	−1.28
1940	−0.71	−0.46	−0.59	−0.58	−1.68	−1.06	−1.35	−1.33
1950	−0.76	−0.51	−0.63	−0.63	−1.64	−1.10	−1.36	−1.34
1960	−0.78	−0.52	−0.65	−0.65	−1.76	−1.27	−1.51	−1.50
1970	−0.77	−0.52	−0.65	−0.64	−1.78	−1.25	−1.50	−1.49
1980	−0.89	−0.60	−0.74	−0.74	−1.82	−1.13	−1.47	−1.45
1990	−0.91	−0.64	−0.78	−0.77	−1.67	−1.07	−1.36	−1.35
2000	−0.71	−0.48	−0.60	−0.60	−1.10	−0.64	−0.87	−0.88
2010	−0.73	−0.40	−0.57	−0.56	−1.04	−0.50	−0.77	−0.78

续表

年份	所有人为源				"by-product"源			
	P10	P90	均值	中值（P50）	P10	P90	均值	中值（P50）
气态 Hg(Ⅱ)								
1850	−1.75	−0.68	−1.17	−1.10	−3.00	−1.34	−2.12	−2.06
1860	−1.64	−0.69	−1.13	−1.07	−2.81	−1.32	−2.02	−1.96
1870	−1.47	−0.50	−0.98	−0.89	−3.00	−1.28	−2.10	−2.02
1880	−1.68	−0.59	−1.10	−1.01	−3.07	−1.28	−2.12	−2.03
1890	−1.46	−0.53	−0.98	−0.92	−2.67	−1.35	−1.98	−1.93
1900	−1.39	−0.56	−0.96	−0.92	−2.42	−1.27	−1.81	−1.76
1910	−1.31	−0.66	−0.97	−0.95	−2.12	−1.12	−1.60	−1.55
1920	−1.21	−0.67	−0.93	−0.91	−1.79	−0.95	−1.35	−1.34
1930	−1.06	−0.65	−0.86	−0.86	−1.50	−0.84	−1.18	−1.19
1940	−1.04	−0.61	−0.83	−0.84	−1.54	−0.78	−1.18	−1.20
1950	−1.30	−0.38	−0.86	−0.90	−1.67	−0.38	−1.07	−1.13
1960	−1.30	−0.62	−0.97	−0.97	−1.69	−0.72	−1.21	−1.23
1970	−1.34	−0.61	−0.99	−1.00	−1.65	−0.69	−1.18	−1.20
1980	−1.38	−0.56	−0.98	−0.99	−1.65	−0.59	−1.13	−1.15
1990	−1.27	−0.48	−0.89	−0.90	−1.59	−0.51	−1.07	−1.09
2000	−1.24	−0.33	−0.81	−0.83	−1.44	−0.31	−0.90	−0.93
2010	−1.26	−0.22	−0.76	−0.78	−1.44	−0.17	−0.82	−0.84
颗粒态 Hg(P)								
1850	−1.76	−0.80	−1.25	−1.19	−2.72	−1.44	−2.04	−1.98
1860	−1.67	−0.84	−1.23	−1.20	−2.54	−1.47	−1.97	−1.92
1870	−1.54	−0.60	−1.05	−0.99	−2.71	−1.45	−2.03	−1.96
1880	−1.72	−0.70	−1.17	−1.10	−2.74	−1.42	−2.04	−1.97
1890	−1.56	−0.67	−1.10	−1.05	−2.41	−1.52	−1.94	−1.90
1900	−1.53	−0.76	−1.14	−1.12	−2.22	−1.50	−1.84	−1.81
1910	−1.54	−0.98	−1.25	−1.24	−2.04	−1.46	−1.75	−1.73
1920	−1.62	−1.15	−1.38	−1.37	−1.95	−1.45	−1.69	−1.68
1930	−1.53	−1.15	−1.34	−1.33	−1.87	−1.42	−1.64	−1.63
1940	−1.46	−1.00	−1.22	−1.21	−1.88	−1.28	−1.57	−1.56
1950	−1.66	−1.15	−1.40	−1.38	−1.96	−1.36	−1.65	−1.63
1960	−1.77	−1.12	−1.44	−1.42	−2.08	−1.28	−1.67	−1.65
1970	−1.84	−1.13	−1.47	−1.45	−2.15	−1.30	−1.72	−1.71
1980	−1.95	−1.09	−1.51	−1.49	−2.18	−1.18	−1.67	−1.65
1990	−1.87	−0.99	−1.42	−1.40	−2.13	−1.09	−1.60	−1.59
2000	−1.69	−0.86	−1.26	−1.23	−1.97	−0.95	−1.45	−1.44
2010	−1.59	−0.80	−1.17	−1.14	−1.92	−0.92	−1.42	−1.40

注：P10、P50 和 P90 分别表示预测值有 10%、20% 和 90% 的概率等于或者小于 P10、P50 和 P90 的值。

几十年来 Hg(P) 的 δ^{202}Hg 值要比总汞低达 0.5‰。与 δ^{202}Hg 的变化趋势相反，Δ^{199}Hg 最初的 100 年间（1850～1950 年）从 –0.03‰ 下降至 –0.10‰，这主要是由于北美和欧洲的燃煤具有非常低的 Δ^{199}Hg 值。之后，Δ^{199}Hg 缓慢增加至 2010 的 –0.06‰，这主要是由于亚洲燃煤的 Δ^{199}Hg 接近于零。大气气态 Hg(0)（–0.02‰～–0.04‰）、气态 Hg(Ⅱ)（–0.01‰～–0.13‰）和颗粒态 Hg(P)（–0.08‰～–0.16‰）的 δ^{202}Hg 值基本上与总汞的变化趋势一致（表 4-7）。

表 4-7 所有人为源以及"by-product"源的总汞和大气形态汞[气态 Hg(0)、气态 Hg(Ⅱ)、颗粒态 Hg(P)]的 Δ^{199}Hg 值　　　　　　　　　　　（单位：‰）

年份	所有人为源				"by-product"源			
	P10	P90	均值	中值（P50）	P10	P90	均值	中值（P50）
大气总汞								
1850	–0.06	0.01	–0.02	–0.02	–0.11	0.04	–0.03	–0.03
1860	–0.05	0.01	–0.02	–0.02	–0.09	0.03	–0.03	–0.03
1870	–0.04	0.01	–0.01	–0.01	–0.07	0.05	–0.01	–0.01
1880	–0.04	0.02	–0.01	–0.01	–0.07	0.05	–0.01	–0.01
1890	–0.05	0.01	–0.02	–0.02	–0.09	0.02	–0.03	–0.03
1900	–0.04	0.01	–0.02	–0.02	–0.08	0.02	–0.03	–0.03
1910	–0.05	0.00	–0.02	–0.02	–0.09	0.00	–0.04	–0.04
1920	–0.05	–0.01	–0.03	–0.03	–0.11	–0.01	–0.06	–0.06
1930	–0.05	–0.02	–0.03	–0.03	–0.10	–0.03	–0.06	–0.06
1940	–0.04	–0.02	–0.03	–0.03	–0.12	–0.03	–0.07	–0.07
1950	–0.06	–0.02	–0.04	–0.04	–0.15	–0.05	–0.10	–0.10
1960	–0.06	–0.02	–0.04	–0.04	–0.14	–0.04	–0.09	–0.09
1970	–0.06	–0.02	–0.04	–0.04	–0.14	–0.04	–0.09	–0.09
1980	–0.07	–0.02	–0.04	–0.04	–0.14	–0.03	–0.08	–0.08
1990	–0.07	–0.02	–0.04	–0.04	–0.12	–0.02	–0.07	–0.07
2000	–0.06	–0.02	–0.04	–0.04	–0.11	–0.02	–0.06	–0.06
2010	–0.07	–0.01	–0.04	–0.04	–0.10	0.00	–0.05	–0.06
气态 Hg(0)								
1850	–0.06	0.01	–0.02	–0.02	–0.10	0.05	–0.03	–0.02
1860	–0.05	0.01	–0.02	–0.02	–0.09	0.04	–0.02	–0.02
1870	–0.04	0.02	–0.01	–0.01	–0.06	0.06	0.00	0.00
1880	–0.04	0.02	–0.01	–0.01	–0.06	0.06	0.00	0.00
1890	–0.04	0.01	–0.02	–0.02	–0.09	0.03	–0.02	–0.02
1900	–0.04	0.01	–0.01	–0.01	–0.06	0.03	–0.01	–0.01
1910	–0.04	0.00	–0.02	–0.02	–0.07	0.03	–0.02	–0.02

续表

年份	所有人为源				"by-product"源			
	P10	P90	均值	中值（P50）	P10	P90	均值	中值（P50）
气态 Hg(0)								
1920	−0.03	0.00	−0.02	−0.02	−0.07	0.02	−0.02	−0.02
1930	−0.03	−0.01	−0.02	−0.02	−0.06	0.00	−0.03	−0.03
1940	−0.03	−0.01	−0.02	−0.02	−0.05	0.00	−0.02	−0.02
1950	−0.03	−0.01	−0.02	−0.02	−0.07	−0.02	−0.04	−0.04
1960	−0.03	−0.01	−0.02	−0.02	−0.07	−0.01	−0.04	−0.04
1970	−0.03	−0.01	−0.02	−0.02	−0.07	−0.01	−0.04	−0.04
1980	−0.03	−0.01	−0.02	−0.02	−0.07	−0.01	−0.04	−0.04
1990	−0.03	−0.01	−0.02	−0.02	−0.07	−0.01	−0.04	−0.04
2000	−0.03	−0.01	−0.02	−0.02	−0.06	−0.01	−0.03	−0.03
2010	−0.04	−0.01	−0.02	−0.02	−0.06	0.00	−0.03	−0.03
气态 Hg(Ⅱ)								
1850	−0.03	0.00	−0.02	−0.02	−0.03	0.00	−0.02	−0.01
1860	−0.05	0.00	−0.02	−0.02	−0.10	0.02	−0.04	−0.04
1870	−0.04	0.01	−0.02	−0.02	−0.08	0.04	−0.02	−0.02
1880	−0.04	0.01	−0.01	−0.02	−0.07	0.04	−0.02	−0.02
1890	−0.05	0.00	−0.02	−0.02	−0.10	0.01	−0.04	−0.04
1900	−0.05	0.00	−0.03	−0.03	−0.09	0.00	−0.04	−0.04
1910	−0.06	−0.01	−0.04	−0.03	−0.11	−0.01	−0.06	−0.06
1920	−0.08	−0.02	−0.05	−0.05	−0.14	−0.03	−0.08	−0.08
1930	−0.10	−0.03	−0.06	−0.06	−0.15	−0.05	−0.10	−0.10
1940	−0.10	−0.04	−0.07	−0.07	−0.18	−0.06	−0.12	−0.11
1950	−0.15	−0.05	−0.10	−0.10	−0.20	−0.07	−0.13	−0.13
1960	−0.15	−0.04	−0.09	−0.09	−0.20	−0.06	−0.13	−0.13
1970	−0.15	−0.05	−0.10	−0.10	−0.20	−0.06	−0.13	−0.13
1980	−0.16	−0.04	−0.10	−0.10	−0.20	−0.05	−0.12	−0.12
1990	−0.13	−0.03	−0.08	−0.08	−0.18	−0.04	−0.11	−0.11
2000	−0.12	−0.03	−0.08	−0.07	−0.15	−0.03	−0.09	−0.09
2010	−0.12	−0.01	−0.06	−0.06	−0.14	0.00	−0.07	−0.07
颗粒态 Hg(P)								
1850	−0.09	−0.01	−0.05	−0.05	−0.16	−0.01	−0.08	−0.08
1860	−0.10	−0.02	−0.05	−0.05	−0.16	−0.02	−0.09	−0.09
1870	−0.07	−0.01	−0.04	−0.04	−0.14	−0.01	−0.07	−0.07
1880	−0.07	0.00	−0.04	−0.04	−0.12	0.00	−0.06	−0.06

续表

年份	所有人为源				"by-product" 源			
	P10	P90	均值	中值（P50）	P10	P90	均值	中值（P50）
	颗粒态 Hg(P)							
1890	−0.09	−0.02	−0.05	−0.05	−0.15	−0.03	−0.09	−0.09
1900	−0.10	−0.02	−0.06	−0.06	−0.16	−0.04	−0.10	−0.10
1910	−0.12	−0.04	−0.08	−0.08	−0.18	−0.05	−0.12	−0.11
1920	−0.16	−0.05	−0.11	−0.10	−0.21	−0.07	−0.14	−0.14
1930	−0.16	−0.06	−0.11	−0.11	−0.21	−0.08	−0.14	−0.14
1940	−0.16	−0.06	−0.11	−0.11	−0.22	−0.08	−0.15	−0.15
1950	−0.19	−0.07	−0.13	−0.13	−0.24	−0.08	−0.16	−0.16
1960	−0.18	−0.05	−0.11	−0.11	−0.21	−0.06	−0.13	−0.13
1970	−0.19	−0.05	−0.12	−0.11	−0.22	−0.06	−0.14	−0.14
1980	−0.18	−0.03	−0.10	−0.10	−0.21	−0.03	−0.11	−0.11
1990	−0.14	0.00	−0.07	−0.07	−0.16	0.00	−0.08	−0.08
2000	−0.11	0.01	−0.05	−0.05	−0.13	0.02	−0.06	−0.06
2010	−0.09	0.01	−0.04	−0.04	−0.12	0.02	−0.05	−0.05

注：P10、P50 和 P90 分别表示预测值有 10%、20% 和 90% 的概率等于或者小于 P10、P50 和 P90 的值。

由于历史上 "intentional Hg use" 汞排放约占所有人为源汞排放的 50%～80%，而 "intentional Hg use" 汞排放的同位素组成被认为是不变的，所以所有人为源总汞排放的 δ^{202}Hg 和 Δ^{199}Hg 值在 1850～2010 年间的变化较小，分别为 −1.1‰～−0.7‰ 和 −0.02‰～−0.04‰（表 4-6 和表 4-7）。大气气态 Hg(0) 的 δ^{202}Hg 和 Δ^{199}Hg 的变化趋势基本上与总汞保持一致，因为约 90% "intentional Hg use" 排放的汞是以 Hg(0) 的形式排放的。大气气态 Hg(Ⅱ) 的 δ^{202}Hg 的变化范围较小（−1.1‰～−0.8‰）且没有明显的趋势；大气颗粒态 Hg(P) 的 δ^{202}Hg 从 1850 年（−1.2‰）到 1980 年（−1.5‰）呈现下降的趋势，但在 2010 年升至 −1.1‰。因为 "intentional Hg use" 汞排放的大气气态 Hg(Ⅱ) 和 Hg(P) 只占所有人为源汞排放的 10% 左右，所有人为源大气气态 Hg(Ⅱ)（−0.02‰～−0.10‰）和 Hg(P)（−0.05‰～−0.13‰）的 Δ^{199}Hg 的变化趋势与两者在 "by-product" 源的变化趋势基本一致。

4.3.4 大气形态汞转化过程中的同位素分馏效应

自然源和人为源的汞释放到大气中，三种形态的汞可以相互转化，主要包括：①气态 Hg(0) 均相氧化成气态 Hg(Ⅱ)；②气态 Hg(0) 非均相氧化成颗粒态 Hg(P)；③气态 Hg(Ⅱ) 溶解至云雾中形成溶解态 Hg(Ⅱ)；④气态 Hg(Ⅱ) 吸附至颗粒物中转化为颗粒态 Hg(P)；⑤含水颗粒态 Hg(P) 或者云雾中的溶解态 Hg(Ⅱ) 还原成气态

Hg(0)。自然界观测到不同地区的大气汞同位素组成(图4-3、图4-4)是不同源区(自然源、人为源)混合并叠加大气形态汞相互转化过程中同位素分馏效应的结果。由大气气态汞、大气降水和大气颗粒态汞非质量分馏的 $\varDelta^{199}$Hg $vs.$ $\varDelta^{201}$Hg 图解(图4-5)

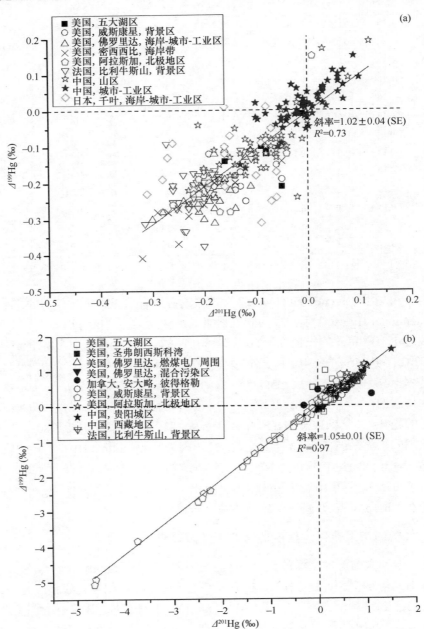

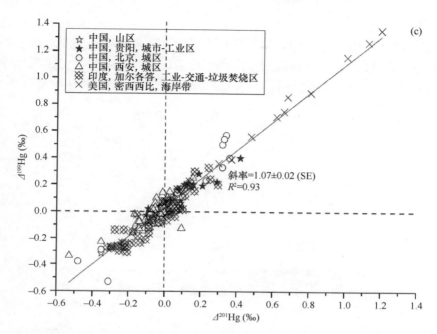

图 4-5　世界不同地区的大气气态汞（a）、大气降水汞（b）以及大气颗粒态汞（c）奇数同位素的非质量分馏值 Δ^{199}Hg 与 Δ^{201}Hg 线性相关关系图

可以看出它们的斜率分别为 1.02、1.05 和 1.07，非常接近于 1.0，表明无机 Hg(Ⅱ) 的光致还原是造成大气样品汞同位素分馏的主要原因。

图 4-6 显示了去除特殊采样位点（北极地区、直接燃煤汞排放、密西西比海岸带）并剔除异常值后，大气气态汞的 δ^{202}Hg $vs.$ Δ^{199}Hg(a)、Δ^{199}Hg $vs.$ Δ^{200}Hg(b)图解。由于人为源远远大于自然源排放的汞通量，且自然源汞排放的同位素组成报道较少（Sun et al., 2016b; Zambardi et al., 2009），图 4-6 只列出 Sun 等（2016b）模拟计算的当今人为源大气形态汞排放的 δ^{202}Hg 和 Δ^{199}Hg 值（Δ^{200}Hg 没有明显异常，参见 4.3.3 节）。通过对比可以发现：①人为源排放的气态 Hg(0)、气态 Hg(Ⅱ) 以及颗粒态 Hg(P) 的 δ^{202}Hg 均值与观测到的大气气态汞、大气降水以及大气颗粒态汞的 δ^{202}Hg 均值都呈递减的趋势，但前者相对后者分别偏移了大约–0.7‰、–0.35‰ 和–0.15‰；②人为源排放大气汞的 Δ^{199}Hg 值主要为负，与观测到的明显偏正的大气降水 Δ^{199}Hg 值不一致；③人为源排放大气汞的 Δ^{200}Hg 值没有显著异常，与观测到的明显偏正的大气降水和偏负的大气气态汞的 Δ^{200}Hg 值不一致。

由此可以得出以下结论：

（1）大气颗粒态汞同位素组成（δ^{202}Hg、Δ^{199}Hg 和 Δ^{200}Hg）估算值与观测值间无显著差别，反映了大气颗粒态 Hg(P) 可能主要来源于人为排放源且没有经历明显的大气分馏过程。

（2）由于大气颗粒态 Hg(P)的所有同位素值相对于大气降水[包括溶解态 Hg(Ⅱ)和颗粒态 Hg(P)]要低，溶解态 Hg(Ⅱ)的同位素值相对于大气降水要高；Yuan 等（2018）对贵阳市雨水的溶解态 Hg(Ⅱ)和颗粒态 Hg(P)的实测数据也支持这个推论。

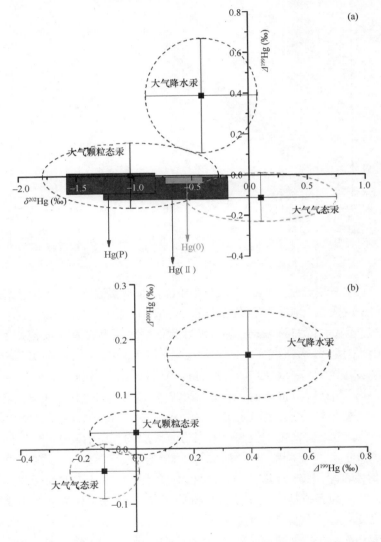

图 4-6　已剔除特殊采样位点和异常值的世界不同地区大气气态汞、大气降水汞、大气颗粒态汞的质量分馏值（δ^{202}Hg）与奇数同位素非质量分馏（Δ^{199}Hg）图解（a）以及奇数同位素非质量分馏值（Δ^{199}Hg）与偶数同位素非质量分馏（Δ^{200}Hg）图解（b）。图中的椭圆代表不同形态汞的均值±1SD 的范围；矩形代表人为源大气形态汞排放的同位素变化区间（Sun et al.，2016b）

(3) 相对于人为源排放的 Hg(0) 和 Hg(Ⅱ)，大气 Hg(0) 和大气降水 [或者大气溶解态 Hg(Ⅱ)] 的 δ^{202}Hg 要高出许多，指示着可能存在其他过程（陆-气交换、海-气交换）造成大气汞富集重的同位素。

(4) 大气气态 Hg(0) 和大气降水 [或者大气溶解态 Hg(Ⅱ)] 的非质量分馏值（Δ^{199}Hg 和 Δ^{200}Hg）分别低于和高于人为源排放的大气形态汞的非质量分馏值，说明大气 Hg(0) 和大气降水 [或者大气溶解态 Hg(Ⅱ)] 是同位素组成互补的两个大气汞储库。这点可以通过大气汞 Δ^{199}Hg 和 Δ^{200}Hg 的平衡方程验证：

$$\Delta^{199}\text{Hg}(\text{Ⅱ}) \times f_1 + \Delta^{199}\text{Hg}(0) \times f_2 + \Delta^{199}\text{Hg}(\text{P}) \times f_3 = \Delta^{199}\text{Hg}^{\text{tot}} \quad (4\text{-}16)$$

$$\Delta^{200}\text{Hg}(\text{Ⅱ}) \times f_1 + \Delta^{200}\text{Hg}(0) \times f_2 + \Delta^{200}\text{Hg}(\text{P}) \times f_3 = \Delta^{200}\text{Hg}^{\text{tot}} \quad (4\text{-}17)$$

$$f_{\text{Hg}}(\text{Ⅱ}) + f_{\text{Hg}}(0) + f_{\text{Hg}}(\text{P}) = 1 \quad (4\text{-}18)$$

利用大气形态汞（Δ^{xxx}Hg(0)：气态零价汞由大气气态汞代表；Δ^{xxx}Hg(Ⅱ)：气态二价汞由大气降水代表；Δ^{xxx}Hg(P)：大气颗粒态汞）以及人为源汞排放总汞（Δ^{xxx}Hg^{tot}）的非质量分馏值，可以计算出大气气态 Hg(0) 的比例在 80%~90% 之间。由于大气气态 Hg(Ⅱ) [即溶解态 Hg(Ⅱ)] 要大于大气降水的 Δ^{xxx}Hg 值，可得出大气气态 Hg(0) 的比例可能会在 90% 以上，与实际观测的大气形态汞的分布类似。

由 4.2.1 节所阐述的"汞在无机转化过程中的同位素分馏规律"可知，通过植物叶片气孔作用所造成的陆地-大气界面汞交换能够产生非常大的质量分馏，使大气汞相对于叶片富集 2‰~3‰（δ^{202}Hg）的重同位素（Obrist et al., 2017; Enrico et al., 2016; Yu et al., 2016; Demers et al., 2013）。该过程很好地解释了为什么观测的大气气态汞的 δ^{202}Hg 值要远高于人为源排放气态汞的 δ^{202}Hg 值。另外，由于大气 Hg(0) 相对于大气降水富集重的汞同位素且大气 Hg(Ⅱ) 的还原反应使得产物 Hg(0) 更富集轻的同位素，说明大气 Hg(0) 的氧化反应必须使得反应物 Hg(0) 富集重的同位素。然而，实验室控制实验却发现 Hg(0) 在被主要氧化物（Br 原子）氧化的过程中富集轻的同位素（Sun et al., 2016a），与观测到的大气形态汞的同位素组成特征不一致。因此，作者建议开展更多的大气 Hg(0) 氧化同位素分馏实验，以厘清大气汞的同位素分馏机理。

由表 4-3 可知，虽然 Hg(Ⅱ) 的还原和 Hg(0) 的氧化反应都使得反应物富集奇数汞同位素，但是前者所产生的奇数同位素非质量分馏的幅度（E^{199}Hg）要远大于后者。这与观测到的大气降水和大气 Hg(0) 的 Δ^{199}Hg 值的相对大小较为一致。除了奇数同位素的非质量分馏，Hg(0) 氧化实验还发现了较为明显的偶数同位素的非质量分馏现象 [Hg(0) 相对缺失偶数同位素 ^{200}Hg]（Sun et al., 2016a）。虽然实验观测的分馏幅度难以解释实地测试的大气降水中较大的 Δ^{200}Hg 值，但其正确预测了大气形态汞 Δ^{200}Hg 值的方向 [即大气 Hg(0) 具有负的 Δ^{200}Hg 值而大气 Hg(Ⅱ) 具有

正的 $\Delta^{200}Hg$ 值]。

4.3.5 大气形态汞同位素组成对汞来源的解析

大气不同形态汞之间的转化可能造成它们所携带的源同位素信号改变，这在一定程度上为解析大气汞跨区域传输带来了困难。但是，一些研究者还是成功地利用汞同位素对大气汞进行了区域来源的解析（Yuan et al., 2018; Xu et al., 2017; Das et al., 2016; Huang et al., 2016; Demers et al., 2015; Chen et al., 2012）。这主要是因为：①大气气态 Hg(0)的大气滞留时间比较长，化学性质不活泼，且其比例较大，短期内其源同位素信号不易被大气过程显著改变；②大气颗粒态 Hg(P)的源示踪研究主要集中在城市区域，采样点与排放源距离较近且样品的汞浓度较高，样品的同位素信号在短期内能够保持源同位素的特征；③一些经过远距离传输的大气氧化汞具有特征的同位素分馏信号，与近距离传输的大气氧化汞相比存在显著的区别。

1. 大气气态（零价）汞

一般而言，人为源直接影响的大气气态 Hg(0)具有较低的 $\delta^{202}Hg$ 值和不明显的非质量分馏值，而自然背景区的大气 Hg(0)由于与陆地植被和土壤进行了较为充分的汞交换而具有较高的 $\delta^{202}Hg$ 值和偏负的非质量分馏值。Demers（2015）通过测试美国佛罗里达州彭萨科拉（Pensacola）海岸带的一座燃煤电厂周围大气气态汞的同位素组成发现其具有中度偏正的 $\delta^{202}Hg$ 值（0.39‰～1.43‰）；结合采样期间的气象参数，作者成功识别出四个不同大气气态汞的来源及其特征 $\delta^{202}Hg$ 值：①具有最低值的城市和工业源大气汞排放；②具有中间值的受海洋影响的大气汞传输；③具有中间值的来自于北部-西北部大气混合层的陆源大气汞传输；④具有最高值的来自于北部-东北部高海拔背景区域（水平面上 2000 m）的陆源大气汞传输。Fu 等（2016）对法国比利牛斯山 Pic du Midi 区域（水平面上 2860 m，自由对流层）进行了连续一年的大气气态汞的采集与同位素组成测试，发现其 $\delta^{202}Hg$ 值主要受控于欧洲大陆输送的具有低 $\delta^{202}Hg$ 值的大气气态汞和大西洋方向输送的具有高 $\delta^{202}Hg$ 值的大气气态汞。Yu 等（2016）发现中国的大气气态汞的同位素组成也存在类似的规律：接近人为污染源大气气态汞具有偏负的 $\delta^{202}Hg$ 值且无 $\Delta^{199}Hg$ 的异常，而背景区域充分混合的大气气态汞具有偏正的 $\delta^{202}Hg$ 值和偏负的 $\Delta^{199}Hg$ 值。

2. 大气降水

溶解在云雾水滴中的二价汞在大气传输的过程中很容易发生光还原及其他物

理化学反应而改变其原始源区汞同位素信号。但是基于汞同位素分馏的规律性，即使源区的汞同位素组成发生了较大的分馏，利用大气降水的汞同位素组成依然能够解析出分馏后的源区汞同位素特征。例如，Sherman 等（2012）发现佛罗里达州 Crystal River 地区一燃煤电厂周围大气降水的 δ^{202}Hg 值极度偏负（均值为 –2.56‰），而远离电厂的大气降水的 δ^{202}Hg 值接近于 0‰。通过对比电厂周围大气降水与原煤的同位素组成发现，前者与后者的 δ^{202}Hg 和 Δ^{199}Hg 存在显著差别，但这并不妨碍利用汞同位素来区分燃煤排放的汞和其他来源的汞。另外，Chen 等（2012）利用大气降水的 δ^{202}Hg vs. 1/[Hg] 图解，结合气团后轨迹模型解析出加拿大多伦多东北部彼得伯勒（Peterborough）地区大气降水中汞的可能来源：南部经过中等距离传输的美国北部及五大湖区域的排放汞（约–0.3‰）、西部经过长距离传输的混合来源（北美大陆中西部和亚洲）的大气汞（接近–2‰）以及北部输入的极地区域的大气汞（约–1.35‰）。值得提出的是，长距离输送的溶解态氧化汞由于经历了较长时间的大气分馏过程，一般具有明显偏正的 Δ^{199}Hg 和 Δ^{200}Hg。例如，Wang 等（2015）和 Yuan 等（2015）认为我国贵阳和拉萨地区部分具有较高 Δ^{199}Hg 和 Δ^{200}Hg 值的雨水样品中的汞可能来自于非本地排放的长距离输送的汞。

3. 大气颗粒态汞

目前报道的大气颗粒态 Hg(P) 同位素组成主要来自于汞含量较高的城市大气，其汞含量主要受本地人为排放源的影响。大气颗粒物中的汞可能来自于源直接排放一次颗粒物和大气 Hg(0)/Hg(Ⅱ) 的转化。如果大气颗粒物中的汞主要来自于前者，同时排放到大气后同位素分馏较小，同位素手段能够有效地解析出具有不同同位素组成的颗粒物来源。如果大气颗粒物中的汞主要来自于后者，源-汇关系的解析就会变得比较复杂。Xu 等（2017）发现西安市大气颗粒物的同位素组成与估算的当地人为源（电厂燃煤、水泥生产）汞排放的同位素组成非常类似，表明大气颗粒物在大气传输的过程中并未发生显著的分馏。但是，Das 等（2015）通过分析不同污染模式下大气颗粒物中的汞同位素组成发现：①垃圾焚烧点附近采集的大气颗粒态汞同位素组成基本上保持了源同位素信号；②工业污染源附近采集的大气颗粒物的 δ^{202}Hg 与燃煤的 δ^{202}Hg 类似，但前者的 Δ^{199}Hg 相对于后者偏移了约 –0.15‰；③交通繁忙路段采集大气颗粒物的 Δ^{199}Hg 相对于源区可能发生了 –0.3‰~0.3‰ 的偏移。这表明源排放的颗粒态汞在大气传输过程中发生了一定程度的同位素分馏。该分馏可能是由于大气颗粒物上的溶解态的 Hg(Ⅱ) 发生了光致还原，使得反应后的颗粒物上的汞富集［(+)MIE］或者缺失［(–)MIE］奇数汞同位素。Huang 等（2016）对北京城区四个季度颗粒态汞同位素的研究发现：生物质燃烧可能是造成秋季颗粒物 Δ^{199}Hg 偏负的主要原因，而春夏季颗粒物上偏正的

Δ^{199}Hg 主要是由于颗粒物长距离传输过程中发生了汞光致还原反应。

4.4 基于汞同位素的全球汞循环传输模型

目前汞同位素主要应用于局部和区域的汞循环示踪的研究，如何把汞同位素扩展到全球尺度来研究全球的汞循环问题是汞同位素发展的重要方向。当前汞循环模型的构建主要基于汞含量和汞形态的实地测试和经验估算，大气-陆地-海洋各储库之间和内部关键过程的汞通量（如大气汞的干/湿沉降、大气汞的氧化和还原、海洋-大气汞交换、陆地-大气汞交换）幅度存在极大的不确定性（Agnan et al.，2016；Obrist et al.，2014；Streets et al.，2011；Holmes et al.，2010；Lindberg et al.，2007）。物理、化学和生物过程驱动着全球的汞循环，同时也使得汞同位素沿着特定的轨迹进行着质量和非质量分馏（Blum et al.，2014）。因此，汞的循环与循环过程中的汞同位素分馏密切相关，两者共同决定了地球不同储库的汞同位素组成。

Sonke（2011）最早在汞同位素模型建立方面进行了初步尝试；他通过在已有全球汞循环模型的基础上加入过程汞同位素分馏系数去解释自然界所观测到的奇数同位素非质量分馏数据。在假设所采用的汞循环模型正确基础上，Sonke（2011）发现奇数同位素的非质量分馏可能发生在表层海洋、陆地和大气环境中且分馏机理不尽相同。由于当时汞同位素研究数据较少，且过程同位素分馏机理不太明确，该研究只是利用全球汞循环模型去理解自然界的汞同位素数据和同位素分馏过程。近年来，随着汞同位素源排放数据、地球各储库数据和过程汞同位素分馏系数的逐渐增加，汞同位素有望用来定量研究汞的全球收支与平衡。

要构建基于汞同位素的全球汞循环模型，必须依赖三个必要条件：①全球汞循环模型的基本框架；②人为/自然源汞排放的同位素数据库；③汞在循环（迁移和转化）过程中的同位素分馏系数。对于条件①，国际上的研究已较为完善，理论比较成熟，广泛认可的全球汞循环模型有大气三维汞化学传递模型 GEOS-Chem（Selin et al.，2008）、海洋汞循环模型 MITgcm 和 OFFTRA（Zhang et al.，2015；2014）、陆地汞循环模型 GTMM（Smith-Downey et al.，2010）以及多箱体全耦合汞循环模型（Amos et al.，2014；2013）。对于条件②，4.3.3 节已经对自然源和人为源的汞排放特征做了详细的描述。对于条件③，国际上已报道了诸多物理、化学和生物过程（如氧化、还原、挥发、扩散、吸附）的汞同位素分馏系数（表 4-3）。现在的关键问题是如何增加条件①的代表性以及减少条件②和条件③的不确定度，使得汞同位素能够准确地定量全球的汞循环过程和收支通量。

目前来看，比较简单的箱体模型最适合构建汞同位素的循环模型。箱体模型的基本理念是把大气、陆地以及海洋中各种形态的汞储库看作是相互联系的箱体；

箱体之间进行着汞的迁移与转化以及汞同位素的交换。汞的迁移与转化受到箱体间元素迁移速率的控制，而同位素交换受到箱体间同位素分馏系数的控制。箱体模型的基本方程可表示为如下质量平衡方程和同位素平衡方程：

$$\frac{dM_i}{dt}=\sum_{j\neq i}\left(k_{j\rightarrow i}\times M_j\right)-\sum_{i\neq j}\left(k_{i\rightarrow j}\times M_i\right)+S \quad (4-19)$$

$$\begin{aligned}\frac{d\left(M_i\times\delta^{202}Hg_i\right)}{dt}=&\sum_{j\neq i}\left[k_{j\rightarrow i}\times M_j\times\left(\delta^{202}Hg_j+\varepsilon^{202}Hg_{(j\rightarrow i)}\right)\right]\\&-\sum_{i\neq j}\left[k_{i\rightarrow j}\times M_i\times\left(\delta^{202}Hg_i+\varepsilon^{202}Hg_{(i\rightarrow j)}\right)\right]+S\times\delta^{202}Hg_{(S)}\end{aligned} \quad (4-20)$$

式中，M(Mg)是各箱体（储库）的质量；$k_{j\rightarrow i}=F_{j\rightarrow i}/M_j$ 和 $k_{i\rightarrow j}=F_{i\rightarrow j}/M_i$ 是一级速率常数，分别由箱体 j 到箱体 i 的汞迁移/转化通量 $F_{j\rightarrow i}$(Mg/a)和箱体 i 到箱体 j 的汞迁移/转化通量 $F_{i\rightarrow j}$(Mg/a)除以输出箱体的质量计算所得；S(Mg/a)是源排放输入的汞通量；$\delta^{202}Hg$ 是拟模拟箱体的质量分馏值，可以替代为非质量分馏值 $\Delta^{xxx}Hg$；$\varepsilon^{202}Hg=1000\times\ln(\alpha^{202}Hg)$ 是质量分馏富集系数，可以替换为非质量分馏富集系数 $E^{xxx}Hg=1000\times\ln(\alpha^{xxx}Hg_{MIF})$，两者分别由同位素质量分馏系数 $\alpha^{202}Hg$ 和非质量分馏系数 $\alpha^{xxx}Hg_{MIF}$ 计算所得；$\delta^{202}Hg_S$ 为源排放的同位素质量分馏值，可替换为源排放的非质量分馏值 $\Delta^{xxx}Hg_S$。

方程（4-19）和方程（4-20）包含有上述提到的条件①～③，其中方程（4-19）即是条件①所提到的汞循环模型的数学表达式。因此，如果对汞同位素的认知（即条件②和条件③）比较深入，上述方程就可以用来约束条件①中所包含的汞迁移/转化速率，从而定量认清汞的生物地球化学循环。

参 考 文 献

Agnan Y, Le Dantec T, Moore C W, Edwards G C, Obrist D, 2016. New constraints on terrestrial surface-atmosphere fluxes of gaseous elemental mercury using a global database. Environmental Science & Technology, 50: 507-524.

Amos H M, Jacob D J, Kocman D, Horowitz H M, Zhang Y, Dutkiewicz S, Horvat M, Corbitt E S, Krabbenhoft D P, Sunderland E M, 2014. Global biogeochemical implications of mercury discharges from rivers and sediment burial. Environmental Science & Technology, 48: 9514-9522.

Amos H M, Jacob D J, Streets D G, Sunderland E M, 2013. Legacy impacts of all-time anthropogenic emissions on the global mercury cycle. Global Biogeochemical Cycles, 27: 410-421.

Angeli I, 2004. A consistent set of nuclear rms charge radii: Properties of the radius surface R (N,Z). At. Data and Nuclear Data Tables, 87: 185-206.

Bergquist B A, Blum J D, 2007. Mass-dependent and -independent fractionation of Hg isotopes by photoreduction in aquatic systems. Science, 318: 417-420.

Bigeleisen J, 1949. The relative reaction velocities of isotopic molecules. The Journal of Chemical Physics, 17: 675-678.

Bigeleisen J, 1996a. Nuclear size and shape effects in chemical reactions. Isotope chemistry of the heavy elements. Journal of the American Chemical Society, 118: 3676-3680.

Bigeleisen J, 1996b. Temperature dependence of the isotope chemistry of the heavy elements. Proceedings of the National Academy of Sciences of the United States of America, 93: 9393-9396.

Blum J D, Bergquist B A, 2007. Reporting of variations in the natural isotopic composition of mercury. Analytical and Bioanalytical Chemistry, 388: 353-359.

Blum J D, Popp B N, Drazen J C, Anela Choy C, Johnson M W, 2013. Methylmercury production below the mixed layer in the North Pacific Ocean. Nature Geosci, 6: 879-884.

Blum J D, Sherman L S, Johnson M W, 2014. Mercury isotopes in Earth and environmental sciences. Annual Review of Earth and Planetary Sciences, 42: 249-269.

Bonsignore M, Tamburrino S, Oliveri E, Marchetti A, Durante C, Berni A, Quinci E, Sprovieri M, 2015. Tracing mercury pathways in Augusta Bay (southern Italy) by total concentration and isotope determination. Environmental Pollution, 205: 178-185.

Buchachenko A L, 2001. Magnetic isotope effect: Nuclear spin control of chemical reactions. Journal of Physical Chemistry A, 105: 9995-10011.

Buchachenko A L, Kouznetsov D A, Shishkov A V, 2004. Spin biochemistry: Magnetic isotope effect in the reaction of creatine kinase with CH_3HgCl. Journal of Physical Chemistry A, 108: 707-710.

Cai H, Chen J, 2015. Mass-independent fractionation of even mercury isotopes. Science Bulletin, 61: 116-124.

Chandan P, Ghosh S, Bergquist B A, 2015. Mercury isotope fractionation during aqueous photoreduction of monomethylmercury in the presence of dissolved organic matter. Environmental Science & Technology, 49: 259-267.

Chen J, Hintelmann H, Feng X, Dimock B, 2012. Unusual fractionation of both odd and even mercury isotopes in precipitation from Peterborough, ON, Canada. Geochimica et Cosmochimica Acta, 90: 33-46.

Cohen E R, Cvitas T, Frey J G, Holmström B, Kuchitsu K, Marquardt R, Mills I, Pavese F, Quack M, Stohner J, Strauss H L, Takami M, Thor A J, 2008. Quantities, units and symbols in physical chemistry, IUPAC Green Book, 3rd Edition, 2nd Printing. Cambridge: IUPAC & RSC Publishing.

Das R, Salters V J M, Odom A L, 2009. A case for *in vivo* mass-independent fractionation of mercury isotopes in fish. Geochemistry, Geophysics, Geosystems, 10: Q11012.

Das R, Wang X, Khezri B, Webster R D, Sikdar P K, Datta S, 2016. Mercury isotopes of atmospheric particle bound mercury for source apportionment study in urban Kolkata, India. Elementa: Science of the Anthropocene, 4: 000098.

Demers J D, Blum J D, Zak D R, 2013. Mercury isotopes in a forested ecosystem: Implications for air-surface exchange dynamics and the global mercury cycle. Global Biogeochemical Cycles, 27: 222-238.

Demers J D, Sherman L S, Blum J D, Marsik F J, Dvonch J T, 2015. Coupling atmospheric mercury isotope ratios and meteorology to identify sources of mercury impacting a coastal urban-industrial region near Pensacola, Florida, USA. Global Biogeochemical Cycles, 29: 1689-1705.

Donovan P M, Blum J D, Yee D, Gehrke G E, Singer M B, 2013. An isotopic record of mercury in San Francisco Bay sediment. Chemical Geology, 349-350: 87-98.

Driscoll C T, Mason R P, Chan H M, Jacob D J, Pirrone N, 2013. Mercury as a global pollutant: Dources, pathways, and effects. Environmental Science & Technology, 47: 4967-4983.

Du B, Feng X, Li P, Yin R, Yu B, Sonke J E, Guinot B, Anderson C W N, Maurice L, 2018. Use of mercury isotopes to quantify mercury exposure sources in inland populations, China. Environmental Science & Technology, 52: 5407-5416.

Enrico M, Le Roux G, Heimbürger L-E, Van Beek P, Souhaut M, Chmeleff J, Sonke J E, 2017. Holocene atmospheric mercury levels reconstructed from peat bog mercury stable isotopes. Environmental Science & Technology, 51: 5899-5906.

Enrico M, Roux G L, Marusczak N, Heimbürger L-E, Claustres A, Fu X, Sun R, Sonke J E, 2016. Atmospheric mercury transfer to peat bogs dominated by gaseous elemental mercury dry deposition. Environmental Science & Technology, 50: 2405-2412.

Estrade N, Carignan J, Sonke J E, Donard O F X, 2009. Mercury isotope fractionation during liquid-vapor evaporation experiments. Geochimica et Cosmochimica Acta, 73: 2693-2711.

Feng C, Pedrero Z, Li P, Du B, Feng X, Monperrus M, Tessier E, Berail S, Amouroux D, 2016. Investigation of Hg uptake and transport between paddy soil and rice seeds combining Hg isotopic composition and speciation. Elementa: Science of the Anthropocene, 4: 000087.

Fu X, Heimburger L-E, Sonke J E, 2014. Collection of atmospheric gaseous mercury for stable isotope analysis using iodine- and chlorine-impregnated activated carbon traps. Journal of Analytical Atomic Spectrometry, 29: 841-852.

Fu X, Marusczak N, Wang X, Gheusi F, Sonke J E, 2016. Isotopic composition of gaseous elemental mercury in the free troposphere of the Pic du Midi Observatory, France. Environmental Science & Technology, 50: 5641-5650.

Fu X W, Zhang H, Yu B, Wang X, Lin C J, Feng X B, 2015. Observations of atmospheric mercury in China: A critical review. Atmospheric Chemistry and Physics, 15: 9455-9476.

Ghosh S, Schauble E A, Lacrampe Couloume G, Blum J D, Bergquist B A, 2013. Estimation of nuclear volume dependent fractionation of mercury isotopes in equilibrium liquid-vapor evaporation experiments. Chemical Geology, 336: 5-12.

Ghosh S, Xu Y, Humayun M, Odom L, 2008. Mass-independent fractionation of mercury isotopes in the environment. Geochemistry, Geophysics, Geosystems, 9: Q03004.

Gratz L E, Keeler G J, Blum J D, Sherman L S, 2010. Isotopic composition and fractionation of mercury in Great Lakes precipitation and ambient air. Environmental Science & Technology, 44: 7764-7770.

Hahn A A, Miller J P, Powers R J, Zehnder A, Rushton A M, Welsh R E, Kunselman A R, Roberson P, Walter H K, 1979. An experimental study of muonic X-ray transitions in mercury isotopes. Nuclear Physics A, 314: 361-386.

Holmes C D, Jacob D J, Corbitt E S, Mao J, Yang X, Talbot R, Slemr F, 2010. Global atmospheric model for mercury including oxidation by bromine atoms. Atmospheric Chemistry and Physics, 10: 12037-12057.

Horowitz H M, Jacob D J, Zhang Y, Dibble T S, Slemr F, Amos H M, Schmidt J A, Corbitt E S, Marais E A, Sunderland E M, 2017. A new mechanism for atmospheric mercury redox chemistry: implications for the global mercury budget. Atmospheric Chemistry and Physics, 17: 6353-6371.

Huang Q, Chen J, Huang W, Fu P, Guinot B, Feng X, Shang L, Wang Z, Wang Z, Yuan S, Cai H, Wei L, Yu B, 2016. Isotopic composition for source identification of mercury in atmospheric fine particles. Atmospheric Chemistry and Physics, 16: 11773-11786.

Huang Q, Liu Y L, Chen J B, Feng X B, Huang W L, Yuan S L, Cai H M, Fu X W, 2015. An improved dual-stage protocol to pre-concentrate mercury from airborne particles for precise isotopic measurement. Journal of Analytical Atomic Spectrometry, 30: 957-966.

Jackson T A, 2015. Evidence for mass-independent fractionation of mercury isotopes by microbial activities linked to geographically and temporally varying climatic conditions in Arctic and Subarctic Lakes. Geomicrobiology Journal, 32: 799-826.

Jackson T A, Whittle D M, Evans M S, Muir D C G, 2008. Evidence for mass-independent and mass-dependent fractionation of the stable isotopes of mercury by natural processes in aquatic ecosystems. Applied Geochemistry, 23: 547-571.

Janssen S E, Schaefer J K, Barkay T, Reinfelder J R, 2016. Fractionation of mercury stable isotopes during microbial methylmercury production by iron- and sulfate-reducing bacteria. Environmental Science & Technology, 50: 8077-8083.

Jiménez-Moreno M, Perrot V, Epov V N, Monperrus M, Amouroux D, 2013. Chemical kinetic isotope fractionation of mercury during abiotic methylation of Hg(II) by methylcobalamin in aqueous chloride media. Chemical Geology, 336: 26-36.

Jiskra M, Wiederhold J G, Bourdon B, Kretzschmar R, 2012. Solution speciation controls mercury isotope fractionation of Hg(II) sorption to goethite. Environmental Science & Technology, 46: 6654-6662.

Jiskra M, Wiederhold J G, Skyllberg U, Kronberg R-M, Hajdas I, Kretzschmar R, 2015. Mercury deposition and re-emission pathways in Boreal Forest soils investigated with Hg isotope signatures. Environmental Science & Technology, 49: 7188-7196.

Koster van Groos P G, Esser B K, Williams R W, Hunt J R, 2013. Isotope effect of mercury diffusion in air. Environmental Science & Technology, 48: 227-233.

Kritee K, Barkay T, Blum J D, 2009. Mass dependent stable isotope fractionation of mercury during mer mediated microbial degradation of monomethylmercury. Geochimica et Cosmochimica Acta, 73: 1285-1296.

Kritee K, Blum J D, Barkay T, 2008. Mercury stable isotope fractionation during reduction of Hg(II) by different microbial pathways. Environmental Science & Technology, 42: 9171-9177.

Kritee K, Blum J D, Johnson M W, Bergquist B A, Barkay T, 2007. Mercury stable isotope fractionation during reduction of Hg(II) to Hg(0) by mercury resistant microorganisms. Environmental Science & Technology, 41: 1889-1895.

Kritee K, Blum J D, Reinfelder J R, Barkay T, 2013. Microbial stable isotope fractionation of mercury: A synthesis of present understanding and future directions. Chemical Geology, 336: 13-25.

Kritee K, Motta L C, Blum J D, Tsui M T-K, Reinfelder J R, 2018. Photomicrobial visible light-induced magnetic mass independent fractionation of mercury in a *Marine microalga*. ACS Earth and Space Chemistry, 2: 432-440.

Kwon S Y, Blum J D, Carvan M J, Basu N, Head J A, Madenjian C P, David S R, 2012. Absence of fractionation of mercury isotopes during trophic transfer of methylmercury to freshwater fish in captivity. Environmental Science & Technology, 46: 7527-7534.

Laffont L, Sonke J E, Maurice L, Hintelmann H, Pouilly M, Sánchez Bacarreza Y, Perez T, Behra P, 2009. Anomalous mercury isotopic compositions of fish and human hair in the Bolivian Amazon. Environmental Science & Technology, 43: 8985-8990.

Laffont L, Sonke J E, Maurice L, Monrroy S L, Chincheros J, Amouroux D, Behra P, 2011. Hg speciation and stable isotope signatures in human hair as a tracer for dietary and occupational exposure to mercury. Environmental Science & Technology, 45: 9910-9916.

Li M, Sherman L S, Blum J D, Grandjean P, Mikkelsen B, Weihe P, Sunderland E M, Shine J P, 2014. Assessing sources of human methylmercury exposure using stable mercury isotopes. Environmental Science & Technology, 48: 8800-8806.

Li P, Du B, Maurice L, Laffont L, Lagane C, Point D, Sonke J E, Yin R, Lin C-J, Feng X, 2017. Mercury isotope signatures of methylmercury in rice samples from the wanshan mercury mining srea, China: Environmental implications. Environmental Science & Technology, 51: 12321-12328.

Lindberg S, Bullock R, Ebinghaus R, Engstrom D, Feng X, Fitzgerald W, Pirrone N, Prestbo E, Seigneur C, 2007. A synthesis of progress and uncertainties in attributing the sources of mercury in deposition. AMBIO: A Journal of the Human Environment, 36: 19-33.

Malinovsky D, Latruwe K, Moens L, Vanhaecke F, 2010. Experimental study of mass-independence of Hg isotope fractionation during photodecomposition of dissolved methylmercury. Journal of Analytical Atomic Spectrometry, 25: 950-956.

Malinovsky D, Vanhaecke F, 2011. Mercury isotope fractionation during abiotic transmethylation reactions. International Journal of Mass Spectrometry, 307: 214-224.

Mead C, Lyons J R, Johnson T M, Anbar A D, 2013. Unique Hg stable isotope signatures of compact fluorescent lamp-sourced Hg. Environmental Science & Technology, 47: 2542-2547.

Meija J, Coplen Tyler B, Berglund M, Brand Willi A, De Bièvre P, Gröning M, Holden Norman E, Irrgeher J, Loss Robert D, Walczyk T, Prohaska T, 2016. Isotopic compositions of the elements 2013 (IUPAC Technical Report). Pure and Applied Chemistry, 293-306.

Nadjakov E G, Marinova K P, Gangrsky Y P, 1994. Systematics of nuclear charge radii. Atomic Data and Nuclear Data Tables, 56: 133-157.

Obrist D, Agnan Y, Jiskra M, Olson C L, Colegrove D P, Hueber J, Moore C W, Sonke J E, Helmig D, 2017. Tundra uptake of atmospheric elemental mercury drives Arctic mercury pollution. Nature, 547: 201-204.

Obrist D, Pokharel A K, Moore C, 2014. Vertical Profile Measurements of soil air suggest immobilization of gaseous elemental mercury in mineral soil. Environmental Science & Technology, 48: 2242-2252.

Perrot V, Bridou R, Pedrero Z, Guyoneaud R, Monperrus M, Amouroux D, 2015. Identical Hg isotope mass dependent fractionation signature during methylation by sulfate-reducing bacteria in sulfate and sulfate-free environment. Environmental Science & Technology, 49: 1365-1373.

Perrot V, Jimenez-Moreno M, Berail S, Epov V N, Monperrus M, Amouroux D, 2013. Successive methylation and demethylation of methylated mercury species (MeHg and DMeHg) induce mass dependent fractionation of mercury isotopes. Chemical Geology, 355: 153-162.

Rodríguez-González P, Epov V N, Bridou R, Tessier E, Guyoneaud R, Monperrus M, Amouroux D, 2009. Species-specific stable isotope fractionation of mercury during Hg(II) methylation by an anaerobic bacteria (*Desulfobulbus propionicus*) under Dark Conditions. Environmental Science & Technology, 43: 9183-9188.

Rolison J M, Landing W M, Luke W, Cohen M, Salters V J M, 2013. Isotopic composition of species-specific atmospheric Hg in a coastal environment. Chemical Geology, 336: 37-49.

Rose C H, Ghosh S, Blum J D, Bergquist B A, 2015. Effects of ultraviolet radiation on mercury isotope fractionation during photo-reduction for inorganic and organic mercury species. Chemical Geology, 405: 102-111.

Rothenberg S E, Yin R, Hurley J P, Krabbenhoft D P, Ismawati Y, Hong C, Donohue A, 2017. Stable mercury isotopes in polished rice (*Oryza sativa* L.) and hair from rice consumers. Environmental Science & Technology, 51: 6480-6488.

Schauble E A, 2007. Role of nuclear volume in driving equilibrium stable isotope fractionation of mercury, thallium, and other very heavy elements. Geochimica et Cosmochimica Acta, 71: 2170-2189.

Schauble E A, 2013. Modeling nuclear volume isotope effects in crystals. Proceedings of the National Academy of Sciences, 110: 17714-17719.

Selin N E, Jacob D J, Yantosca R M, Strode S, Jaeglé L, Sunderland E M, 2008. Global 3-D land-ocean-atmosphere model for mercury: Present-day versus preindustrial cycles and anthropogenic enrichment factors for deposition. Global Biogeochemical Cycles, 22: GB2011.

Sherman L S, Blum J D, Douglas T A, Steffen A, 2012a. Frost flowers growing in the Arctic ocean-atmosphere-sea ice-snow interface: 2. Mercury exchange between the atmosphere, snow, and frost flowers. Journal of Geophysical Research: Atmospheres, 117(D14).

Sherman L S, Blum J D, Dvonch J T, Gratz L E, Landis M S, 2015. The use of Pb, Sr, and Hg isotopes in Great Lakes precipitation as a tool for pollution source attribution. Science of the Total Environment, 502: 362-374.

Sherman L S, Blum J D, Franzblau A, Basu N, 2013. New insight into biomarkers of human mercury exposure using naturally occurring mercury stable isotopes. Environmental Science & Technology, 47: 3403-3409.

Sherman L S, Blum J D, Johnson K P, Keeler G J, Barres J A, Douglas T A, 2010. Mass-independent fractionation of mercury isotopes in Arctic snow driven by sunlight. Nature Geosci, 3: 173-177.

Sherman L S, Blum J D, Keeler G J, Demers J D, Dvonch J T, 2012b. Investigation of local mercury deposition from a coal-fired power plant using mercury isotopes. Environmental Science & Technology, 46: 382-390.

Smith-Downey N V, Sunderland E M, Jacob D J, 2010. Anthropogenic impacts on global storage and emissions of mercury from terrestrial soils: Insights from a new global model. Journal of Geophysical Research-Atmospheres, 115(G3): G03008.

Sonke J E, 2011. A global model of mass independent mercury stable isotope fractionation. Geochimica et Cosmochimica Acta, 75: 4577-4590.

Streets D G, Devane M K, Lu Z, Bond T C, Sunderland E M, Jacob D J, 2011. All-time releases of mercury to the atmosphere from human activities. Environmental Science & Technology, 45: 10485-10491.

Štrok M, Baya P A, Hintelmann H, 2015. The mercury isotope composition of Arctic coastal seawater. Comptes Rendus Geoscience, 347: 368-376.

Štrok M, Hintelmann H, Dimock B, 2014. Development of pre-concentration procedure for the determination of Hg isotope ratios in seawater samples. Analytica Chimica Acta, 851: 57-63.

Sun G, Sommar J, Feng X, Lin C-J, Ge M, Wang W, Yin R, Fu X, Shang L, 2016a. Mass-dependent and -independent fractionation of mercury isotope during gas-phase oxidation of elemental mercury vapor by atomic Cl and Br. Environmental Science & Technology, 50: 9232-9241.

Sun R, Enrico M, Heimbürger L-E, Scott C, Sonke J E, 2013. A double-stage tube furnace—Acid-trapping protocol for the pre-concentration of mercury from solid samples for isotopic analysis. Analytical and Bioanalytical Chemistry, 405: 6771-6781.

Sun R, Streets D G, Horowitz H M, Amos H M, Liu G, Perrot V, Toutain J-P, Hintelmann H,

Sunderland E M, Sonke J E, 2016b. Historical (1850–2010) mercury stable isotope inventory from anthropogenic sources to the atmosphere. Elementa: Science of the Anthropocene, 4(1): 000091.

Ulm G, Bhattacherjee S K, Dabkiewicz P, Huber G, Kluge H J, Kühl T, Lochmann H, Otten E W, Wendt K, Ahmad S A, Klempt W, Neugart R, 1987. Isotope shift of ^{182}Hg and an update of nuclear moments and charge radii in the isotope range ^{181}Hg-^{206}Hg Zeitschrift für Physik A: Atomic Nuclei, 325: 247-259.

Urey H C, 1947. The thermodynamic properties of isotopic substances. Journal of the Chemical Society, 0: 562-581.

Wang Z, Chen J, Feng X, Hintelmann H, Yuan S, Cai H, Huang Q, Wang S, Wang F, 2015. Mass-dependent and mass-independent fractionation of mercury isotopes in precipitation from Guiyang, SW China. Comptes Rendus Geoscience, 347: 358-367.

Wiederhold J G, Cramer C J, Daniel K, Infante I, Bourdon B, Kretzschmar R, 2010. Equilibrium mercury isotope fractionation between dissolved Hg(Ⅱ) species and thiol-bound Hg. Environmental Science & Technology, 44: 4191-4197.

Xu H, Sonke J E, Guinot B, Fu X, Sun R, Lanzanova A, Candaudap F, Shen Z, Cao J, 2017. Seasonal and annual variations in atmospheric Hg and Pb isotopes in Xi'an, China. Environmental Science & Technology 51: 3759-3766.

Yamakawa A, Moriya K, Yoshinaga J, 2017. Determination of isotopic composition of atmospheric mercury in urban-industrial and coastal regions of Chiba, Japan, using cold vapor multicollector inductively coupled plasma mass spectrometry. Chemical Geology, 448: 84-92.

Yang L, Sturgeon R, 2009. Isotopic fractionation of mercury induced by reduction and ethylation. Analytical and Bioanalytical Chemistry, 393: 377-385.

Young E D, Galy A, Nagahara H, 2002. Kinetic and equilibrium mass-dependent isotope fractionation laws in nature and their geochemical and cosmochemical significance. Geochimica et Cosmochimica Acta, 66: 1095-1104.

Yu B, Fu X, Yin R, Zhang H, Wang X, Lin C-J, Wu C, Zhang Y, He N, Fu P, Wang Z, Shang L, Sommar J, Sonke J E, Maurice L, Guinot B, Feng X, 2016. Isotopic composition of atmospheric mercury in China: New evidence for sources and transformation processes in air and in vegetation. Environmental Science & Technology, 50: 9262-9269.

Yuan S, Chen J, Cai H, Yuan W, Wang Z, Huang Q, Liu Y, Wu X, 2018. Sequential samples reveal significant variation of mercury isotope ratios during single rainfall events. Science of the Total Environment, 624: 133-144.

Yuan S, Zhang Y, Chen J, Kang S, Zhang J, Feng X, Cai H, Wang Z, Wang Z, Huang Q, 2015. Large variation of mercury isotope composition during a single precipitation event at Lhasa City, Tibetan Plateau, China. Procedia Earth and Planetary Science, 13: 282-286.

Zambardi T, Sonke J E, Toutain J P, Sortino F, Shinohara H, 2009. Mercury emissions and stable isotopic compositions at Vulcano Island (Italy). Earth and Planetary Science Letters, 277: 236-243.

Zhang Y, Jacob D J, Dutkiewicz S, Amos H M, Long M S, Sunderland E M, 2015. Biogeochemical drivers of the fate of riverine mercury discharged to the global and Arctic oceans. Global Biogeochemical Cycles, 29: 854-864.

Zhang Y, Jaeglé L, Thompson L, Streets D G, 2014. Six centuries of changing oceanic mercury. Global Biogeochemical Cycles, 28(11): 2014GB004939.

Zheng W, Foucher D, Hintelmann H, 2007. Mercury isotope fractionation during volatilization of Hg(0) from solution into the gas phase. Journal of Analytical Atomic Spectrometry, 22: 1097-1104.

Zheng W, Hintelmann H, 2009. Mercury isotope fractionation during photoreduction in natural water is controlled by its Hg/DOC ratio. Geochimica et Cosmochimica Acta, 73: 6704-6715.

Zheng W, Hintelmann H, 2010a. Isotope fractionation of mercury during its photochemical reduction by low-molecular-weight organic compounds. Journal of Physical Chemistry A, 114: 4246-4253.

Zheng W, Hintelmann H, 2010b. Nuclear Field Shift Effect in Isotope Fractionation of Mercury during Abiotic Reduction in the Absence of Light. Journal of Physical Chemistry A, 114: 4238-4245.

Zheng W, Obrist D, Weis D, Bergquist B A, 2016. Mercury isotope compositions across North American forests. Global Biogeochemical Cycles, 30: 1475-1492.

第 5 章 大气汞的长距离传输

> **本章导读**
> - 简要介绍大气汞的主要形态分类，重点介绍不同形态大气汞的物理化学性质及不同形态大气汞之间的转化过程和控制因素，介绍不同形态大气汞的分析测试方法及其分析误差。
> - 简要介绍全球大气汞的主要污染来源，详细介绍全球和我国不同形态大气汞的分布特征和影响因素。
> - 详细介绍大气汞长距离传输的研究方法，包括源汇关系模型、全球和区域大气汞循环模型，重点介绍国内外大气汞长距离传输的研究进展。

5.1 大气汞形态和理化性质

5.1.1 大气汞形态分类

大气汞的常规分类主要包括气态单质汞（gaseous elemental mercury, Hg(0), GEM）、活性气态汞（reactive gaseous mercury, RGM）和颗粒态汞（particulate bound mercury, PBM）（Schroeder and Munthe, 1998; Gustin and Jaffe, 2010）。此外，大气汞还包括一些痕量的单甲基汞（monomethyl mercury, MMHg）和二甲基汞（dimethyl mercury, DMHg）（Bloom and Fitzgerald, 1988; Brosset and Lord, 1995）。气态单质汞是指存在于大气气相的汞蒸气。活性气态汞主要指存在于大气气相的一价和二价汞化合物，包括 HgO、$HgCl_2$、$HgBr_2$、$Hg(OH)_2$、$HgSO_4$、$Hg(NO_2)_2$ 等（Lindberg and Stratton, 1998; Feng et al., 2000; Lindberg et al., 2000; Gustin et al., 2013; Huang et al., 2013）。由于目前还没有测定不同类型大气活性汞的分析仪器，文献中有关大气活性汞的分类主要是一种操作性定义。大气颗粒态汞指的是赋存于大气颗粒物上汞的总称，包括单质汞、活性汞和一些痕量的有机形态汞。气态单质汞和活性气态汞统称为大气气态总汞（total gaseous mercury, TGM）。

5.1.2 大气汞物理化学性质和形态转化

大气汞的物理化学性质对汞在大气中的传输具有重要影响作用。通常来讲，化学性质较为稳定和沉降速率较低的大气汞形态具有较长的大气居留时间，进而能随大气环流进行长距离传输；而化学性质较为活跃、沉降速率较高的大气汞形态在大气中容易转化为其他形态汞或者很快沉降到地表和水生生态系统，通常不能随大气环流进行长距离迁移，被认为是一种区域性污染物。

1. 大气气态单质汞

气态单质汞是大气汞中性质最为稳定的形态。传统理论认为，大气气态单质汞具有极低的水溶性和干沉降速率，且具有很强的化学反应惰性，因此气态单质汞在大气中比较稳定，具有较长的大气居留时间，能够随大气环流进行全球性的迁移（Schroeder and Munthe，1998）。研究指出（Schroeder and Munthe，1998），常温下气态单质汞的水溶性约为 4.9×10^{-5} g/L，远低于其他大气活性汞和有机形态汞（例如 $HgCl$、HgO、$(CH_3)_2Hg$、CH_3HgCl 等：5.3×10^{-2}~66 g/L），也远低于其他常规大气污染物的水溶性，比如二氧化碳（CO_2: 1.45 g/L）、二氧化硫（SO_2: 110 g/L）、一氧化碳（CO: 2.6×10^{-4} g/L）、臭氧（O_3: 2.1 g/L）等。

大气气态单质汞的干沉降速率（V_d）主要受到地表基质和植被覆盖条件的影响，通常变化范围较大。研究结果表明，大气气态单质汞在水体和土壤界面的干沉降速率通常比较低，如水体界面大气气态单质汞的干沉降速率通常在 0.003~0.012 cm/s，而土壤界面大气气态单质汞的干沉降速率通常在 0.002~0.03 cm/s（Xiao et al.，1991；Kim et al.，1995；Lindberg and Stratton，1998；Gårdfeldt et al.，2001；Xin and Gustin，2007；Kuiken et al.，2008）。受植物对大气气态单质汞吸收作用的影响，一些有植被覆盖的区域大气气态单质汞的地区干沉降速率会明显升高，如森林地区大气气态单质汞干沉降速率为 0.05~1.9 cm/s（Lindberg and Stratton，1998；Bash et al.，2004；Fu et al.，2016b），农田生态系统大气气态单质汞干沉降速率为 0.05~0.26 cm/s（Kim et al.，2003；Cobbett et al.，2007）、草地和湿地生态系统大气气态单质汞的干沉降速率为 0.02~0.19 cm/s（Lindberg et al.，2002b；Poissant et al.，2004；Zhang et al.，2005；Fritsche et al.，2008）。

大气汞的物理化学转化是全球大气汞研究的一个重要研究领域。通过对北极地区加拿大阿勒特（Alert）站大气气态单质汞长期监测，研究人员发现北极地区大气气态单质汞在每年春季（3~6月）都会出现显著的"亏损"现象（Schroeder et al.，1998）：在春季，大气气态单质汞会从全年平均值的约 1.5 ng/m^3 突然降低到 0.5 ng/m^3 甚至是仪器检出限以下（0.1 ng/m^3）。这种现象在夜晚与白天均有发生，有时可持续一星期的时间。在南极的春季（8～12 月）也发现了类似的大气气态单质汞"亏损"现象（Ebinghaus et al.，2002b）。之后，研究人员陆续在其他南极和北极地区或者靠近极地的地区发现了类似的大气气态单质汞"亏损"现象（Lindberg et al.，2002a；Berg et al.，2003；Skov et al.，2004；Steffen et al.，2005），从而证明了极地区域大气气态单质汞"亏损"是一个普遍的现象。通过对不同形态大气汞（气态单质汞、颗粒态汞和活性气态汞）和大气氧化物的联合研究，Lindberg 等（2002a）揭示了极地地区大气气态单质汞通过气相和冰晶等颗粒界面氧化转化为活性气态汞和颗粒态汞是导致极地地区大气气态单质汞"亏损"的原因。经过系列的研究，研究人员基本确定了这一现象的内在机制，即极地地区春季的冰盖融化导致大量卤族元素被释放到大气中，这些卤族元素被紫外线激发形成卤族自由基和卤族化合物（如：$Br^{\cdot}$、$Cl^{\cdot}$、$BrCl$、$BrO^{\cdot}$、$ClO^{\cdot}$等），能够快速氧化大气气态单质汞形成活性气态汞和颗粒态汞，进而造成气态单质汞的"亏损"（Steffen et al.，2008）。这种大气气态单质汞的氧化现象可能主要集中在极地地区的行星边界层（指地面/海平面至其上部 1000 m），每年可以导致大量气态单质汞（约 300 t）沉降到极地地区（Banic et al.，2003；Skov et al.，2004），给偏远的极地生态环境造成了一定的汞污染风险。

除南北极地区外，大气气态单质汞的异常"亏损"也广泛分布在自由对流层、海洋边界层和森林边界层。Talbot 等（2007）通过飞机航测观测到北美和东太平洋自由对流层上部和平流层存在明显的大气气态单质汞"亏损"，通常会降低到 0.5 ng/m^3 以下，推测高空大气中较高的臭氧和卤族氧化物导致了大量的大气气态单质汞被氧化为活性气态汞和颗粒态汞。随后，研究人员通过多次的欧洲-亚洲、欧洲-南美和欧洲-北美等航线的航空观测发现全球自由对流层顶部附近的确存在显著的大气气态单质汞"亏损"现象，并证实氧化作用是导致自由对流层顶气态单质汞"亏损"的重要原因（Slemr et al.，2009；Lyman and Jaffe，2012；Slemr et al.，2016）。此外，研究人员也开展了自由对流层中部和下部大气气态单质汞"亏损"现象的研究，如在太平洋上自由对流层中部和下部的大气气态单质汞和活性气态汞研究中，尽管没有发现显著的大气气态单质汞"亏损"现象，但发现自由对流层中部和下部存在明显的大气气态单质汞浓度降低和活性气态汞浓度升高现象，其观测到的活性气态汞浓度约是大陆边界层平均浓度的 10～50 倍，证实了海洋上空自由对流层中部和下部均存在明显的大气气态单质汞氧化现象（Swartzendruber et al.，2009）。

在高海拔山顶地区同样发现自由对流层中部和下部存在显著的大气气态单质汞氧化现象。比如，美国学士山观测站（Mount Bachelor Observatory）（海拔 2700 m）发现最高约有 40%的大气气态单质汞被氧化，美国风暴峰天文台（Storm Peak Laboratory）（海拔 3220 m）最高约有 10%的大气气态单质汞被氧化，中国台湾鹿林山（Mt. Lulin）（海拔 2862 m）最高约有 15%的大气气态单质汞被氧化，法国比利牛斯山 Pic du Midi 天文台（海拔 2877 m）最高可有 20%的大气气态单质汞被氧化，进一步证明了这些氧化主要发生在海洋上空自由对流层的中部和下部，极有可能与海洋释放的卤族氧化物有关（Swartzendruber et al.，2006；Fain et al.，2009；Sheu et al.，2010；Timonen et al.，2013；Fu et al.，2016c）。

海洋边界层也发现了明显的大气气态单质汞"亏损"现象。早期 Laurier 等（2003）在北太平洋的研究发现太平洋海洋边界层存在微弱的大气气态单质汞"亏损"现象，同时监测到了大气活性气态汞浓度升高的现象，证实海洋边界层存在一定程度的大气气态单质汞氧化现象。在南非好望角（Cape Point）全球大气基准站也发现了显著的大气气态单质汞"亏损"现象，其大气气态单质汞能够在很短时间从背景值（1.0 ng/m³）下降到 0.2 ng/m³ 以下，气态单质汞的"亏损"能够达到 80%以上，而且这种"亏损"现象主要发生在午后和下午（12:00～18:00），持续时间 3～6 个小时（Brunke et al.，2010）。研究表明，海洋边界层大气气态单质汞的氧化"亏损"现象极有可能与卤族氧化物的氧化作用有关，如以色列死海的大气气态单质汞氧化主要是由 BrO 导致的（Obrist et al.，2011）。然而，对于其他海洋边界层，其他卤族元素（如碘自由基）、二氧化氮（NO_2）、羟基自由基以及 HO_2•的作用不容忽视（Wang et al.，2014a）。模型研究指出，全球海洋边界层大气气态单质汞的居留时间在 10 天左右（Hedgecock and Pirrone，2004），远远低于传统认识的 0.5～1.0 年（Lindberg et al.，2007；Holmes et al.，2010）。

除南北极、自由对流层和海洋边界层，森林边界层大气气态单质汞的"亏损"现象是近年来发现的另一个全球性普遍现象。早在 2008 年，研究人员通过对美国新英格兰州多个森林监测站大气气态单质汞的长期监测发现：大气气态单质汞在每年的夏季均存在显著的"亏损"，其含量从背景浓度的 1.5 ng/m³ 能够降低到 0.2 ng/m³ 以下，而且这种"亏损"主要出现在夜间形成稳定边界层的情况下（Mao et al.，2008）。通过进一步的模型估算，研究推测森林地区大气气态单质汞"亏损"30%是由大气氧化造成的，而 70%是由大气气态单质汞的干沉降造成的（Mao et al.，2008）。近年来，我国研究人员结合大气汞形态分析、植物/大气交换通量监测、汞同位素技术和模型分析等手段对我国长白山森林地区大气气态单质汞"亏损"的机理进行了系统的研究，揭示了森林边界层的大气气态单质汞"亏损"主要是由植物的吸收作用造成的，而大气氧化作用所起到的作用非常小，从而证实了植物吸收作用也是全球大气气态单质汞一个非常重要的汇（Fu et al.，2016b）。结合全球森林凋落物汞含量和凋落物沉降量的观测数据，研究人员估算了全球森林每年吸收大气气态单质汞的通量，约为 1180 t±710 t（Wang et al.，2016a），约占全球大气汞总沉降通量的 15%（Holmes et al.，2010），证实了森林地区植物吸收大气气态单质汞在全球大气汞循环中起着重要作用。

尽管大气气态单质汞通常被认为是一种具有较长大气居留时间和可进行全球性传输的污染物，近年来发现的区域性大气气态单质汞"亏损"现象给大气汞的全球循环提出了新的问题。从上述大气气态单质汞"亏损"现象可以看出，南北极、海洋边界层以及森林边界层由于存在大气气态单质汞的快速氧化或沉降作用，大气自由对流层和平流层气态单质汞在这些地区极有可能具有非常短的大气居留时间（如1天至1个星期），这么短的大气居留时间很可能限制了大气气态单质汞进行全球性或者跨大洲的传输，这会进一步给大气汞长距离传输研究带来一定的不确定性。尽管氧化后或沉降后的大气汞还可能通过一系列大气光化学还原作用再次转化为气态单质汞，这部分汞由于经过了复杂的化学转化过程，其原有的指纹特征（比如形态比值、汞和其他常规大气污染物比值、同位素指纹特征等）均会发生改变，原来单一的来源也会被其他来源干扰，从而给大气汞长距离传输的研究带来一定的误差。

2. 活性气态汞

活性气态汞是一种具有不稳定性质的大气汞形态。理论认为，活性气态汞具有较高的水溶性和干沉降速率，而且很容易被大气气溶胶吸附或进一步发生光致还原反应。因此，活性气态汞具有很短的大气居留时间（几个小时至1个星期，取决于大气环境如水汽溶度、气溶胶浓度和环境界面等），通常认为不参与汞的长距离大气传输。不过，在一些特殊的大气环境下，比如干燥的自由对流层大气中，活性气态汞的大气居留时间会达到10天左右（Selin et al.，2007）。研究指出，常温下活性气态汞的水溶性约为74 g/L，接近于二氧化硫（SO_2：110 g/L）和一氧化氮（NO：47 g/L）的水体溶解度。

和大气气态单质汞类似，活性气态汞的干沉降速率主要受到气象条件（如风速、大气温度和相对湿度等）、地表基质和植被覆盖条件的影响。森林生态系统大气活性气态汞的干沉降速率通常为0.1~6.0 cm/s，草地生态系统活性气态汞的干沉降速率为0.1~1.7 cm/s，湿地生态系统活性气态汞的干沉降速率为0.021~7.6 cm/s，而北极冰雪区活性气态汞的干沉降速率为1.0~5.2 cm/s（Zhang et al.，2009）。需要指出的是：活性气态汞干沉降速率研究通常是在测定大气活性气态汞浓度的基础上，采用动态通量箱法、微气象梯度法、涡度相关法等通量方法计算得到的。目前活性气态汞干沉降通量的测试方法有所不同，且不同方法之间的可比性较差。另外，活性气态汞浓度目前的测试方法（如镀KCl扩散管）还存在一定的问题（Lyman et al.，2010；Gustin et al.，2013；Huang et al.，2013；McClure et al.，2014；Gustin et al.，2015），可能会导致以往研究低估了活性气态汞的干沉降速率。另外，干沉降速率测定方法本身亦存在测试误差，这些会导致以往监测到的活性气体汞干沉降速率存在很大不确定性。Zhang 等（2012）基于界面双向阻抗模型，模拟了美国大陆活性气态汞的干沉降速率，结果为 0.4~2.0 cm/s。总体来说，活性气态汞是一种非常容易干沉降到地表生态系统的大气汞形态，因此其在边界层的大气居留时间通常比较短。因此，在低空大气中，活性气态汞通常是不参与长距离大气传输的。但在高空自由对流层特别是干燥的空气中，由于缺少相关的清除机制（吸附和干沉降），活性气态汞可能具有较长的大气居留时间（如 1 个星期）（Selin et al.，2007；Timonen et al.，2013）。

活性气态汞的物理化学转化过程比较复杂，一方面是因为活性气态汞包括多种氧化汞形态，不同汞形态转化速率会有所不同。另外，活性气态汞的反应还受到气象条件、氧化物和颗粒物的影响。目前的理论认为（Ariya et al.，2015），活性气态汞能够由气态单质汞氧化生成，同时活性气态汞又可以通过还原反应转化为气态单质汞，这种氧化和还原过程可能受到大气氧化物、大气温度、光照强度和大气相对湿度等的影响，低温和高辐射环境通常有利于气态单质汞氧化为活性气态汞，而高温则有利于活性气态汞的还原。另外，活性气态汞和大气颗粒态汞之间还经常发生吸附和解吸附的过程，该过程主要受到大气温度、光照强度和颗粒物浓度的影响（Lindberg et al.，2002a；Amos et al.，2012；Fu et al.，2016c）。在大气温度较高的环境下，活性汞主要存在于气相（＞90%），而在低温和颗粒物浓度较高的环境下，活性汞主要赋存于颗粒相（＞90%）（Amos et al.，2012）。另外，较大的辐射强度也会导致赋存在颗粒相的活性汞转化为活性气态汞。这一点在两极地区的大气汞形态转化中特别显著，在冬季末和春季初，北极圈通过大气气态单质汞氧化生成的活性汞很快通过吸附作用转化为颗粒态汞，而随着春季气温和光照强度的升高，北极圈大气颗粒态汞浓度快速降低，而活性气态汞浓度快速升高（Cobbett et al.，2007）。

因此，在近地表和海洋边界层，通常不存在活性气态汞的长距离传输，这些地区观测的活性气态汞主要来自于近距离的人为源汞排放和本地大气气态单质汞的氧化作用。而一些高海拔地区或者自由对流层，活性气态汞进行长距离传输的影响则较为普遍，不过这些区域的活性气态汞会经常被大气气溶胶和云团捕获，最终通过降水进入地表生态系统。

3. 颗粒态汞

颗粒态汞是一种稳定性介于大气气态单质汞和活性气态汞之间的大气汞形态。通常认为，颗粒态汞具有比大气气态单质汞高而比活性气态汞低的干沉降速率，而且较易通过云雾捕捉的降雨冲刷作用经由湿沉降输入到地表生态系统。一一般来说，颗粒态汞的大气居留时间为几个小时至几个星期。和其他大气汞形态类似，颗粒态汞的大气居留时间主要受到颗粒物粒径、气象条件和地表界面的影响。在自由对流层的上部或者平流层，由于气温极低而且缺少云雾捕捉作用，颗粒态汞具有较长的大气居留时间，会在这些区域出现一定的富集（Murphy et al.，2006）。而在边界层或者自由对流层下部，颗粒态汞比较容易被云雾和降水作用清除，或通过干沉降快速沉降到地表，亦或被光解/光还原为气态单质汞和活性气态汞，在大气中的居留时间略短。因此，颗粒态汞通常被认为是一种区域性污染物。

云雾和降水对大气颗粒态汞的清除作用主要取决于颗粒态汞本身的性质，如粒径、理化参数等因素。一般来说，颗粒态汞的云雾和降水清除速率通常为 1.3×10^{-5}，略低于活性气态汞的清除速率（Lee et al.，2001）。大气颗粒态汞的干沉降速率主要取决于颗粒物的粒径大小、气象和地表覆盖等因素。通常来讲，较大和较小粒径以及较高的风速条件都能加速颗粒态汞的沉降（Zhang et al.，2001）。野外监测结果显示，一些森林和湿地地区的颗粒态汞干沉降速率为 1.2～2.1 cm/s，而河流和湖泊地区的干沉降速率通常较低（0.02 cm/s）（Zhang et al.，2009）。

颗粒态汞的物理化学转化过程主要包括吸附/解吸附、氧化/光还原，吸附和氧化能够生成颗粒态汞，而解吸附和光还原能够使颗粒态汞转化为气态单质汞和活性气态汞。吸附/解吸附主要发生在颗粒态汞和活性气态汞之间，低温和低光照的气象条件能够促进活性气态汞吸附从而加速颗粒态汞的形成，而高温和较强光照气象条件下颗粒态汞的解吸附作用增强，会促进颗粒态汞向活性气态汞转化。颗粒物对气态单质汞的氧化也较为普遍，尤其容易发生在南北极的冬末、春初的平流层中（Lindberg et al.，2002a；Murphy et al.，2006）。颗粒态汞的光还原也是一种全球性普遍现象，但目前这方面的研究还比较少。

5.1.3 不同形态大气汞测定方法

1. 气态总汞

早期气态总汞的采样和分析方法采用的是镀金石英砂预富集结合冷原子荧光光谱法测定（Brosset，1987）。这种方法较为方便灵活，通常的采样流速为 0.3 L/min，样品采集的时间为几个小时，方法检出限随采样时间变化，通常最低检出限为 0.1 ng/m³，采集的气态总汞可以通过二次金汞齐法消除基质干扰，之后气态总汞热解为气态单质汞由冷原子荧光光谱仪或者冷原子吸收光谱法测定。金砂捕汞管长期（如 2 个星期）采样容易造成捕汞管的钝化，特别是在一些森林和海洋边界层环境。目前并不清楚钝化因素，但配置碱性干燥剂能够有效降低钝化现象（Fu et al.，2014）。化学处理活性炭（如碘化、氯化和硫化活性炭）也是一种非常有效的气态汞预富集方法，其结合消解/热解法和冷原子荧光光谱法也能精确测定大气气态总汞浓度（Bloom et al.，1995；Fu et al.，2014）。这种方法本身具有一定的空白，在短期采样中容易造成本底干扰，不过这种方法特别适合于大气气态总汞的长期采样（Fu et al.，2014）。

基于金对汞高效吸附特性，目前已开发了一系列基于金捕汞管预富集的在线大气气态总汞分析仪，如加拿大 Tekran 公司生产的 Tekran 2537A/B/X 系列、立陶宛 Tikslieji Prietaisai Ltd 公司生产的 Gardis 系列、日本 Nihon 公司生产的 Mercury/AM-3 和德国 Mercury Instruments GmbH 公司生产的 UT-3000 等（Urba et al.，1995；Tekran，1998；Osawa et al.，2007）。这些测汞仪均是基于金管预富集、热解吸和冷原子荧光光谱法联用技术，能够实现大气气态总汞高精度的监测，其仪器检出限通常能达到 0.1 ng/m^3 以下，时间分辨率为 5 min～1 h。这些基于预富集技术的测汞仪可以通过内部汞源和外部汞源进行校正，确保大气气态总汞的分析精度。另外，俄罗斯 Lumex 公司基于塞曼原子吸收技术开发了 RA-915A 系列测汞仪（Kim et al.，2006），该系列测汞仪利用高频调制偏振光的塞曼原子吸收技术来进行分析，不需要样品的预富集，因此具有相对较高的时间分辨率（如 1～10 s）。但仪器的检出限（0.3～1.0 ng/m^3）高于基于预富集技术的测汞仪（Ci et al.，2011）。

2. 活性气态汞

活性气态汞在大气中的浓度极低，因此目前的采样方法主要是通过预富集、解吸/提取和冷原子荧光光谱法测定。早期活性汞的预富集方法包括多级滤膜法（离子交换膜）、回流喷雾箱和镀 KCl 扩散管法（Xiao et al.，1997；Feng et al.，2000；Sheu and Mason，2001），预富集的活性气态汞可通过稀盐酸洗脱或者热解吸法结合冷原子荧光光谱法测定。Landis 等（2002）开发了镀 KCl 的环形扩散管预富集活性气态汞的方法，提高了活性气态汞的采样流速，另外结合二次热解法和 Tekran 2537A/B/X 的冷原子荧光光谱技术，开发了 Tekran 1130+Tekran537A/B/X 大气活性气态汞自动分析仪，实现了高时间分辨率（1～2 h）的活性气态汞在线连续监测，检出限为 0.5～6 pg/m^3（Landis et al.，2002）。此外，还可以通过 Tekran 2537A/B/X 测定大气中气态总汞和气态单质汞的差值来计算活性气态汞的浓度，其中气态单质汞测定首先采用多根镀 KCl 环形扩散管、石英棉或阳离子交换膜过滤掉活性气态汞，然后用 Tekran 2537A/B/X 测定气态单质汞浓度（Swartzendruber et al.，2009；Lyman and Jaffe，2012）。这种方法的检出限（约 80 pg/m^3）要高于 Tekran 1130+Tekran 2537A/B/X 活性气态汞分析系统，不适合测定低含量的大气活性气态汞（Ambrose et al.，2013）。

3. 颗粒态汞

大气颗粒态汞分析通常采用滤膜进行采集。目前用于采集颗粒态汞的滤膜有石英纤维滤膜、特氟龙滤膜、石英棉、玻璃纤维滤膜和醋酸乙酯滤膜等（Keeler et al., 1995; Lu et al., 1998; Ebinghaus et al., 1999; Munthe et al., 2001），其中石英纤维滤膜和特氟龙滤膜的空白容易控制，是比较常用的两种滤膜。采集有颗粒态汞的滤膜通常采用酸消解法和热解吸法进行处理（Lynam and Keeler, 2002），之后采用冷原子荧光光谱仪测定。加拿大 Tekran 公司基于石英纤维滤膜法、二次热解吸和 Tekran 2537A/B/X 汞蒸气分析仪，开发了 Tekran 1135 + Tekran 2537A/B/X 大气颗粒态汞在线分析系统。这套系统具有较高的时间分辨率（低至 1 h）和较低的仪器检出限（低至 1 pg/m^3），是目前国际上大气颗粒态汞监测使用最普遍的分析系统。

4. 不同形态大气汞的分析误差

大气气态总汞的测定目前采用最多的是加拿大 Tekran 公司生成的 Tekran 2537A/B/X 自动汞分析仪。Ebinghaus 等（1999）对比了 Tekran 2537A/B/X 自动汞分析仪和一些常见的大气气态总汞手动分析方法，发现不同方法之间具有较好的一致性（误差通常小于 10%）。另外，在城市地区较高的大气气态总汞环境下，Tekran 2537A/B/X 自动汞分析仪与 Lumex RA-915AM 测汞仪的监测结果也比较一致，两者之间的含量差异一般在 5%以下（付学吾等，2011）。

活性气态汞的分析方法是目前存在争议比较多的研究领域。一些研究认为，镀 KCl 扩散管在采集大气活性气态汞过程中会存在一定的活性气态汞流失现象，特别是在一些大气臭氧浓度和大气相对湿度较高的条件下（Lyman et al., 2010; McClure et al., 2014）。最近，有研究对比了 Tekran 2537 + Tekran 1130 + Tekran 1135 大气汞形态分析仪和其他活性气态汞分析方法（Gustin et al., 2013; Huang et al., 2013; Gustin et al., 2015; Huang and Gustin, 2015），发现镀 KCl 的扩散管法显著低估了大气活性气态汞浓度，而且不同的大气活性气态汞形态会有较大差异。由于目前对大气活性气态汞的具体化学形态并不是很清楚，对不同采样方法所采集的活性气态汞形态也不完全了解，因此尚不清楚其他形态的大气汞是否会对活性气态汞的采集产生影响，有关大气活性气态汞分析方法的争议还比较大。

与大气气态单质汞和活性气态汞研究相比，有关大气颗粒态汞分析方法误差的研究相对较少。Lynam 和 Keeler（2002）对比了滤膜酸消解法和热解吸法对大气颗粒态汞的影响，发现酸消解法测定的浓度通常会高于（约 30%）热解法测定的浓度。这可能会因为热解法采用的一次金汞齐方法产生的酸性气体影响到了金捕汞管的吸附效率。另外，颗粒态汞滤膜在采样期间有可能会吸附少量的活性气态汞，进而导致颗粒态汞浓度的高估（Malcolm and Keeler，2007；Rutter and Schauer，2007）。采样时间的增加也有可能导致测到的颗粒态汞浓度降低（Malcolm and Keeler，2007）。

5.2 大气汞的来源和分布特征

5.2.1 大气汞的来源

大气汞的来源包括人为源和自然源。其中人为源主要包括化石燃料的燃烧、金属和非金属材料冶炼、水泥生产、氯碱制造、水银和金的生产以及垃圾焚烧活动等，而自然源主要包括火山活动、土壤和水体排放、森林火灾和大气的物理化学转化过程。目前的全球人为源清单显示，2005～2010 年期间人为源全球年排汞量为 1875～2320 t（Streets et al.，2009；Pacyna et al.，2010；Pirrone et al.，2010；AMAP/UNEP，2013）。从形态分布来讲，2000 年人为源每年排放的大气气态单质汞、活性气态汞和颗粒态汞量分别为 1278 t、720 t 和 192 t（Pacyna et al.，2006b；Wilson et al.，2006）。全球人为源大气汞排放具有非常显著的区域分布特征，一些发展中国家或者欠发达国家和地区（如中国、印度、东南亚和中南美洲）是全球人为源大气汞排放比较集中的地区。

大气汞的自然源来源研究具有很大的难度，相关研究给出的自然源排放量估算结果一般具有很大的误差。目前多数有关全球自然源排放估算的结果主要是基于全球大气汞质量平衡的研究，普遍接受的一个范围为 4000～6000 t，其中大部分来自于海洋水体和陆地土壤的排放（Lamborg et al.，2002；Mason and Sheu，2002；Seigneur et al.，2004；Strode et al.，2007；Holmes et al.，2010；Pirrone et al.，2010）。与人为源不同，自然源排放的汞主要是气态单质汞。

近年来，一些研究采用界面交换模型估算了部分区域的自然源排汞量。如北美地区的研究结果显示，北美及近海海域的年均自然源排汞量为 95～150 t（Hartman et al.，2009；Wang et al.，2014c）。而亚洲的研究结果表明，东亚地区（主要包括中国、日本、韩国和部分南亚和东南亚地区）的自然源年均排放量为 834 t，远远高于北美地区，其中一个重要原因是亚洲人为源排汞沉降后的再释放（Shetty et al.，2008）。我国自然源排汞清单的研究显示，我国大陆自然源年均排放量在 574 t 左右，其中绝大部分来自于裸露土壤、戈壁和草地区域的排放（Wang et al.，2016b）。然而，基于全球陆地地表/大气汞交换通量的野外监测分析结果获得的全球陆地地表年均大气汞排放量约为 607 t（Agnan et al.，2016），远低于传统的认识，这主要是因为发展中和欠发达国家的野外监测数据比较缺乏，而欧美国家和地区的研究数据（通量通常比较低）整体上低估了全球陆地不同地表平均汞排放通量。例如，同样是基于我国陆地地表/大气排放通量监测结果的估算，认为我国自然源的年排汞量达到了 528 t，而其他亚洲地区如南亚、中南半岛和中亚的年均自然源汞排放量分别为 240 t、113 t 和 220 t（Fu et al.，2015a），这一结果要远远高于 Agnan 等（2016）对全球陆地排放量的估算，说明发展中和欠发达国家由于受到人为源再释放（包括人为源沉降后再释放、人为源排放到水体和土壤后的再释放）的影响，其自然源汞排放量和人为源一样要高于全球其他地区。

全球地表自然源的排汞主要以大气气态单质汞为主，而一些大气物理化学转化则是大气活性气态汞和颗粒态汞的重要来源。例如，模型研究发现全球每年约有 8000～15000 t 的气态单质汞通过大气氧化过程转化为活性气态汞和颗粒态汞，而这些转化过程主要发生在自由对流层、海洋边界层、平流层和两极地区（Selin et al.，2007；Holmes et al.，2010；Horowitz et al.，2017）。尽管这些研究结果目前还缺乏有效的证据，但可以看出，自由对流层可能是全球活性气态汞和颗粒态汞的一个主要来源区域，对全球大气汞的长距离传输和地表大气汞的沉降起着至关重要的影响。

5.2.2　全球大气汞区域分布特征

受到区域性大气汞排放的影响，大气汞分布具有非常明显的区域性分布特征。全球背景区大气气态单质汞的平均浓度为（1.74±0.71）ng/m^3（n = 93），而城市地区大气气态单质汞平均浓度为（4.33±3.60）ng/m^3（n = 34），约是全球背景区平均浓度的 2.5 倍（图 5-1）。北美背景区大气气态单质汞的平均浓度为（1.65±0.38）ng/m^3（n = 37），欧洲背景区大气气态单质汞的平均浓度为（1.66±0.27）ng/m^3（n = 20），南美洲背景区大气气态单质汞的平均浓度为（1.40±0.79）ng/m^3（n = 5），亚洲背景区大气气态单质汞的平均浓度为（2.57±1.06）ng/m^3（n = 18），而两极地区大气气态单质汞的平均浓度为（1.29±0.31）ng/m^3（n = 13）（图 5-1）。欧洲和北美城市地区大气气态单质汞平均浓度为（2.74±1.51）ng/m^3（n = 19），而亚洲城市地区大气气态单质汞平均浓度为（6.34±4.46）ng/m^3（n = 14）（图 5-1）。全球大气气态单质汞分布还具有明显的纬度差异，比如南半球背景区大气气态单质汞平均浓度为（0.94±0.09）ng/m^3（n = 9），赤道背景区大气气态单质汞平均浓度为（1.43±0.62）ng/m^3（n = 7），而北半球背景区大气气态单质汞平均浓度为（1.87±0.70）ng/m^3（n = 77）（图 5-1）。气态单质汞的区域分布基本上与全球人为源大气汞排放强度的区域分布吻合，说明大气气态单质汞浓度主要受人为源排放的影响。由于大气气态单质汞是一种具有较长大气居留时间的污染物，大气气态单质汞浓度较高的区域（比如亚洲）通常是污染源净输出区域，而浓度较低的区域（如南半球和两极地区）则主要表现为大气气态单质汞的汇。

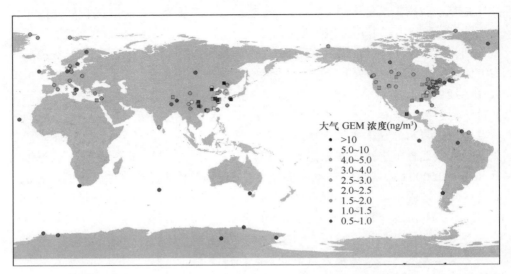

图 5-1　全球地表大气气态单质汞（GEM）浓度的分布特征（圆形表示背景区大气气态单质汞浓度，方形表示城市大气气态单质汞浓度）(Amouroux et al., 1999; Wangberg et al., 2001; Lindberg et al., 2002a; Liu et al., 2002; Berg et al., 2003; Kellerhals et al., 2003; Malcolm et al., 2003; Munthe et al., 2003; Temme et al., 2003; Weiss-Penzias et al., 2003; Fang et al., 2004; Han et al., 2004; Gabriel et al., 2005; Lynam and Keeler, 2005; Poissant et al., 2005; Zielonka et al., 2005; Swartzendruber et al., 2006; Laurier and Mason, 2007; Nguyen et al., 2007; Valente et al., 2007; Wang et al., 2007; Chand et al., 2008; Choi et al., 2008; Fu et al., 2008a; Fain et al., 2009; Yang et al., 2009; Fu et al., 2010; Liu et al., 2010; Sheu et al., 2010; Brooks et al., 2011; Ci et al., 2011; Friedli et al., 2011; Fu et al., 2011; Nguyen et al., 2011; Steen et al., 2011; Cheng et al., 2012; Fu et al., 2012b; Lan et al., 2012; Li, 2012; Zhu et al., 2012; Chen et al., 2013; Zhang et al., 2013; Wang et al., 2014a; Slemr et al., 2015; Xu et al., 2015; Yu et al., 2015; Zhang et al., 2015a; Fu et al., 2016c; Liu et al., 2016; Sprovieri et al., 2016; Zhang et al., 2016a; 李舒等，2016)

大气活性气态汞的研究主要集中于北美、欧洲、亚洲和南北极地区。全球背景区大气活性气态汞的浓度平均值为 (14.8 ± 21.3) pg/m^3（$n = 51$），而城市地区大气活性气态汞的浓度平均值为 (27.5 ± 27.8) pg/m^3（$n=21$）（图 5-2）。全球背景区大气活性气态汞的分布具有明显的区域分布特征，两极地区的浓度最高，为 (42.4 ± 46.7) pg/m^3（$n = 6$）；其次是欧洲 $[(20.9\pm10.4)$ pg/m$^3]$、北美 $[(7.3\pm9.8)$ pg/m$^3]$ 和亚洲 $[(5.7\pm2.9)$ pg/m$^3]$（图 5-2）。这与全球背景区大气气态单质汞的区域分布规律基本是相反的。南北极地区大气活性气态汞偏高主要与春季这些地区大气汞的氧化作用有关，而亚洲大气活性气态汞偏低有可能与这些地区较高的大气颗粒物浓度有关，其可能促进了大气活性气态汞向颗粒态汞转化（van Donkelaar et al.，2010；Fu et al.，2016a）。另外，一些高海拔地区的大气活性气态汞平均浓度为（25.5±13.1）pg/m^3（$n=4$）（图 5-2），明显高于全球背景区的整体平均值，主要受到自由对流层大气汞氧化作用的影响。尽管南北极地区大气活性气态汞的浓度明显高于全球其他地区，但这些大气活性气态汞可能主要集中于行星边界层内，且活性气态汞极容易沉降到地表，因此，南北极地区（边界层）很少存在大气活性气态汞的输出。不过，自由对流层通常具有较低的大气温度和相对湿度，大气活性气态汞具有相对较长的大气居留时间。因此，自由对流层富大气活性汞气团向低海拔地区的入侵通常是地表大气活性气态汞的一个重要来源（Gratz et al.，2015；Weiss-Penzias et al.，2015；Shah et al.，2016）。

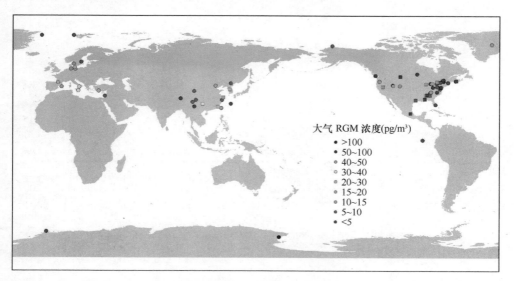

图 5-2　全球大气活性气态汞（RGM）浓度分布特征（圆形表示背景区活性气态汞浓度，方形表示城市活性气态汞浓度）（Sprovieri et al.，2002；Berg et al.，2003；Munthe et al.，2003；Temme et al.，2003；Weiss-Penzias et al.，2003；Han et al.，2004；Gabriel et al.，2005；Poissant et al.，2005；Swartzendruber et al.，2006；Laurier and Mason，2007；Chand et al.，2008；Fu et al.，2008b；Choi et al.，2009；Fain et al.，2009；Liu et al.，2010；Sheu et al.，2010；Brooks et al.，2011；Fu et al.，2011；Steen et al.，2011；Cheng et al.，2012；Fu et al.，2012a；Lan et al.，2012；Zhang et al.，2013；Wang et al.，2014a；Fu et al.，2015b；Xu et al.，2015；Yu et al.，2015；Zhang et al.，2015a；de Foy et al.，2016；Fu et al.，2016c；Liu et al.，2016；Zhang et al.，2016a；李舒等，2016）

全球背景区大气颗粒态汞浓度平均值为（21.9±32.5）pg/m^3（$n=48$），而城市地区大气颗粒态汞的浓度平均值为（147±251）pg/m^3（$n=24$）（图 5-3）。全球背景区大气颗粒态汞具有明显的区域分布特征，其中亚洲背景区平均浓度最高，为（49.0±52.7）pg/m³（$n=6$），其次是欧洲[（25.1±28.5）pg/m³]和北美[（8.4±10）pg/m³]（图5-3）。全球城市地区大气颗粒态汞的区域分布特征和背景区类似，亚洲城市地区的平均值为（314±324）pg/m³，北美城市地区的平均值为（28.4±49.3）pg/m³（图5-3）。全球大气颗粒态汞的区域分布特征与全球人为源排汞强度分布特征较为类似，表明人为源直接排放是影响全球地表大气颗粒态汞分布的重要因素，这与大气气态单质汞的分布较为一致。

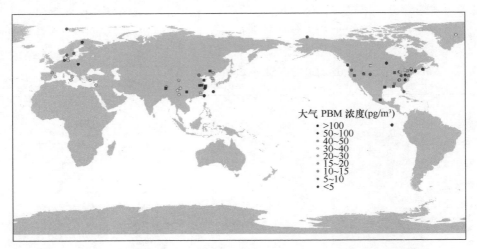

图5-3 全球大气颗粒态汞（PBM）浓度的分布特征（圆形表示背景区颗粒态汞浓度，方形表示城市颗粒态汞浓度）（Fang et al.，2001；Lindberg et al.，2002a；Malcolm et al.，2003；Munthe et al.，2003；Weiss-Penzias et al.，2003；Gabriel et al.，2005；Poissant et al.，2005；Swartzendruber et al.，2006；Valente et al.，2007；Chand et al.，2008；Choi et al.，2008；Fu et al.，2008b；Fain et al.，2009；Liu et al.，2010；Sheu et al.，2010；Brooks et al.，2011；Fu et al.，2011；Steen et al.，2011；Cheng et al.，2012；Fu et al.，2012a；Lan et al.，2012；Xu et al.，2013；Zhang et al.，2013；Wang et al.，2014a；Zhu et al.，2014；Fu et al.，2015b；Xu et al.，2015；Yu et al.，2015；Zhang et al.，2015a；Zhang et al.，2015c；de Foy et al.，2016；Fu et al.，2016c；Huang et al.，2016；Zhang et al.，2016a；李舒等，2016）

5.2.3 我国大气汞区域分布特征

受到我国较高人为源直接排放和再释放（人为源排汞沉降后的再释放）的影响，我国的大气气态单质汞和颗粒态汞浓度要高于全球平均值（参见图5-1和图5-3）。我国背景区大气气态单质汞的平均浓度为（2.74±1.05）ng/m³（$n=13$，图5-4），城市地区的气态单质汞平均值为（6.24±4.55）ng/m³（$n=13$，图5-4）；而全球其他地区背景区大气气态单质汞的平均浓度为（1.58±0.48）ng/m³（$n=81$，图5-1），城市地区的气态单质汞平均值为（3.15±2.26）ng/m³（n = 21，图 5-1）。我国背景区大气颗粒态汞的平均浓度为（54.1±53.2）pg/m³（n = 9，图 5-5），城市地区的颗粒态汞平均值为（346±326）pg/m³（n = 9，图 5-5）；而全球其他地区背景区大气颗粒态汞的平均浓度为（14.4±19.7）pg/m³（n = 39，图 5-3），城市地区的颗粒态汞平均值为（28.1±47.5）pg/m³（n = 15，图 5-3）。

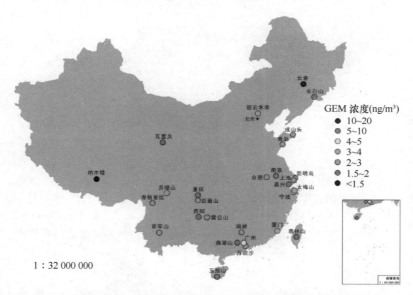

图 5-4　我国城市和背景区地表大气气态单质汞浓度分布特征（Liu et al.，2002；Fang et al.，2004；Wang et al.，2007；Fu et al.，2008a；Yang et al.，2009；Fu et al.，2010；Sheu et al.，2010；Ci et al.，2011；Friedli et al.，2011；Fu et al.，2011；Nguyen et al.，2011；Fu et al.，2012b；Li，2012；Zhu et al.，2012；Chen et al.，2013；Zhang et al.，2013；Xu et al.，2015；Yu et al.，2015；Zhang et al.，2015a；Liu et al.，2016；Zhang et al.，2016a；李舒等，2016）

与大气气态单质汞和颗粒态汞不同，我国大气活性气态浓度和其他地区相比并未出现升高的现象。我国城市和背景区的大气活性气态汞平均浓度分别为（23.2±11.4）pg/m³ 和（5.8±3.0）pg/m³（图 5-6），而全球其他城市和背景区大气活性气态汞平均浓度分别为（28.2±29.9）pg/m³ 和（16.4±22.6）pg/m³（图 5-2）。由于大气活性气态汞具有很高的干沉降速率和水溶性，人为源排放的活性气态汞通常会在排放源附近发生沉降，因此对野外监测点的影响比较有限。此外，我国大气颗粒物的浓度普遍偏高，会导致更多的活性汞向颗粒态汞进行转化。需要注意的是，由于目前活性气态汞监测数据多存在一定误差，区域性的分布差异可能存在很大的不确定性（Fu et al.，2015b）。

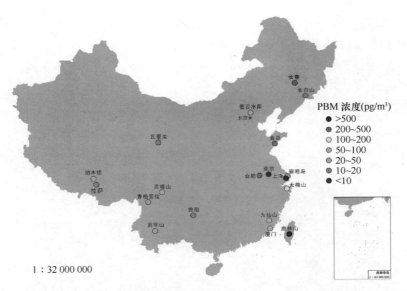

图 5-5 我国大气颗粒态汞（PBM）浓度分布特征（Fang et al., 2001; Fu et al., 2008b; Sheu et al., 2010; Fu et al., 2011; Fu et al., 2012a; Xu et al., 2013; Zhang et al., 2013; Zhu et al., 2014; Fu et al., 2015b; Xu et al., 2015; Yu et al., 2015; Zhang et al., 2015a; Zhang et al., 2015c; de Foy et al., 2016; Huang et al., 2016; Zhang et al., 2016a; 李舒等, 2016）

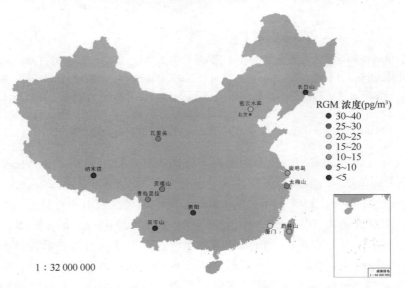

图 5-6 中国大气活性气态汞（RGM）浓度分布特征（Sheu et al., 2010; Fu et al., 2011; Fu et al., 2012a; Zhang et al., 2013; Fu et al., 2015b; Xu et al., 2015; Yu et al., 2015; Zhang et al., 2015a; de Foy et al., 2016; Zhang et al., 2016a; 李舒等, 2016）

5.3 大气汞长距离传输研究方法

5.3.1 基于监测数据的源汇关系模型

目前的源汇关系模型主要包括两种类型：一种是以确定污染源类型和贡献份额为主的受体模型，另一种是以污染源分布和传输过程为主的扩散模型。受体模型是通过测定受体样品和排放源样品的物理化学特征，基于受体和污染源之间的污染物质量平衡关系，识别和解析受体处大气污染物不同来源及其贡献率的数值模式和方法。受体模型主要用于研究排放源对受体的贡献，不考虑排放源的排放条件、气象条件和大气传输过程，因此不能用于示踪大气污染物的长距离迁移过程和污染源的空间分布特征。扩散模型重点用于研究特定污染源排放污染物的传输过程及其对受体污染物含量变化的影响，或者用于示踪受体处大气污染物的主要污染来源区域和主要传输过程。

1. 受体模型

受体模型的出现至今已经过半个世纪的历程，早期主要用于示踪大气颗粒物的污染来源（Henry et al., 1984）。目前的主要研究方法包括物理法和化学统计法，其中化学统计法是应用最为广泛、发展也比较成熟的研究方法。基于化学统计法的受体模型包括主成分分析法（PCA）、正交矩阵因子分析法（PMF）、富集因子法（EF）、化学质量平衡法（CMB）等，其中主成分分析法和正交矩阵因子分析法在大气汞的源解析中应用较为广泛。

主成分分析法（principal component analysis，PCA）的核心是运用降维的思想将研究对象的多个相关变量转化为几个不相关的变量，并且要求这些不相关的变量能够最大化反映原变量提供的数据信息（Thurston and Spengler, 1985）。在大气汞的PCA源解析研究中，通过对多种大气污染物浓度和化学组分、气象参数和汞浓度等相关数据的分析，利用正交旋转法来估算大气汞的来源。不同的大气污染物和污染物化学组分可以用来指示特定的污染源。例如，元素碳（elemental carbon，EC）主要用于示踪一次燃烧源；有机碳（organic carbon，OC）和元素碳比值（OC/EC），以及 Ba、Ca、Na、Pb、CO 和 NO_x 用于示踪机动车排放；$^{13}C/^{14}C$ 同位素用于示踪生物质源；As、Se 和 SO_2 主要用于示踪燃煤电厂烟气排放；Ni 和 V 主要用于示踪燃油排放；Ca 和 Fe 主要用于示踪水泥生产排放；Zn、Pb、Cu 和 Cl 主要用于示踪垃圾焚烧排放；V、Cr、Mn 和 Fe 主要用于示踪金属制造排放；K、有机碳和左旋葡萄糖主要用于示踪生物质燃烧排放；Na、Ca、Al 和 Fe 主要用于示踪土壤和地壳物质排放；Na 和 Cl 主要用于示踪海盐气溶胶等排放（Keeler et al.,

2006; Lee and Hopke, 2006; Lynam and Keeler, 2006; Watson et al., 2008; Zhang et al., 2008; 刘娜等, 2010)。

正交矩阵因子分析（positive matrix factorization, PMF）法早期是由 Paatero 和 Tapper (1994) 提出的, 目前可以通过美国环境保护署 (USEPA) 官方网站下载。PMF 首先利用权重确定出污染物化学组分的误差, 之后采用最小二乘法来估算出污染物的主要污染源及其贡献率。正交矩阵因子分析法不需要直接测定特定污染源的指纹谱, 因此可用于研究未知污染源的贡献。当然, 对潜在污染源指纹谱的认知有助于解释模型预测结果 (Watson et al., 2008)。PMF 模型目前主要应用于大气颗粒物和挥发性有机物源示踪研究 (Lee and Hopke, 2006; Song et al., 2008), 而在大气汞源示踪中的研究相对较少 (Liu et al., 2003; Keeler et al., 2006; Cheng et al., 2009; Wang et al., 2013)。PMF 模型通常需要获得大量的大气污染物含量和化学组分数据。在大气汞的源示踪研究中, 采用的数据包括不同形态大气汞（气态单质汞、活性气态汞和颗粒态汞）、常规大气污染物（CO、NO_x、O_3 和 SO_2）、重金属、$PM_{2.5}$ 数浓度和碳（如黑炭）等 (Liu et al., 2003; Keeler et al., 2006; Cheng et al., 2009; Wang et al., 2013)。

2. 扩散模型

大气汞的扩散模型主要用于研究污染物在大气中的传输过程。扩散模型需要计算受体大气的后向三维传输轨迹和排放源大气前进方向的三维传输轨迹, 据此分析不同区域来源的大气对受体大气汞含量变化的影响, 以及污染源排放的大气汞可能影响区域。大气后向和前向的三维传输轨迹可以由美国国家海洋和大气管理局空气资源实验室提供的混合单粒子拉格朗日轨迹（hybrid single particle Lagrangian integrated trajectory, HYSPLIT）模型进行计算（Draxler and Rolph, 2014; Stein et al., 2015）。进行 HYSPLIT 模型需要输入起始点的三维位置信息、开始计算时间、大气传输的方向（后向或前向）、大气传输的时间和轨迹计算的频次等信息。HYSPLIT 模型能够计算出起始点大气在之前或之后某个时间段的三维位置、空气温度、位温、下垫面海拔高度、小时降雨量、混合层高度、空气相对湿度和辐射强度等相关信息。根据起始点监测的大气汞和其他大气污染物浓度, 以及 HYSPLIT 模型模拟的大气前向和后向传输轨迹便可以分析大气传输和气象要素对大气汞或其他大气污染物浓度变化影响。通过对比起始点监测到的气象资料和 HYSPLIT 模型模拟的气象资料, 还可以分析 HYSPLIT 模型模拟大气传输轨迹的不确定性。为了降低模型对大气传输轨迹的不确定性, HYSPLIT 模型还提供了轨迹矩阵式起始点（trajectory endpoint matrix）、轨迹集合（trajectory ensemble）和轨迹概率（trajectory frequency）等集成化功能。

在对大气传输轨迹计算的基础上，相关源汇关系模型通过分析大气汞和其他大气污染物监测数据与大气传输轨迹相互关系，能够估算大气汞和其他大气污染物的潜在污染来源。这些模型主要包括轨迹网格化概率分布（trajectory gridded frequency distribution，GFD）、潜在污染源贡献函数（potential source contribution function，PSCF）和浓度加权轨迹法（concentration-weighted trajectory，CWT）等。

1) GFD 模型

GFD 模型是一种多起始点轨迹计算方法，目前主要应用于一些大气汞形态和活性气态汞沉降通量研究（Weiss-Penzias et al.，2009；Weiss-Penzias et al.，2011；Gustin et al.，2012）。这些研究选取典型的大气汞污染事件（如大气汞含量高于某个特定值或者在某个特定区间），通过计算这些大气汞污染事件期间监测点周边多个点位（如监测点周边 36 个点位均匀分布在监测点周边 0.5°×0.5°和 500 m、1000 m、1500 m 和 2000 m 的空间内）的大气气团的来源，然后对大气气团的来源进行统计分析，得出大气汞污染事件期间大气气团来源的区域概率分布图，从而能够解析这些大气汞污染事件的污染来源（Gustin et al.，2012）。GFD 模型的优点是通过计算监测点周边多个起始点的大气传输轨迹，从而降低了单一起始点轨迹计算的不确定性。不过，由于 GFD 模型通常只选取少数几个大气汞污染事件的大气传输过程进行研究，这些污染片段在监测点大气汞的分布中不一定具有很强代表性，因此在解析监测点大气汞整体污染来源方面可能具有一定局限性。另外，目前的 GFD 模型研究主要集中在大气活性气态汞的来源解析（Weiss-Penzias et al.，2009；Weiss-Penzias et al.，2011；Gustin et al.，2012），由于活性气态汞在边界层内通常不参与长距离传输，且活性气态汞的监测分析误差普遍较大（Gustin et al.，2013；Huang et al.，2013；Gustin et al.，2015），因此有关活性气态汞来源解析的可靠性会比较低。

2) PSCF 模型

PSCF 模型主要用于解析监测点污染源来源的分布情况。其主要是通过计算落入某个空间区域的高于某个浓度阈值相关大气轨迹节点与全部监测浓度相关大气轨迹节点比值来确定该空间区域是否是受体监测点大气汞的重点来源区域。其模型计算公式为（Hopke，2003）

$$\mathrm{PSCF}_{ij} = \frac{M_{ij}}{N_{ij}} W_{ij} \tag{5-1}$$

式中，PSCF_{ij} 为某个空间区域经纬度栅格（i，j）的潜在污染源指数（无量纲，范围在 0~1 之间）；M_{ij} 为落入某个空间区域经纬度栅格（i，j）而且高于某个浓度阈值大气轨迹节点的数量；N_{ij} 为落入某个空间区域经纬度栅格（i，j）全部大气轨迹节点的数量；W_{ij} 为某个空间区域经纬度栅格 (i,j) 权重系数，W_{ij} 可由公式（5-2）计算（Polissar et al.，2001）：

$$W_{ij} = \begin{cases} 1.0 & N_{ij} > 3N_{ave} \\ 0.7 & 3N_{ij} > N_{ij} > 1.5N_{ave} \\ 0.4 & 1.5N_{ij} > N_{ij} > N_{ave} \\ 0.2 & N_{ij} < N_{ave} \end{cases} \quad (5\text{-}2)$$

式中，N_{ave} 为研究区域所有空间区域经纬度栅格 (i,j) 大气轨迹节点数量的平均值。

PSCF 模型的运算首先需要采用 HYSPLIT 模型计算大量的后向大气传输轨迹。后向大气轨迹的传输时间的设定通常与污染物的大气居留时间有关，如大气气态单质汞的传输时间通常为 72～240 h（Choi et al.，2008；Xu and Akhtar，2010；Fu et al.，2011；Fu et al.，2012a；Fu et al.，2012b；Fu et al.，2016c），而大气颗粒态汞和活性气态汞的传输时间通常为 24～48 h（Han et al.，2005；Choi et al.，2008）。为了降低大气传输轨迹的空间不确定性，通常需要计算受体监测点处地面以上 500 m、1000 m 和 1500 m 三个高度的大气后向轨迹（Fu et al.，2011；Fu et al.，2012b）。另外，还可以通过加密轨迹计算频次来降低大气后向轨迹的不确定性，比如选择每 2～4 个小时计算一条大气后向轨迹（Fu et al.，2012b；Fu et al.，2016c）。空间区域经纬度栅格 (i,j) 的划分也会影响 PSCF 模型模拟结果。对于全球、洲际或者其他大尺度的长距离传输研究来说，栅格 (i,j) 的大小通常为 1°×1°或者 2°×2°（Zhang et al.，2015a；Fu et al.，2016c）；而对一些区域性的污染源解析来说，栅格 (i,j) 的大小通常为 0.25°×0.25°或者 0.5°×0.5°（Abbott et al.，2008；Fu et al.，2011）。

PSCF 模型的优点是能够描绘受体大气汞污染来源的空间分布情况，通过对比大气汞排放清单能进一步确定主要污染源的类型和分布情况。同时，PSCF 模型运算不需要添加辅助的大气污染物监测数据。当然，PSCF 模型也具有一些缺陷。比如 PSCF 模型没有考虑到大气汞的物理化学转化和干湿沉降过程。另外，尽管大量的后向轨迹计算能够降低一定的不确定性，但大气后向轨迹本身仍具有很大空间不确定性。PSCF 模型还不适合解析监测点近距离的排放源，由于受到大气后向轨迹"拖尾效应"的影响，真实排放源上风向或下风向的区域也可能会被识别为主要的污染源区域。PSCF 模型结果还受到浓度阈值的影响，设定的浓度阈值通常为监测浓度的平均值、75%位值或 97.5%位值（Weiss-Penzias et al.，2009；Weiss-Penzias et al.，2011；Fu et al.，2016c），较高的浓度阈值可能会导致部分污染源区的缺失，而较低的阈值则会将非污染源区预测为潜在的污染源区域。

3）CWT 模型

与 PSCF 模型类似，CWT 模型也是通过对大气汞监测数据和大气后向传输轨迹的综合分析来确定大气汞的潜在污染来源区域（Rutter et al., 2009; de Foy et al., 2012; Cheng et al., 2013b; Zhang et al., 2016a）。和 PSCF 模型不同的是，CWT 模型采用大气汞形态监测浓度来校正落入某个空间区域经纬度栅格（i, j）大气后向轨迹节点，其模型运行公式为

$$P_{ij} = \sum_{l=1}^{L} C_t \tau_{ijt} \bigg/ \sum_{l=1}^{L} \tau_{ijt} \qquad (5\text{-}3)$$

式中，P_{ij} 为经纬度栅格（i, j）污染源贡献指数，C_t 为大气汞监测浓度，τ_{ijt} 为大气汞监测浓度 C_t 的后向大气传输轨迹（t）落入纬度栅格（i, j）的节点，L 为大气汞监测期间计算的大气后向传输轨迹的数量。

CWT 模型模拟首先需要采用 HYSPLIT 模型和 Flexpart 模型计算大量的后向大气传输轨迹（Stohl et al., 2005; Zhang et al., 2016a），之后将模型计算的每条大气后向传输轨迹和大气汞监测浓度匹配。和 PSCF 模型类似，计算的大气后向轨迹的传输时间、计算频次和单个经纬度栅格（i, j）的划分主要依据大气汞形态的大气居留时间、模型精度和研究区域的范围来进行确定。

与 PSCF 模型采用单一的大气汞浓度阈值相比，CWT 模型采用全部的大气汞监测浓度对大气后向轨迹进行加权分析，能够更精确解析大气汞污染来源的区域分布（Han et al., 2007; Cheng et al., 2015）。CWT 模型采用某个空间栅格（i, j）的轨迹节点数量来校准污染源区域贡献指数，克服了 PSCF 模型通常将受体附近区域计算为重点来源区域的缺点。CWT 模型的缺点主要与大气后向轨迹的空间不确定性有关。另外，与 GFD 和 PSCF 模型类似，CWT 模型缺乏与大气汞排放清单的有效联系，是一种单向性源解析方法。

5.3.2 区域大气汞循环模型

前面介绍的受体模型通常用于解析大气汞的重点污染来源类型，而扩散模型只能模拟大气汞污染来源的区域分布特征。这两种方法主要基于野外监测数据和大气后向轨迹模拟，没有和大气汞排放清单形成紧密的联系，因此通常被认为是一种定性化的源解析方法。而区域和全球大气汞循环模型一定程度上克服了受体模型和扩散模型的缺点，能够定量描述不同污染来源和不同污染源排放区域对大气汞浓度变化的贡献份额。

现有的大气汞循环模型主要分为区域尺度和全球尺度模型两种类型。其中区域尺度模型主要包括 CMAQ-Hg（community multiscale air quality-mercury）模型、GRAHM 模型（global/regional atmospheric heavy metal model）、MSCE-HM 模型（meteorological synthesizing center east heavy metal model）、DEHM 模型（Danish

Eulerian hemispheric model)、STEM-Hg 模型（sulfur transport and deposition model-mercury）和 ADOM 模型（acid deposition and oxidant model）（Bullock，2000；Lee et al.，2001；Petersen et al.，2001；Christensen et al.，2004；Dastoor and Larocque，2004；Ryaboshapko et al.，2007；Pan et al.，2008；Lin et al.，2010）。

近年来，CMAQ 模型成为区域汞循环研究应用最为广泛的一种模型（Bullock，2000；Bullock and Brehme，2002；Gbor et al.，2006；Lin et al.，2006；Lin et al.，2010）。该模型是美国 EPA 为仿真分析各种复杂大气污染物物理化学过程开发的大气模式，已更新至第三代（Models-3）。CMAQ 模型突破了传统模型只针对单一化合物或单相化合物的局限性，考虑了实际大气中不同化合物之间的相互转化和影响，包括多层网格嵌套、可供选择的多种用途、多空间尺度和化学机制。CMAQ-Hg 模型在 2002 年由 Bullock 和 Brehme（2002）提出，之后不断改进，新版本的 CMAQ-Hg 模型修正了大气气态单质汞与 H_2O_2 和•OH 等大气氧化物反应的化学动力学参数以及气态单质汞氧化产物的分配等。

5.3.3 全球大气汞循环模型

全球尺度的大气汞循环模型包括 CTM-Hg 模型（global chemical transport model）、DEHM 系统模型（Danish Eulerian hemispheric model system）、GRAHM 模型（global/regional atmospheric heavy metal model）、MSCE-HM（hemispheric model of mercury airborne transport and deposition）模型和 GEOS-Chem（Goddard earth observing system-Chem）模型等（Seigneur et al.，2001；Mason and Sheu，2002；Christensen et al.，2004；Dastoor and Larocque，2004；Seigneur et al.，2004；Travnikov，2005；Seigneur and Lohman，2008；Holmes et al.，2010；Zhang et al.，2016b；Horowitz et al.，2017）。这些模型是具有自洽封闭边界的全球尺度大气汞循环模型。

全球大气汞循环模型和区域性大气汞循环模型需要导入大气汞排放清单，标准化的气象数据，全球地形、地貌和土地利用类型等数据以及大气汞转化反应的化学动力和大气汞干湿沉降参数。全球气象数据以及全球地形、地貌和土地利用类型等数据目前已经非常精确，而大气汞排放清单、大气汞物理化学转化过程和参数以及大气汞干湿沉降参数是影响全球和区域汞循环模型精度的关键因子。目前，发达国家人为源大气汞排放清单已经比较完善（Pacyna et al.，2010；Pirrone et al.，2010；Streets et al.，2011），我国的人为源大气汞清单经过近 10 年的研究也得到了大幅度的提升（Wu et al.，2006；Zhang et al.，2015b；Wu et al.，2016）。但是，非洲、南美、东南亚等地区仍缺乏可靠的大气汞排放因子，其大气汞排放清单具有很大不确定性（Brunke et al.，2012；Fu et al.，2015a）。对全球自然源特别是海洋水体和陆地地表与大气间的汞交换通量尚认识不足。此外，模型通常需要加入大量的大气汞转化过程和反应动力学参数（Horowitz et al., 2017），而绝大多数的大气汞物理化学转化过程主要集中于实验室研究，且动力学反应系数通常具有较大的变化范围，能否代表实际大气中汞的转化过程还存在一定疑问。对一些特殊环境条件下大气汞物理化学转化过程的研究还比较缺乏，如汞在人为源排放烟气刚刚进入大气中的反应过程、海洋上空边界层和自由对流层汞的反应过程、部分地区复合污染状况和复杂气象条件下汞的迁移和转化过程等。大气汞形态的干湿沉降速率也是影响模型精度的另一个关键因素，如何在模型中建立干沉降速率与环境气象条件之间的联系也是一个重要课题。大气汞的湿沉降通常只考虑活性气态汞和颗粒态汞，有关活性气态汞和颗粒态汞的降雨清除速率的研究相对较少，多数模型将活性气态汞和颗粒态汞的降雨清除速率分别等同于其他易溶的大气化学物质（如 HNO_3）和水溶态气溶胶（Lee et al., 2001；Selin et al., 2007）。同时，不同形态大气汞干沉降速率的确定也比较关键，目前研究估算的大气活性汞和颗粒态汞干沉降速率具有较大的变化范围。此外，一些模型研究忽略了大气气态单质汞的干沉降过程（Seigneur et al., 2004；Travnikov, 2005；Selin et al., 2007），然而近些年的研究发现植物生态系统地区大气气态单质汞的干沉降对全球汞的循环起着重要的作用（Mao et al., 2008；Fu et al., 2016b；Wang et al., 2016a），这会显著影响到模型预测结果的精度。

5.4　国内外大气汞长距离传输研究进展

5.4.1　我国大气汞传输过程和特征

我国大气汞的长距离传输和模型研究开始于 21 世纪初，近些年来出现了快速的发展，显著提升了对我国大气汞的跨省、跨区域和跨境传输以及我国大气汞的质量平衡等相关科学认知，为我国的汞污染防治、国际汞公约谈判和履行国际汞公约提供了重要的科技支撑。

1. 我国城市地区大气汞的传输过程

我国城市地区通常分布有密集的工业活动，再加上居民生活活动（如早期的散煤燃烧）、机动车排放和土壤二次汞排放的影响（Feng et al., 2004；Feng et al., 2005；Tang et al., 2007；Zhu et al., 2013），城市区域内大气汞的排放强度比较高。研究表明，我国城市地区的大气气态单质汞浓度与区域人为源排放强度之间的相关性不太明显，而我国偏远地区大气气态单质汞浓度则和区域人为源大气汞排放强度呈显著的正相关，说明我国城市地区大气气态单质汞主要受城市自身的大气汞排放影响，而偏远地区的大气气态单质汞主要与区域性汞排放的大气传输有关

(Fu et al., 2015b)。

贵阳市是我国西南地区经济发展最快的城市之一。其大气气态单质汞、颗粒态汞和活性气态汞的平均浓度分别为（9.72±7.06）ng/m³、（368±676）pg/m³ 和（35.7±43.9）pg/m³（Fu et al.，2011），约是全球背景值浓度的 5~50 倍（Valente et al.，2007）。PSCF 模型结果表明，贵阳市大气气态单质汞的重点来源地区是贵阳市辖区和贵阳市以北的几个城镇区域（如修文和息烽），其主要工业活动包括燃煤电厂、水泥厂、钢铁厂和小型燃煤锅炉，是西南地区人为源大气汞排放的重点区域（Fu et al.，2011；Zhang et al.，2015b）。此外，贵阳市主要受西南季风的影响，西南季风从东南亚向我国西南地区传输的过程中会捕获东南亚工业和生物质燃烧排放的汞以及我国西南地区工业排放的汞，进而给贵阳市的大气带来附加的汞污染。这一结论也得到了贵阳市大气气态单质汞长期监测结果的印证，贵阳市 2009/2010 年度大气气态单质汞的几何平均值（8.88 ng/m³±7.06 ng/m³）比 2001/2002 年度的几何平均值（7.45 ng/m³±12.8 ng/m³）升高了约 19%。与 2001/2002 年度相比，2009/2010 年度大气气态单质汞浓度的升高主要在白天，升高幅度达到了 33%，显著高于夜间大气气态单质汞的升高幅度（7%）。受到夜间城市大气逆温层的影响，贵阳市 2009/2010 年度夜间大气气态单质汞浓度的升高可能主要来自于贵阳市辖区的排放，而白天大气气态单质汞浓度的显著升高则表明西南地区大气汞的排放和长距离传输作用起到重要作用，其中来自于我国中东部、云贵高原和中南半岛大气汞排放的贡献较大（付学吾和冯新斌，2015）。

厦门市是典型的海滨城市，其大气气态单质汞（3.50 ng/m³±1.21 ng/m³）和颗粒态汞（174 pg/m³±281 pg/m³）平均浓度明显低于贵阳市，不过其大气活性气态汞平均浓度（61.0 pg/m³±69.4 pg/m³）约是贵阳市的 2 倍，表明海洋大气环境条件促进了城市地区大气汞的氧化过程（Xu et al.，2015）。PCA 模型研究发现，厦门市的大气汞浓度主要受到大气汞光化学过程、燃煤电厂汞排放、机动车汞排放、生物质燃烧和边界层条件变化等因素影响（Xu et al.，2015）。这些影响因素可能来自于城市辖区，也可能来自于周边区域和长距离大气传输作用。

南京市位于长江三角洲城市群，是我国人为源大气汞排放的重点区域（Zhang et al.，2015b）。南京市大气气态总汞的年均浓度为（7.9±7.0）ng/m³，略低于贵阳市但明显高于厦门市。通过对南京市一年大气后向轨迹的聚类分析发现，来自于南京市本地大气的大气气态总汞浓度最高，达到了 11.9 ng/m³。其次是来自于我国东南部的大气（平均浓度：9.82 ng/m³）、我国东海的大气（平均浓度：8.3 ng/m³）、我国北部沿海地区大气（平均浓度：7.6 ng/m³）、我国华北和蒙古国的大气（平均浓度：5.9 ng/m³）以及我国东北和西伯利亚的大气（平均浓度：4.7 ng/m³）（Zhu et al.，2012）。结合这些大气轨迹聚类的出现频率，可以推测南京市本地汞排放和来自于我国东南地区的大气汞长距离传输是南京市大气气态总汞的两个关键来源（Zhu et al., 2012）。

2. 我国青藏高原大气汞的传输过程

青藏高原是被称为世界的"第三极"，当地人为活动排放的大气汞较少，是大气环境整体较为洁净的区域。然而，与青藏高原毗邻的南亚、我国中东部地区、东南亚和中亚地区是全球大气汞排放最集中的区域（Pacyna et al., 2010）。在青藏高原季风环流的作用下，这些地区排放的大气汞能够通过长距离的跨界和跨境传输对青藏高原的生态环境造成潜在风险。瓦里关全球大气基准站位于青藏高原东北缘，其大气气态单质汞、颗粒态汞和活性气态汞分别为（1.98±0.98）ng/m³、（19.4±18.1）pg/m³ 和（7.4±4.8）pg/m³，略高于北半球背景值（Fu et al., 2012a）。PSCF 模型分析的结果显示，我国西北工业和城市地区（如西宁-兰州-西安-银川一带）排放的大气气态单质汞经长距离大气传输是瓦里关大气气态单质汞最重要的污染来源，另外，南亚特别是印度西北工业地区排放的大气气态单质汞也可以"翻越"喜马拉雅山进入青藏高原，进而影响到青藏高原的东北缘并进一步向我国国内进行输送（Fu et al., 2012a）。瓦里关大气颗粒态汞的潜在污染源区域主要为四川西北部、青海东部和甘肃西部，这些地区分布有大面积的黄土和零散的隔壁沙漠，形成的沙尘有利于气态总汞向颗粒态汞转化。另外与大气气态总汞类似，印度西北部的大气颗粒态汞也可以通过长距离的大气传输作用影响到瓦里关（Fu et al., 2012a）。瓦里关大气活性气态汞的来源主要分布在瓦里关山南部的低海拔地区，这些地区分布有零散的居民活动，其排放的活性气态汞能够在"谷风"作用下向瓦里关山爬升（Fu et al., 2012a）。

3. 我国华北和东北背景区大气汞的传输过程

密云水库背景站和长白山背景站分别位于我国华北地区和东北地区，其中密云水库的大气气态总汞年均值为（3.22±1.74）ng/m³，约是北半球背景区的 2.1 倍；而长白山大气气态总汞年均浓度为（1.60±0.51）ng/m³，和北半球基本相当（Fu et al., 2012b；Zhang et al., 2013）。密云水库大气活性气态汞有典型的日变化特征，其中活性气态汞的高值主要出现在午后，说明本地的大气汞氧化过程是大气活性气态汞的最主要来源（Zhang et al., 2013）。密云水库春季和夏季 PBM/PM$_{2.5}$ 比值偏高，说明该地的大气颗粒态汞主要来自于区域性的人为源汞排放，来源区域包括密云水库南部的华北平原，其中包括北京市及其周边城市群（Zhang et al., 2013）。密云水库大气气态总汞的污染源主要来自于华北地区和华东地区，包括北京南部、山东西部、河南东部和安徽北部，这些地区是我国人为源汞排放最大的地区之一

(Zhang et al., 2013)。受区域性主导风向和季风气候的影响，密云水库大气气态总汞的来源具有明显的区域分布规律。其中冬季的主要污染源来自于密云水库以东的辽东半岛，春季污染源主要来自于北京南部、山东西部、河南东部和安徽北部等地，夏季污染源主要分布在北京和山东，秋季的污染源主要分布在山西和陕西的北部（Zhang et al., 2013）。密云水库冬季污染源类型主要为区域家庭燃煤汞排放，这一排放源具有较低的 TGM/CO 比值，夏季、春季和秋季密云水库的 TGM/CO 比值略高，而且 TGM 和 CO 回归曲线的截距比冬季明显升高，表明这些季节自然源直接和再释放的贡献比例增加（Zhang et al., 2013）。

长白山大气气态总汞浓度和我国其他背景区相比普遍偏低（Fu et al., 2015b），其中一个重要原因是东北地区的森林生态系统在夏季植物生长阶段吸收了大量的大气气态总汞，从而造成该地区夏季大气气态总汞的"亏损"（Fu et al., 2016b）。长白山 PSCF 的模型结果显示，长白山地区的大气气态总汞主要受长距离大气传输作用的影响，其中辽宁南部、河北、北京和山西东部是国内最主要的污染来源；另外，朝鲜中部人为源排放的大气汞也可以通过长距离的大气跨境传输影响长白山地区（Fu et al., 2012b）。长白山不同季节大气气态总汞的来源区域也有一定差异，其中春季和冬季主要来自于华北地区，而秋季来自于辽宁和内蒙古中部的污染源增加，夏季来源主要集中在朝鲜中部，主要受到了西太平洋气团主导大气传输作用的影响（Fu et al., 2012b）。

4. 我国西南背景区大气汞的传输过程

我国西南地区是经济相对欠发达的区域，不过由于这些地区原煤汞含量偏高且有大量的金属和非金属冶炼活动，这一地区也是我国大气汞排放的重点区域（Feng et al., 2002；Zhang et al., 2015b）。四川贡嘎山、贵州雷公山、重庆四面山、云南香格里拉和哀牢山等背景区的大气气态总汞浓度分别为（3.98±1.62）ng/m³、（2.80±1.51）ng/m³、（2.88±1.54）ng/m³、（2.55±0.73）ng/m³ 和（2.09±0.63）ng/m³（Fu et al., 2008a；Fu et al., 2010；Zhang et al., 2015a；Zhang et al., 2016a；刘伟明等，2016），约是北半球背景区平均浓度 1.5 ng/m³ 的 1.4~2.6 倍。我国西南背景区大气气态总汞浓度的升高与区域的大气排汞有关，比如四川贡嘎山的大气气态总汞主要污染源为贡嘎山较低海拔地区城镇的汞排放，这些城镇分布有零散的居民燃煤锅炉和小规模的锌和铅冶炼工厂，是重要的区域大气汞排放源。在"谷风"作用下，这些地区排放的大气汞通过不断地向上爬升，能够对贡嘎山的高海拔地区造成显著的大气汞污染（Fu et al., 2009）。这种区域性污染物的爬升作用也从侧面反映了四川盆地排放的污染物翻越贡嘎山山系向青藏高原高海拔地区扩散的过程。

雷公山位于贵州省的东部，PSCF 模型结果显示（Fu et al.，2015b），雷公山的大气气态总汞污染主要来自于贵阳市及其周边城市和工业中心；此外，云南昆明及其周边的工业和城市区域排放的大气汞能够通过长距离的大气传输影响雷公山大气汞；东南亚和南亚排放的大气汞（如工业排放和生物质燃烧）在西南季风作用下也可以深入我国西南部进而影响到雷公山。

重庆四面山位于重庆市、成都市和贵阳市等城市群中间。大气后向轨迹的聚类分析表明季风控制的大气汞区域/长距离迁移和传输对四面山大气汞分布有较大的影响（刘伟明等，2016）。比如，四面山春季大气气态总汞主要受到贵州省北部和重庆市南部区域大气汞扩散的影响，夏季主要受到我国西北地区和华北地区大气汞长距离传输的影响，秋季主要受到来自于我国西北部甘肃和陕西长距离大气传输的影响，冬季主要受到区域大气汞扩散的影响，主要污染源区域为重庆市和贵阳市及其周边工业和城市中心。综合分析表明，重庆市主城区汞排放极有可能是四面山最主要的污染来源。

香格里拉位于青藏高原东南缘，是西南季风和东南季风进入青藏高原的一个门户。研究结果显示（Zhang et al.，2015a）：香格里拉季风期大气气态总汞浓度明显高于非季风期，表明西南季风作用下南亚和东南亚排汞的大气长距离传输是香格里拉大气气态总汞的一个重要污染源。大气后向轨迹聚类分析发现（Zhang et al.，2015a），起源于我国西北部经过四川西部的大气长距离传输携带的大气气态总汞浓度最高，达到了 3.9 ng/m^3，约是香格里拉全年平均值的 1.5 倍。不过这个传输过程在全年的监测时间段内占的比重较小（4%），因此对香格里拉大气气态总汞污染的影响贡献相对较小。值得指出的是，来自于东南亚的大气云团平均大气气态总汞浓度达到了 2.6 ng/m^3，明显高于来自于青藏高原和南亚大气云团平均汞浓度（2.3~2.4 ng/m^3），且来自于东南亚的大气约占到香格里拉大气来源的一半以上，因此东南亚排放大气气态总汞的长距离传输是影响我国青藏高原东南缘的大气气态总汞污染的重要因素（Zhang et al.，2015a）。PSCF 模型结果显示，香格里拉主要污染源区域集中于缅甸的北部，可能与这些地区春季严重的生物质燃烧排汞有关（Zhang et al.，2015a）。

云南哀牢山大气汞受西南季风大气长距离传输更加显著（Zhang et al.，2016a）。哀牢山季风期大气气态总汞的平均浓度为（2.22±0.58）ng/m^3，约是非季风期平均浓度（1.99 ng/m^3±0.66 ng/m^3）的 1.12 倍。哀牢山季风期大气颗粒态汞的平均浓度为（36.3±30.6）pg/m^3，约是非季风期平均浓度（27.4 pg/m^3±26.1 pg/m^3）的 1.34 倍。大气气态总汞和颗粒态汞的风玫瑰图显示，季风期大气气态总汞、颗粒态汞和活性气态汞的污染片段主要分布于西南风向。通过对典型大气气态总汞污染事件的分析，发现来自于东南亚大气汞的污染事件与这些地区的生物质燃烧密切相关，证明了东南亚生物质燃烧汞排放是我国西南地区大气汞的重要污染来源。通过对东南亚生物质燃烧排放气态汞和 CO 浓度的相关性分析，估算的东南亚生物质燃烧年排汞量为（11.4±2.1）t（Wang et al., 2015）。PSCF 模型结果显示（Zhang et al., 2016a）：季风期来自于缅甸、泰国和老挝北部以及我国云南中东部（昆明城市集群）和贵州西部大气汞的长距离传输作用是哀牢山大气汞的重要污染来源，而非季风期哀牢山大气汞的污染来源主要集中在印度的中部和北部以及缅甸，这些地区是亚洲大气汞排放的重点区域，在青藏高原南支绕流的作用下，存在着向我国西南地区跨国界输送过程。

5. 我国东部沿海背景区大气汞的传输过程

我国的中东部地区是我国乃至全球大气人为源汞排放强度最大的区域之一（Zhang et al., 2015b）。宁波大梅山和山东成山头分别位于浙江省和山东省的沿海背景区，其年均大气气态总汞浓度分别为（3.3±1.4）ng/m^3 和（2.31±0.74）ng/m^3，约是北半球背景值的 1.5~2.2 倍（Ci et al., 2011；Yu et al., 2015）。大梅山的研究显示，区域性的人为源汞排放和长距离的大气传输是影响该地大气汞的两个重要因素。大气后向轨迹聚类分析表明：春季与来自于长江三角洲的区域大气汞扩散有关的大气气态总汞浓度最高，约是海洋来源体的 1.2 倍；夏季来自于东南沿海的气团大气气态总汞浓度最高，约是海洋来源气团 1.3 倍；秋季则以来自于我国华北地区的长距离传输云团气态总汞含量最高，约是海洋来源气团的 1.3 倍；冬季则以华东地区来源的区域性气团气态总汞浓度最高，约是海洋和东北亚来源气团的 1.4 倍（Yu et al., 2015）。通过 PSCF 模型分析，确定了江苏南部、安徽东部和浙江北部为大梅山大气气态总汞的重点来源区域（Yu et al., 2015）。成山头的研究分析了该地典型的汞污染事件和低含量大气汞事件大气后向轨迹。结果发现（Ci et al., 2011）：汞污染事件的大气主要来自于我国华北地区和东部沿海地区，包括内蒙古、山西、河南、河北、辽宁、江苏、安徽和山东等省（自治区）以及北京和天津等大型城市；另一方面，低含量的大气气态总汞主要来自于我国东北和西伯利亚地区。另外，研究结果发现，韩国和日本的人为源汞排放也可以通过长距离的传输影响到成山头（Ci et al., 2011）。

6. 我国华南背景区大气汞的传输过程

华南地区特别是珠江三角洲地区是我国大气汞排放较为集中的一个地区。这些地区主要受到东南季风和冬季风的影响，人为源排放的不均一性和季风的双重作用对华南背景区的大气汞分布具有重要影响。广东鼎湖山和南岭的大气气态总汞浓度分别为（5.07±2.89）ng/m^3 和（2.53±0.93）ng/m^3，约是北半球背景值的

3.4 倍和 1.7 倍；海南省五指山的大气气态总汞平均浓度为（1.58±0.71）ng/m³，和北半球背景值基本相当（Chen et al.，2013；Liu et al.，2016；高志强等，2016）。鼎湖山大气气态总汞浓度与 SO_2 和 NO_2 等污染物浓度具有显著的正相关关系，表明珠江三角洲核心区域（广州市及周边城市群）排放大气汞的区域性传输是影响鼎湖山大气汞的最重要因素（Chen et al.，2013）。南岭位于广东省北部，PSCF 模型结果表明：春季大气气态总汞的污染主要来自于珠江三角洲地区，夏季的污染主要来自于湖南和江西南部，秋季的污染主要来自于湖南南部、广西东部和广东西部，冬季的污染主要来自于湖南南部和广西北部，上述地区的排放源主要包括有色金属冶炼和燃煤排放（高志强等，2016）。五指山位于海南省南部。通过对典型大气气态总汞污染事件的分析发现，五指山多数（73%）的大气气态总汞污染事件与来自于我国中南部的大气长距离传输有关，其余的污染事件则主要来自于中南半岛（Liu et al.，2016）。CWT 模型结果显示：五指山夏季大气汞的污染来源主要来自于越南的北部，而冬季的污染来源主要来自于珠江三角洲和华南南部的长距离大气传输（Liu et al.，2016）。

7. 我国大气汞的质量平衡

在建立我国人为源和自然源汞排放清单的基础上，王训（2016）采用 CMAQ-Hg 模型构建了我国大气汞的区域循环模型。模型结果表明，我国区域内边界层大气气态总汞的平均浓度为 2.9 ng/m³，与我国背景区大气汞监测数据的平均值（2.74 ng/m³±1.05 ng/m³）较为一致（Fu et al.，2015b）；我国边界层大气活性气态汞和颗粒态汞的平均浓度分别为 18 pg/m³ 和 113 pg/m³，略高于我国背景区平均监测浓度，但低于城市地区的平均浓度。2013 年我国大气汞的年排放量约为 1109 t，其中自然源排汞 565 t，人为源排汞 544 t。我国大气汞的年均沉降量为 586 t，其中湿沉降约占 42%，干沉降约占 58%。由此，我国大气汞的净输出量为每年 426 t，其中有 97 t 滞留在我国区域上空的边界层和自由对流层。我国大气汞的输出主要是由我国较高的自然源汞排放造成的，其中我国大气汞净输出的 61% 来自于自然源排放，自然源以排放具有较长大气居留时间的大气气态单质汞为主（王训，2016）。此外，我国大气汞的外源输入为 194 t/a，其中大部分来自于南亚、东南亚和中东地区。该模型结果表明，以前的研究高估了我国大气汞的输出量。另外，我国大气汞的输出可能主要（30%～60%）累积在东亚和西太平洋地区，向全球其他地区的输送量较少。考虑到全球大气中汞的总量为 5000～6000 t，中国的排放可能会导致全球大气汞沉降通量平均增加 10%～15%（王训，2016），这一结果也低于以前发表的数据（Lin et al.，2010）。

GEOS-Chem 模型结果显示，我国人为源汞排放对我国大气汞浓度、干沉降通量和湿沉降通量的贡献量分别为21%、37.5%和36.5%,大气汞的全球/跨境传输和我国自然源汞排放仍然是我国大气汞的最主要来源。我国人为源汞排放对长江三角洲、华中地区和华北地区大气汞的影响最大,其中对大气汞浓度的贡献接近50%,对干湿沉降通量的贡献在60%~70%之间;而一些偏远地区如西北、西南、东北等地,我国人为源汞排放的贡献通常小于20%(Wang et al., 2014b)。燃煤电厂、非金属冶炼和水泥生产是我国3个最大的人为排放源,其对我国大气汞浓度、干沉降通量和湿沉降通量的贡献分别为6.2%、9.7%和7.3%(Wang et al., 2014b)。

我国是全球最大的人为源排放国,国际社会对我国人为源排汞的输出非常关注。Chen 等(2015)开发了第一代 GNAQPMS-Hg 模型(global nested air quality prediction modeling system for Hg),系统研究了中国人为源汞排放对全球大气汞浓度和沉降通量的影响,认为我国人为源排汞对我国大气汞浓度和大气汞沉降量的贡献分别为30%和62%,略高于Wang 等(2014b)的模型结果。受自西向东大气环流的影响,我国人为源排汞沿中国自西向东、自北向南的贡献逐渐降低,朝鲜半岛受我国的影响最大,其大气汞浓度和大气汞沉降量的 11%和 15.2%分别来自于我国人为源汞排放;其次是东南亚,有 10.4%和 8.2%的大气汞浓度和大气汞沉降量来自于我国人为源汞排放;然后是日本,其大气汞浓度和大气汞沉降量的 5.7%和 5.9%分别来自于我国人为源汞排放;其他洲如北美(沉降量的 3.5%来自中国)、欧洲(沉降量的 3.0%来自中国)、南美和大洋洲受我国人为源汞排放的影响较低,普遍低于5%(Chen et al., 2015)。

5.4.2 亚洲其他地区大气汞传输过程和特征

除我国外,亚洲其他国家和地区也是全球大气汞排放的重点区域。全球人为源清单表明,亚洲地区的人为源汞排放占全球一半以上(Pacyna et al., 2010; Pirrone et al., 2010)。根据飞机航测结果,东亚(包括我国东部海域、日本海、西太平洋、韩国和日本等地区)边界层和自由对流层的大气气态总汞浓度变化非常显著,其中靠近中国的海洋地区、日本和韩国本土及其近海海洋上空的大气气态总汞浓度普遍偏高,而远离东亚大陆的远海地区大气汞普遍较低,表明亚洲人为源直接排放和再释放对东亚地区大气汞的分布具有重要影响(Friedli et al., 2004; Pan et al., 2006)。通过对大气气态总汞和二氧化硫、一氧化碳、颗粒物钙和钾含量的相关性分析,发现日本的火山喷发、中国北方和蒙古国沙尘、东亚地区的生物质燃烧及中国沿海地区工业排放是东亚大气汞的主要来源(Friedli et al., 2004)。对上海、青岛和北京等城市的工业中心排放烟气中气态总汞和一氧化碳比值分析结果表明,我国工业烟气的大气气态总汞和一氧化碳比值为 6.4×10^{-6} w/w(质量比),非常接近于我国工业排汞和一氧化碳的比值(7.8×10^{-6} w/w),据此推算我国大气汞的输出量为 600 t/a（Friedli et al.，2004）。综合飞机航测和 STEM 区域大气汞循环模型结果，Pan 等（2006）对典型的污染事件进行了源解析，发现日本的汞排放对日本及其周边区域大气汞的贡献最大，在高于全球大气背景值的地区，其大气汞有 80%以上来自于日本的大气汞排放。

日本冲绳岛（Cape Hedo）观测站位于中国东海和西太平洋之间，距我国大陆的最近距离为 600 km，距日本本土的最近距离约为 500 km。冲绳岛观测站的大气气态单质汞、活性气态汞和颗粒态汞浓度分别为（2.04±0.38）ng/m^3、（4.5±5.4）pg/m^3 和（3.0±2.5）pg/m^3，其中气态单质汞浓度约是北半球背景值的 1.4 倍，而活性气态汞和颗粒态汞则与北半球背景值大体相当（Chand et al.，2008）。大气后向轨迹分析的结果发现，来自我国的大气气态总汞浓度高于来自日本南部的大气气态总汞浓度，表明我国的大气汞传输是冲绳岛大气汞重要来源（Chand et al.，2008）。

韩国济州岛（Jeju Island）位于中国大陆以东约 500 km，韩国南部约 100 km。济州岛大气气态总汞年均浓度为（3.85±0.38）ng/m^3，约是北半球背景值的 2.5 倍，来自中国的气团大气气态总汞平均浓度为 3.35～4.73 ng/m^3、来自于朝鲜半岛的气团大气气态总汞平均浓度为 3.77～4.43 ng/m^3，来自于日本的气团大气气态总汞平均浓度为 5.27 ng/m^3，而来自于日本海和西太平洋的气团大气气态总汞浓度为 3.39～3.62 ng/m^3（Nguyen et al.，2010）。由于来自中国的大气气团占到了济州岛大气来源的 51%，该研究认为我国的大气汞长距离传输是济州岛大气汞的重要来源。需要指出的是，尽管济州岛远离中国、朝鲜半岛和日本，其大气气态总汞平均浓度要明显高于中国沿海城市、韩国和日本城市地区的大气气态总汞监测结果（Kock et al.，2005；Kim et al.，2009；Fu et al.，2015b），因此，该研究把济州岛大气汞的污染归咎于来自于我国或者日本的长距离传输作用缺乏可靠依据。由于受到大气后向轨迹"拖尾效应"的影响，来自于中国和日本的气团在到达济州岛观测站之前极有可能混合了岛内人为源排放的大气汞，从而高估了来自于中国和日本长距离大气汞传输对济州岛的影响。

韩国首尔（Seoul）是亚洲乃至全球最大的城市之一，其大气气态单质汞、活性气态汞和颗粒态汞年均浓度分别为（3.22±2.10）ng/m^3、（27.2±19.2）pg/m^3 和（23.9±19.6）pg/m^3（Kim et al.，2009），均明显低于我国城市地区（Fu et al.，2015b）。通过对首尔 34 个典型大气气态单质汞污染片段的分析发现，我国大气汞长距离传输的贡献占所有污染片段的 73%；日本大气汞长距离传输约占 12%；而来自于东北亚和中国东海的污染事件贡献较少（Kim et al.，2009）。需要说明的是，由于大气后向轨迹分析无法区分本地源和长距离大气传输的相对贡献，特别是对于具有较大人为源汞排放的城市地区，后向轨迹分析揭示的大气长距离传输混合了本地源的汞排放和通过长距离输送的大气汞，因此采用后向轨迹对城市污染来源的解析存在较大误差（Fu et al., 2011）。

作为全球大气汞排放最多的地区，亚洲地区大气汞和沉降量主要受到本地排放的影响。研究表明，亚洲人为源和自然源汞排放对亚洲大气汞沉降量的贡献达到了68%，其余的部分主要来自于欧洲（10%）、海洋水体（10%）、南半球（6%）、北美（4%）和非洲（2%）排汞的长距离大气传输（Travnikov，2005）。亚洲不同地区受本地大气汞排放和长距离大气传输的影响程度有所差异，东亚人为源汞排放对东亚地区大气气态单质汞、活性气态汞和颗粒态汞浓度的贡献分别为41%、64%和71%；南亚人为源汞排放对南亚地区大气气态单质汞、活性气态汞和颗粒态汞浓度的贡献分别为17%、53%和48%；而东南亚和中亚则主要受其他地区大气汞长距离传输的影响，本地人为源对大气气态单质汞、活性气态汞和颗粒态汞的贡献分别为5%~9%、13%~25%和15%~18%（Chen et al., 2014）。东亚人为源汞排放对本地大气汞沉降（包括干沉降和湿沉降）的贡献为50%；南亚人为源汞排放对本地大气汞沉降的贡献为27%，东南亚和中亚人为源汞排放对本地大气汞沉降的贡献分别为10.6%和4.3%（Chen et al., 2014）。

5.4.3 北美大气汞传输过程和特征

1. 北美地区大气汞的污染源解析

北美地区大气汞源解析研究多数采用的是主成分分析法（Lynam and Keeler, 2006; Liu et al., 2007; Temme et al., 2007; Cheng et al., 2009; Cheng et al., 2012），这些研究通常将燃煤排放、工业排放、机动车尾气、生物质燃烧和垃圾焚烧识别为几个较为典型的污染源。此外，也有采用PMF模型研究了北美城市地区和背景区大气汞的重点污染源（Liu et al., 2003; Cheng et al., 2009; Wang et al., 2013）。对加拿大多伦多的研究表明，污水处理是多伦多大气气态单质汞的最主要污染源，对GEM浓度的贡献高达84%，金属冶炼是多伦多大气气态单质汞的第二大污染源（11.6%），其余的部分主要来自于大气的物理化学过程和城市外源输入。对于大气颗粒态汞而言，金属冶炼排放则是多伦多最重要的污染源，贡献达到了78.9%，其余的部分主要来自于污水处理（20.3%）和大气物理化学转化过程（0.8%）。多伦多大气活性气态汞的来源主要为金属冶炼活动，贡献达到56.5%，其余部分主要来自于污水处理（24.1%）、大气物理化学转化过程（10.9%）和城市外源的输入（8.5%）（Cheng et al., 2009）。针对美国罗切斯特的研究确定了6种主要的大气汞污染源，其中与臭氧污染有关的污染源、大气光化学氧化过程和燃煤电厂分别是大气气态单质汞、活性气态汞和颗粒态汞的最主要来源，贡献分别为50%、85%和48%，其他的污染来源还包括生物质燃烧、机动车排放和核电前处理工艺（Wang et al.,

2013)。

2. 北美地区大气汞的长距离传输

阿迪朗达克山（Adirondacks）森林背景区位于美国纽约州的北部，其大气气态单质汞、活性气态汞和颗粒态汞年均浓度分别为（1.4±0.4）ng/m^3、（1.8±2.2）pg/m^3 和（3.2±3.7）pg/m^3（Choi et al.，2008）。PSCF 模型结果显示，阿迪朗达克山森林背景区大气气态单质汞主要来自采样点以西和以南区域人为源排汞的长距离大气传输，其中美国五大湖南部至大西洋的区域（包括宾夕法尼亚州、弗吉尼亚西部、俄亥俄州、肯塔基州和得克萨斯州）是该地区大气气态单质汞重点污染来源区域；阿迪朗达克山森林背景区活动气态汞主要来自于弗吉尼亚州西部、俄亥俄州、印第安纳州和密西里州；该地区大气颗粒态汞的污染来源主要分布在俄亥俄州、弗吉尼亚州西部、印第安纳州和缅因州。这些地区广泛分布有大型的燃煤电厂、生物质燃烧、炼油厂和垃圾焚烧厂，是美国人为源大气汞排放最多的地区（Choi et al.，2008）。

Salmon Falls Creek 水库（SFCR）位于美国爱达荷州中南部，其大气气态单质汞和活性气态汞年均浓度分别为（1.57±0.6）ng/m^3 和（6.8±12）pg/m^3（Abbott et al.，2008）。SFCR 的大气气态单质汞季节性变化非常显著，其中夏季大气气态单质汞的平均浓度为（1.91±0.9）ng/m^3，比冬季的平均大气气态单质汞浓度约偏高1.45 倍，也明显高于北半球背景值。PSCF 模型结果发现区域性大气汞传输是影响SFCR 大气汞的重要因素，其中位于采样点以南内华达州北部和犹他州西部是该地区大气气态单质汞的最主要污染来源区域，这些地区的大气排放源主要为金矿、水泥厂和燃煤电厂。

温莎市（Windsor）是加拿大安大略省西南部的城市，周边（五大湖流域）是北美人为源排汞最大的区域（Pacyna et al.，2010）。温莎市大气气态总汞的年均浓度为（2.02±1.63）ng/m^3，约是北半球背景浓度的 1.34 倍，其中夏季大气气态总汞浓度最高，平均值为（2.48±2.68）ng/m^3，表明区域性大气污染物的传输作用是影响温莎市大气汞的首要因素（Xu and Akhtar，2010）。PSCF 模型结果显示，美国中东部地区排放的大气汞长距离传输作用是影响温莎市大气汞污染重要因素，受北美季风气候的影响，这种长距离传输作用在夏季和春季特别显著，是导致温莎市夏季大气汞浓度升高的首要因素（Xu and Akhtar，2010）。

达特茅斯（Dartmouth）位于加拿大新斯科舍州的东南部海岸，其大气气态单质汞、活性气态汞和颗粒态汞年均浓度分别为（1.67±1.01）ng/m^3、（2.1±3.4）pg/m^3 和（2.3±3.1）pg/m^3（Cheng et al.，2013a），其中大气气态单质汞浓度略高于北半球背景值，而活性气态汞和颗粒态汞则和北半球背景值相当。CWT 模型结果显示，达特茅斯大气气态单质汞的大部分（85%~97%）污染来源区域没有明显的工业排放源，说明面源排放、自然源排放和家庭生活排放是这些地区重要的排放源；而活性气态汞和颗粒态汞的污染来源主要为魁北克和安大略地区的金属冶炼厂（Cheng et al.，2013a）。

墨西哥墨西哥城（Mexico City）是美洲人口最多的都市区，其大气气态单质汞、活性气态汞和颗粒态汞年均浓度分别为（7.2±4.8）ng/m³、（62±64）pg/m³和（187±300）pg/m³（Rutter et al.，2009），显著高于美国城市地区的监测结果，略低于中国贵阳市的大气汞浓度（Valente et al.，2007；Fu et al.，2011）。研究表明，墨西哥城本地以及周边区域的人为源汞（包括水泥厂、燃油厂、造纸厂、化工厂和啤酒厂）排放是造成其大气汞污染的首要因素，其中人为源汞排放对大气活性气态汞和颗粒态汞的贡献约为 93%±3%，对大气气态单质汞的平均贡献为 72%±1%（Rutter et al.，2009）。

3. 自由对流层和平流层入侵对北美大气活性气态汞的影响

和大气气态单质汞主要来自于大陆边界层的人为源直接和再释放不同，大气中的活性气态汞主要来自于大气汞的氧化过程，而这一过程通常发生在自由对流层和平流层等海拔较高的区域（Tan et al.，2000；Swartzendruber et al.，2006；Swartzendruber et al.，2009；Holmes et al.，2010；Lyman and Jaffe，2012）。因此，在大气活性气态汞的来源解析中，自由对流层和平流层大气的长距离传输通常是一个不可忽视的来源。

近年来，研究发现高海拔地区通常具有较高的大气活性气态汞浓度。美国学士山观测站（Mount Bachelor Observatory，海拔 2700 m）、风暴峰天文台（Storm Peak Laboratory，海拔 3200 m）、里诺市（Reno，海拔 1497 m）的大气活性气态汞平均浓度分别为 43 pg/m³、（20±21）pg/m³ 和（18±22）pg/m³，约是美国低海拔背景区活性气态汞浓度的 10~20 倍（Swartzendruber et al.，2006；Fain et al.，2009；Lyman and Gustin，2009）。通过对大气汞形态、一氧化碳和臭氧浓度变化的综合分析，提出了美国西海岸学士山观测站大气活性气态汞的 3 种来源模式（Timonen et al.，2013）：第一种为自由对流层顶部和平流层富活性气态汞大气云团的入侵，由高纬度地区向中纬度地区入侵；第二种为亚洲排放的大气气态单质汞在自由对流层长距离传输过程中被大气氧化物（例如：卤族氧化物和自由基、羟基自由基、臭氧和硝酸根自由基等）氧化为活性气态汞，并最终输送到美国西部的高海拔地区；第三种为太平洋边界层的大气气态单质汞在温暖干燥的环境条件下被大洋卤族氧化物（例如：Br·、Cl·、BrCl、BrO·、ClO· 等）氧化为活性气态汞，并通过向上的爬升作用影响美国西部的高海拔地区。对于落基山脉以东的美国中部地区，自由对流层和平流层富活性气态汞气体的入侵则是主要的活性气态汞来源（Fain et al., 2009; Lyman and Gustin, 2009; Weiss-Penzias et al., 2015）。与美国西海岸相比，美国中部地区还受到太平洋平流层活性气态汞入侵的影响，如飞机航测对得克萨斯州自由对流层（海拔7000 m）活性气态汞的监测发现其活性气态汞平均浓度达到了（266±38）pg/m^3，约是背景区边界层的50~100倍（Gratz et al., 2015）。研究发现，赤道辐合带空气的强烈对流能导致海洋排放的卤族元素传输到平流层，进而发生大规模的大气气态单质汞氧化现象，形成的活性气态汞进而通过长距离传输到美国中部（Gratz et al., 2015）。

4. 跨境传输对北美大气汞分布的影响

北美地区的大气汞浓度通常比北半球其他地区偏低（Lan et al., 2012; Sprovieri et al., 2016），这主要是因为北美地区的人为源和自然源排汞量相对较低（Pacyna et al., 2010; Agnan et al., 2016），因此跨境大气汞传输对北美地区大气汞的影响要普遍高于亚洲和欧洲。Seigneur等（2004）采用CTM-Hg全球汞循环模型分析了全球其他地区汞排放对美国本土大气汞沉降的影响，结果表明北美地区的人为源排放对美国大气汞湿沉降的贡献约为24%，要低于来自亚洲长距离大气汞传输的影响（约为26%），高于欧洲（8%）、南美洲（4%）和非洲（5%）大气汞传输的贡献，说明亚洲大气汞的长距离传输是影响北美大气汞的重要因素。不过，加上北美地区的自然源汞排放，北美本地区汞排放成为北美大气汞沉降的最主要来源，贡献达到了33%，其余的部分主要来自于亚洲（24%）、海洋（18%）、欧洲（14%）、非洲（3%）以及南美洲和大洋洲（二者合计8%）（Travnikov, 2005）。

北美地区大气气态单质汞主要来自于全球自然源的汞排放（68%），其次是全球大气汞背景库（27%），而北美本土的人为源汞排放的贡献仅占5%。北美本土人为源汞排放对其大气活性气态汞和颗粒态汞的贡献高于气态单质汞，分别为20%和41%（Chen et al., 2014）。亚洲大气汞的长距离传输是北美洲大气汞的重要来源，占北美大气气态单质汞的15%以上；其次是欧洲，对北美大气气态单质汞的贡献小于5%。北美大气气态单质汞从西部到东部受亚洲长距离传输的影响程度逐渐降低，美国西海岸落基山脉区域的大气气态单质汞的18%和活性气态汞的18%来自于亚洲地区的人为源汞排放，最高可达41%，而在美国中东部地区这一比值通常会下降到10%以下（Strode et al., 2008）。

5.4.4 欧洲大气汞传输过程和特征

从20世纪的80年代开始，欧洲已经开始大气汞的长期研究（Iverfeldt, 1991; Slemr and Scheel, 1998; Ebinghaus et al., 2002a; Slemr et al., 2003; Slemr et al.,

2011)。受益于对人为源排汞的有效控制,欧洲地区人为源汞排放量已经从 80 年代的 860 t 下降到 2000 年左右的 233 t,下降幅度达到了 73%(Pacyna et al.,2006a),而与此同时全球人为源大气汞排放呈现增加的趋势(约升高了 40%)(Streets et al.,2011)。与之对应,欧洲地区的大气气态总汞浓度从 80 年代的 3.0 ng/m^3 下降到 2000 年左右的 1.5 ng/m^3,约下降了 50%(Slemr et al.,2003)。由此可见,欧洲特别是 20 世纪的大气汞主要受欧洲地区人为源汞排放的影响,但全球其他地区的大气汞排放对欧洲地区大气汞贡献不容忽视。

  Mace Head 全球大气基准站位于爱尔兰岛西部。受到北半球西风带的影响,Mace Head 全球大气基准站大气气态总汞主要受大西洋北部海洋大气长距离传输的影响(Ebinghaus et al.,2011),而这一地区海洋水体的汞排放通量普遍高于全球其他海洋(Soerensen et al.,2010),表明海洋水体排汞是欧洲大气气态总汞的一个重要来源。另一方面,北美地区也是 Mace Head 全球大气基准站大气气态总汞的一个重要来源(Ebinghaus et al.,2011)。

  Pic du Midi 天文台位于法国南部的比利牛斯山山顶,其大气气态单质汞、活性气态汞和颗粒态汞年均浓度分别为(1.86±0.27)ng/m^3、(27±34)pg/m^3 和(14±10)pg/m^3(Fu et al.,2016c),均明显高于全球背景区大气汞的平均浓度。Pic du Midi 天文台大气气态单质汞主要受到法国南部低海拔地区汞排放的影响,这些大气汞通过上升气流影响比利牛斯山的高海拔地区。另一方面,Pic du Midi 天文台大气颗粒态汞和活性气态汞主要受到长距离大气传输的影响,其中颗粒态汞主要来自于北美和北大西洋地区的自由对流层;而活性气态汞则主要来自于低纬度大西洋的自由对流层下部和高纬度地区的自由对流层上部地区,这些地区较高的大气氧化物浓度有利于大气气态单质汞发生氧化(Fu et al.,2016c)。

  全球大气汞循环模型结果显示:欧洲地区大气气态单质汞主要受到全球大气汞传输机制的影响,欧洲本土的人为源汞排放仅贡献其 11%,其余绝大部分来自于全球其他地区的人为源和自然源汞排放(Chen et al.,2014)。相反,欧洲的大气活性气态汞和颗粒态汞主要来自于欧洲的人为源汞排放,贡献份额分别为 52%和 66%,这与亚洲地区相似,但明显高于北美本土人为源对大气活性气态汞和颗粒态汞的贡献份额(Chen et al.,2014)。欧洲大气汞沉降主要来自于本土的人为源和自然源汞排放,占到了 61%(绝大部分来自人为源排放),其次是来自于亚洲的大气汞长距离传输(15%),其他的跨境传输来自于海洋汞排放(12%)、北美(3%)、南半球(4%)和非洲(3%)(Travnikov,2005)。不过,近期的一项模型研究认为跨境传输也是欧洲地区大气汞沉降的最主要来源,约占其沉降总量的 80%以上(Chen et al.,2014)。模型结果之间的差异主要来自于对自然源汞排放、大气汞形态沉降速率以及大气汞转化机制认识的不足。

5.4.5 其他地区大气汞长距离传输过程和特征

非洲、大洋洲、南美洲特别是南北极地区是世界上人为源排汞较低的地区（Pacyna et al.，2010），因此这些地区大气汞通常主要受到全球大气汞传输的影响。模型结果显示，南美洲、非洲和大洋洲本地人为源汞排放分别只占到了其大气气态单质汞来源的7%、3%和2%；而这些地区人为源汞排放占其大气活性气态汞和颗粒态汞以及大气汞沉降量的比例通常小于10%，远低于亚洲、欧洲和北美地区人为源对大气活性气态汞和颗粒态汞的贡献（Chen et al.，2014）。对于南北极地区，其大气汞和汞沉降的来源几乎全部来自于世界上其他地区的长距离输送作用（Travnikov，2005）。

开普敦岛（Cape Point）位于非洲的西南部，其大气气态单质汞的多年平均值约为（0.94 ± 0.16）ng/m^3，远远低于北半球的背景值（Brunke et al.，2010）。研究表明，非洲南部的人为源和生物质燃烧汞排放是开普敦岛大气气态单质汞的重要来源，而其余的大气气态单质汞主要来自于南大西洋（Brunke et al.，2010）。北领地岛（Gunn Point peninsula）位于澳大利亚的北部，其大气气态单质汞年平均值为（0.95 ± 0.12）ng/m^3，和南非开普敦岛相当但低于北半球背景值（Howard et al.，2017）。研究发现，南半球雨季期间北领地岛高含量的大气气态单质汞主要来自于印度尼西亚的西部（如雅加达和万隆城市群），而其干季的大气气态单质汞污染事件主要来自于印度尼西亚的中部和北部以及菲律宾的南部（Howard et al.，2017）。新尼克里是苏里南的沿海城市，位于南美洲东北部，毗邻大西洋，其大气气态单质汞年均浓度为1.40 ng/m^3，接近于北半球背景值（Muller et al.，2012）。后向大气轨迹分析发现，与来自于北半球大气传输有关的平均大气气态单质汞浓度（1.45 ng/m^3）约比来自于南半球的大气（汞浓度：1.32 ng/m^3）高10%；另外，来自于南美洲大陆气团的平均气态单质汞浓度（1.43 ng/m^3）也明显高于来自西大西洋的气团，新尼克里个别大气气态单质汞污染事件与巴西中部的森林火灾汞排放有关（Muller et al.，2012）。

北极是北半球大气环境最为洁净的地区，北极圈内极少有人为排汞源，另外受长期冰雪覆盖的影响，北极圈海洋水体排汞量也明显偏低（Pacyna et al.，2010；Soerensen et al.，2010）。不过与南半球以及南极地区相比，北极地区大气气态单质汞的浓度是普遍偏高的（Steffen et al.，2008；Dommergue et al.，2010）。尽管有研究指出北极地区夏季融雪期水体排汞能导致大气气态单质汞浓度的升高（Sommar et al.，2010），但总的来说，北极地区大气气态单质汞主要来源为大气的长距离传输。Durnford等（2010）采用GRAHM模型分析了北极圈内6个台站大气气态单质汞观测数据，发现北极地区绝大部分的大气气态单质汞污染与来自于北半球大陆地区的汞排放有关。其中来自于亚洲地区的大气长距离传输对北极大气气态单质汞的污染贡献达到43%，其余的污染来自于俄罗斯（27%）、北美（16%）和欧洲（14%）。和气态单质汞不同，北极地区大气活性气态汞和颗粒态汞主要来自于北极边界层大气气态单质汞氧化过程，主要出现于春季（Steffen et al., 2008）。和大气气态单质汞来源近似，北极地区大气汞沉降主要来自于亚洲（33%）、欧洲（22%）和北美洲（10%）（Travnikov，2005）。北极地区大气汞的沉降主要来自于活性气态汞和颗粒态汞的干沉降，每年的沉降量在300 t左右，占大气气态单质汞干湿沉降总量的80%以上，这些活性气态汞和颗粒态汞主要由长距离输送的大气气态单质汞在北极氧化所产生（Skov et al., 2004）。近年来，随着对植物吸收大气气态单质汞认识的不断深入，大气气态单质汞的沉降被认为是北极苔原地区最主要的污染源，占北极土壤汞来源的55%~73%（Obrist et al., 2017）。

5.5 展　　望

近20年来，研究人员对全球大气汞的来源、分布、迁移和转化开展了大量的基础性研究工作，取得了一些重要的进展。然而，随着认识的不断深入，一些制约全球大气汞循环研究的薄弱环节逐渐凸显，急需开展深入的研究工作。

（1）大气汞形态是研究大气汞迁移转化重要环节。目前大气气态单质汞分析测试方法较为成熟，然而有关活性气态汞和颗粒态汞的采样和分析方法还存在较多的争议。尽管近些年来对活性汞和颗粒态汞相关研究方法开展了大量的研究，也发现了不少的问题，但由于活性汞和颗粒态汞含量极低而且性质较为活跃，目前提出的研究方法并未得到普遍认可。因此，今后在大气汞特别是活性汞和颗粒态汞的方法学研究领域急需一些突破性的研究工作。

（2）大气汞形态转化是开展全球汞循环研究的一个关键环节。近年来野外监测工作揭示的大气汞物理化学转化现象得到了实验室相关工作的证实，显著提高了全球汞循环模型研究的精度。然而，目前对大气汞物理化学转化的具体过程还存在认识不足的环节，转化速率测定结果也存在很大不确定性，从而导致模型模拟结果与实际观测值仍存在一定偏差。另外，目前的研究重心多偏向于大气汞的氧化行为，而在其他方面如颗粒态汞还原、雨雾汞还原、气相-液相汞转化、汞沉降速率等方面还缺乏系统性的研究工作。

（3）《水俣公约》已经生效并开始正式实施，其要求对环境中汞及其化合物的含量变化趋势进行长期监测，从而对公约的成效进行有效评估。环境其他介质（如海洋、水生生物体、陆地农田和森林生态系统）汞污染与大气汞含量的变化息息相关。目前欧美在区域大气汞含量的长期变化领域开展了较为系统性的长期研究工作，然而作为全球大气汞排放最多的区域，亚洲在大气汞的变化趋势方面的研究基本上是一个空白，为验证这一地区履行《水俣公约》成效造成了不小的难度。因此，亚洲地区需继续开展大气汞长期变化趋势的研究工作，并在此基础上开展大气汞含量变化与其他环境介质汞污染变化之间的响应关系研究，为降低区域和全球环境汞污染的风险提供理论支撑。

参 考 文 献

付学吾, 冯新斌, 2015. 贵阳市2001/2002和2009/2010两个年度大气气态总汞浓度变化特征及其对区域大气汞排放强度的指示意义. 矿物岩石地球化学通报, 34(2): 242-249.

付学吾, 冯新斌, 张辉, 2011. 贵阳市大气气态总汞:Lumex RA-915AM 与 Tekran 2537A 的对比观测. 生态学杂志, 30(5): 939-943.

高志强, 刘明, 陈来国, 孙家仁, 陈多宏, 黄向峰, 欧劼, 李杰, 许振成, 2016. 广东南岭大气背景点气态元素汞含量变化特征. 中国环境科学, 36(2): 342-348.

李舒, 高伟, 王书肖, 张磊, 李智坚, 王龙, 郝吉明, 2016. 上海崇明地区大气分形态汞污染特征. 环境科学, 37(9): 3290-3299.

李政, 2012. 城市和边远地区大气汞浓度、形态及影响因素. 合肥: 中国科学技术大学.

刘娜, 仇广乐, 冯新斌, 2010. 大气汞源解析受体模型研究进展. 生态学杂志, (4): 798-804.

刘伟明, 马明, 王定勇, 孙涛, 魏世强, 2016. 中亚热带背景区重庆四面山大气气态总汞含量变化特征. 环境科学, 37(5): 1639-1645.

王训, 2016. 中国自然源汞排放清单的建立与大气汞质量平衡的评估. 北京:中国科学院大学.

Abbott M L, Lin C J, Martian P, Einerson J J, 2008. Atmospheric mercury near salmon falls creek reservoir in southern Idaho. Applied Geochemistry, 23: 438-453.

Agnan Y, Le Dantec T, Moore C W, Edwards G C, Obrist D, 2016. New constraints on terrestrial surface atmosphere fluxes of gaseous elemental mercury using a global database. Environmental Science & Technology, 50: 507-524.

AMAP/UNEP, 2013. Technical Background Report for the Global Mercury Assessment 2013. Switzerland: UNEP Division of Technology.

Ambrose J L, Lyman S N, Huang J Y, Gustin M S, Jaffe D A, 2013. Fast time resolution oxidized mercury measurements during the Reno Atmospheric Mercury Intercomparison Experiment (RAMIX). Environmental Science & Technology, 47: 7285-7294.

Amos H M, Jacob D J, Holmes C D, Fisher J A, Wang Q, Yantosca R M, Corbitt E S, Galarneau E, Rutter A P, Gustin M S, Steffen A, Schauer J J, Graydon J A, St Louis V L, Talbot R W, Edgerton E S, Zhang Y, Sunderland E M, 2012. Gas-particle partitioning of atmospheric Hg(II) and its effect on global mercury deposition. Atmospheric Chemistry and Physics, 12: 591-603.

Amouroux D, Wasserman J C, Tessier E, Donard O F X, 1999. Elemental mercury in the atmosphere of a tropical Amazonian forest (French Guiana). Environmental Science & Technology, 33: 3044-3048.

Ariya P A, Amyot M, Dastoor A, Deeds D, Feinberg A, Kos G, Poulain A, Ryjkov A, Semeniuk K, Subir M, Toyota K, 2015. Mercury physicochemical and biogeochemical transformation in the atmosphere and at atmospheric interfaces: A review and future directions. Chemical Reviews, 115:

3760-3802.

Banic C M, Beauchamp S T, Tordon R J, Schroeder W H, Steffen A, Anlauf K A, Wong H K T, 2003. Vertical distribution of gaseous elemental mercury in Canada. Journal of Geophysical Research-Atmospheres, 108(D9): 4264, doi: 10.1029/2002JD002116.

Bash J O, Miller D R, Meyer T H, Bresnahan P A, 2004. Northeast United States and Southeast Canada natural mercury emissions estimated with a surface emission model. Atmospheric Environment, 38: 5683-5692.

Berg T, Sekkesaeter S, Steinnes E, Valdal A K, Wibetoe G, 2003. Springtime depletion of mercury in the European Arctic as observed at Svalbard. Science of the Total Environment, 304: 43-51.

Bloom N, Fitzgerald W F, 1988. Determination of volatile mercury species at the picogram level by low-temperature gas-chromatography with cold-vapor atomic fluorescence detection. Analytica Chimica Acta, 208: 151-161.

Bloom N S, Prestbo E M, Hall B, Vondergeest E J, 1995. Determination of atmospheric Hg by collection on iodated carbon, acid digestion and CVAFS detection. Water Air and Soil Pollution, 80: 1315-1318.

Brooks S, Moore C, Lew D, Lefer B, Huey G, Tanner D, 2011. Temperature and sunlight controls of mercury oxidation and deposition atop the Greenland ice sheet. Atmospheric Chemistry and Physics, 11: 8295-8306.

Brosset C, 1987. The behavior of mercury in the physical-environment. Water Air and Soil Pollution, 34: 145-166.

Brosset C, Lord E, 1995. Methylmercury in ambient air-method of determination and some measurement results. Water Air and Soil Pollution, 82: 739-750.

Brunke E G, Ebinghaus R, Kock H H, Labuschagne C, Slemr F, 2012. Emissions of mercury in southern Africa derived from long-term observations at Cape Point, South Africa. Atmospheric Chemistry and Physics, 12: 7465-7474.

Brunke E G, Labuschagne C, Ebinghaus R, Kock H H, Slemr F, 2010. Gaseous elemental mercury depletion events observed at Cape Point during 2007—2008. Atmospheric Chemistry and Physics, 10: 1121-1131.

Bullock O R, 2000. Modeling assessment of transport and deposition patterns of anthropogenic mercury air emissions in the United States and Canada. Science of the Total Environment, 259: 145-157.

Bullock O R, Brehme K A, 2002. Atmospheric mercury simulation using the CMAQ model: Formulation description and analysis of wet deposition results. Atmospheric Environment, 36: 2135-2146.

Chand D, Jaffe D, Prestbo E, Swartzendruber P C, Hafner W, Weiss-Penzias P, Kato S, Takami A, Hatakeyama S, Kajii Y Z, 2008. Reactive and particulate mercury in the Asian marine boundary layer. Atmospheric Environment, 42: 7988-7996.

Chen H S, Wang Z F, Li J, Tang X, Ge B Z, Wu X L, Wild O, Carmichael G R, 2015. GNAQPMS-Hg v1.0, a global nested atmospheric mercury transport model: Model description, evaluation and application to trans-boundary transport of Chinese anthropogenic emissions. Geoscientific Model Development, 8: 2857-2876.

Chen L, Wang H H, Liu J F, Tong Y D, Ou L B, Zhang W, Hu D, Chen C, Wang X J, 2014. Intercontinental transport and deposition patterns of atmospheric mercury from anthropogenic emissions. Atmospheric Chemistry and Physics, 14: 10163-10176.

Chen L G, Liu M, Xu Z C, Fan R F, Tao J, Chen D H, Zhang D Q, Xie D H, Sun J R, 2013. Variation trends and influencing factors of total gaseous mercury in the Pearl River Delta: A highly industrialised region in South China influenced by seasonal monsoons. Atmospheric Environment, 77: 757-766.

Cheng I, Lu J, Song X J, 2009. Studies of potential sources that contributed to atmospheric mercury in Toronto, Canada. Atmospheric Environment, 43: 6145-6158.

Cheng I, Xu X, Zhang L, 2015. Overview of receptor-based source apportionment studies for speciated atmospheric mercury. Atmospheric Chemistry and Physics, 15: 7877-7895.

Cheng I, Zhang L, Blanchard P, Dalziel J, Tordon R, 2013a. Concentration-weighted trajectory approach to identifying potential sources of speciated atmospheric mercury at an urban coastal site in Nova Scotia, Canada. Atmospheric Chemistry and Physics, 13: 6031-6048.

Cheng I, Zhang L, Blanchard P, Graydon J A, Louis V L S, 2012. Source-receptor relationships for speciated atmospheric mercury at the remote experimental lakes area, northwestern Ontario, Canada. Atmospheric Chemistry and Physics, 12: 1903-1922.

Cheng I, Zhang L M, Blanchard P, Dalziel J, Tordon R, Huang J Y, Holsen T M, 2013b. Comparisons of mercury sources and atmospheric mercury processes between a coastal and inland site. Journal of Geophysical Research-Atmospheres, 118: 2434-2443.

Choi E M, Kim S H, Holsen T M, Yi S M, 2009. Total gaseous concentrations in mercury in Seoul, Korea: Local sources compared to long-range transport from China and Japan. Environmental Pollution, 157: 816-822.

Choi H D, Holsen T M, Hopke P K, 2008. Atmospheric mercury (Hg) in the Adirondacks: Concentrations and sources. Environmental Science & Technology, 42: 5644-5653.

Christensen J H, Brandt J, Frohn L M, Skov H, 2004. Modelling of mercury in the Arctic with the Danish Eulerian Hemispheric Model. Atmospheric Chemistry and Physics, 4: 2251-2257.

Ci Z J, Zhang X S, Wang Z W, Niu Z C, 2011. Atmospheric gaseous elemental mercury (GEM) over a coastal/rural site downwind of East China: Temporal variation and long-range transport. Atmospheric Environment, 45: 2480-2487.

Cobbett F D, Steffen A, Lawson G, Van Heyst B J, 2007. GEM fluxes and atmospheric mercury concentrations (GEM, RGM and Hg-P) in the Canadian Arctic at Alert, Nunavut, Canada (February-June 2005). Atmospheric Environment, 41: 6527-6543.

Dastoor A P, Larocque Y, 2004. Global circulation of atmospheric mercury: A modelling study. Atmospheric Environment, 38: 147-161.

De Foy B, Tong Y D, Yin X F, Zhang W, Kang S C, Zhang Q G, Zhang G S, Wang X J, Schauer J J, 2016. First field-based atmospheric observation of the reduction of reactive mercury driven by sunlight. Atmospheric Environment, 134: 27-39.

De Foy B, Wiedinmyer C, Schauer J J, 2012. Estimation of mercury emissions from forest fires, lakes, regional and local sources using measurements in Milwaukee and an inverse method. Atmospheric Chemistry and Physics, 12: 8993-9011.

Dommergue A, Sprovieri F, Pirrone N, Ebinghaus R, Brooks S, Courteaud J, Ferrari C P, 2010. Overview of mercury measurements in the Antarctic troposphere. Atmospheric Chemistry and Physics, 10: 3309-3319.

Draxler R R, Rolph G D, 2014. HYSPLIT (HYbrid Single-Particle Lagrangian Integrated Trajectory), NOAA Air Resources Laboratory, College Park, MD.

Durnford D, Dastoor A, Figueras-Nieto D, Ryjkov A, 2010. Long range transport of mercury to the

Arctic and across Canada. Atmospheric Chemistry and Physics, 10: 6063-6086.
Ebinghaus R, Jennings S G, Kock H H, Derwent R G, Manning A J, Spain T G, 2011. Decreasing trends in total gaseous mercury observations in baseline air at Mace Head, Ireland from 1996 to 2009. Atmospheric Environment, 45: 3475-3480.
Ebinghaus R, Jennings S G, Schroeder W H, Berg T, Donaghy T, Guentzel J, Kenny C, Kock H H, Kvietkus K, Landing W, Muhleck T, Munthe J, Prestbo E M, Schneeberger D, Slemr F, Sommar J, Urba A, Wallschlager D, Xiao Z, 1999. International field intercomparison measurements of atmospheric mercury species at Mace Head, Ireland. Atmospheric Environment, 33: 3063-3073.
Ebinghaus R, Kock H H, Coggins A M, Spain T G, Jennings S G, Temme C, 2002a. Long-term measurements of atmospheric mercury at Mace Head, Irish west coast, between 1995 and 2001. Atmospheric Environment, 36: 5267-5276.
Ebinghaus R, Kock H H, Temme C, Einax J W, Lowe A G, Richter A, Burrows J P, Schroeder W H, 2002b. Antarctic springtime depletion of atmospheric mercury. Environmental Science & Technology, 36: 1238-1244.
Fain X, Obrist D, Hallar A G, Mccubbin I, Rahn T, 2009. High levels of reactive gaseous mercury observed at a high elevation research laboratory in the Rocky Mountains. Atmospheric Chemistry and Physics, 9: 8049-8060.
Fang F M, Wang Q C, Li J F, 2004. Urban environmental mercury in Changchun, a metropolitan city in Northeastern China: Source, cycle, and fate. Science of the Total Environment, 330: 159-170.
Fang F M, Wang Q C, Liu R H, Ma Z W, Hao Q J, 2001. Atmospheric particulate mercury in Changchun City, China. Atmospheric Environment, 35: 4265-4272.
Feng X B, Shang L H, Wang S F, Tang S L, Zheng W, 2004. Temporal variation of total gaseous mercury in the air of Guiyang, China. Journal of Geophysical Research-Atmospheres, 109, D03303, doi: 10.1029/2003JD004159.
Feng X B, Sommar J, Gardfeldt K, Lindqvist O, 2000. Improved determination of gaseous divalent mercury in ambient air using KCl coated denuders. Fresenius Journal of Analytical Chemistry, 366: 423-428.
Feng X B, Sommar J, Lindqvist O, Hong Y T, 2002. Occurrence, emissions and deposition of mercury during coal combustion in the province Guizhou, China. Water Air and Soil Pollution, 139: 311-324.
Feng X B, Wang S F, Qiu G A, Hou Y M, Tang S L, 2005. Total gaseous mercury emissions from soil in Guiyang, Guizhou, China. Journal of Geophysical Research-Atmospheres, 110, D14306, doi: 10.1029/2004JD005643.
Friedli H R, Arellano A F, Geng F, Cai C, Pan L, 2011. Measurements of atmospheric mercury in Shanghai during september 2009. Atmospheric Chemistry and Physics, 11: 3781-3788.
Friedli H R, Radke L F, Prescott R, Li P, Woo J H, Carmichael G R, 2004. Mercury in the atmosphere around Japan, Korea, and China as observed during the 2001 ACE-Asia field campaign: Measurements, distributions, sources, and implications. Journal of Geophysical Research-Atmospheres, 109, D19S25, doi: 10.1029/2003JD004244.
Fritsche J, Obrist D, Zeeman M J, Conen F, Eugster W, Alewell C, 2008. Elemental mercury fluxes over a sub-alpine grassland determined with two micrometeorological methods. Atmospheric Environment, 42: 2922-2933.
Fu X, Yang X, Lang X, Zhou J, Zhang H, Yu B, Yan H, Lin C J, Feng X, 2016a. Atmospheric wet and litterfall mercury deposition at urban and rural sites in China. Atmospheric Chemistry and Physics,

16: 11547-11562.

Fu X, Zhu W, Zhang H, Sommar J, Yu B, Yang X, Wang X, Lin C J, Feng X, 2016b. Depletion of atmospheric gaseous elemental mercury by plant uptake at Mt. Changbai, Northeast China. Atmospheric Chemistry and Physics, 16: 12861-12873.

Fu X W, Feng X, Dong Z Q, Yin R S, Wang J X, Yang Z R, Zhang H, 2010. Atmospheric gaseous elemental mercury (GEM) concentrations and mercury depositions at a high-altitude mountain peak in south China. Atmospheric Chemistry and Physics, 10: 2425-2437.

Fu X W, Feng X, Liang P, Deliger, Zhang H, Ji J, Liu P, 2012a. Temporal trend and sources of speciated atmospheric mercury at Waliguan GAW station, Northwestern China. Atmospheric Chemistry and Physics, 12: 1951-1964.

Fu X W, Feng X, Shang L H, Wang S F, Zhang H, 2012b. Two years of measurements of atmospheric total gaseous mercury (TGM) at a remote site in Mt. Changbai area, Northeastern China. Atmospheric Chemistry and Physics, 12: 4215-4226.

Fu X W, Feng X B, Qiu G L, Shang L H, Zhang H, 2011. Speciated atmospheric mercury and its potential source in Guiyang, China. Atmospheric Environment, 45: 4205-4212.

Fu X W, Feng X B, Wang S F, Rothenberg S, Shang L H, Li Z G, Qiu G L, 2009. Temporal and spatial distributions of total gaseous mercury concentrations in ambient air in a mountainous area in southwestern China: Implications for industrial and domestic mercury emissions in remote areas in China. Science of the Total Environment, 407: 2306-2314.

Fu X W, Feng X B, Zhu W Z, Wang S F, Lu J L, 2008a. Total gaseous mercury concentrations in ambient air in the eastern slope of Mt. Gongga, South-Eastern fringe of the Tibetan plateau, China. Atmospheric Environment, 42: 970-979.

Fu X W, Feng X B, Zhu W Z, Zheng W, Wang S F, Lu J Y, 2008b. Total particulate and reactive gaseous mercury in ambient air on the eastern slope of the Mt. Gongga area, China. Applied Geochemistry, 23: 408-418.

Fu X W, Heimburger L E, Sonke J E, 2014. Collection of atmospheric gaseous mercury for stable isotope analysis using iodine- and chlorine-impregnated activated carbon traps. Journal of Analytical Atomic Spectrometry, 29: 841-852.

Fu X W, Marusczak N, Heimburger L E, Sauvage B, Gheusi F, Prestbo E M, Sonke J E, 2016c. Atmospheric mercury speciation dynamics at the high-altitude Pic du Midi Observatory, southern France. Atmospheric Chemistry and Physics, 16: 5623-5639.

Fu X W, Zhang H, Lin C J, Feng X B, Zhou L X, Fang S X, 2015a. Correlation slopes of GEM/CO, GEM/CO_2, and GEM/CH_4 and estimated mercury emissions in China, South Asia, the Indochinese Peninsula, and Central Asia derived from observations in northwestern and southwestern China. Atmospheric Chemistry and Physics, 15: 1013-1028.

Fu X W, Zhang H, Yu B, Wang X, Lin C J, Feng X B, 2015b. Observations of atmospheric mercury in China: A critical review. Atmospheric Chemistry and Physics, 15: 9455-9476.

Gabriel M C, Williamson D G, Brooks S, Lindberg S, 2005. Atmospheric speciation of Southeastern mercury in two contrasting US airsheds. Atmospheric Environment, 39: 4947-4958.

Gårdfeldt K, Feng X B, Sommar J, Lindqvist O, 2001. Total gaseous mercury exchange between air and water at river and sea surfaces in Swedish coastal regions. Atmospheric Environment, 35: 3027-3038.

Gbor P K, Wen D Y, Meng F, Yang F Q, Zhang B N, Sloan J J, 2006. Improved model for mercury emission, transport and deposition. Atmospheric Environment, 40: 973-983.

Gratz L E, Ambrose J L, Jaffe D A, Shah V, Jaegle L, Stutz J, Festa J, Spolaor M, Tsai C, Selin N E, Song S, Zhou X, Weinheimer A J, Knapp D J, Montzka D D, Flocke F M, Campos T L, Apel E, Hornbrook R, Blake N J, Hall S, Tyndall G S, Reeves M, Stechman D, Stell M, 2015. Oxidation of mercury by bromine in the subtropical Pacific free troposphere. Geophysical Research Letters, 42: 10, 494-510, 502.

Gustin M, Jaffe D, 2010. Reducing the uncertainty in measurement and understanding of mercury in the atmosphere. Environmental Science & Technology, 44: 2222-2227.

Gustin M S, Amos H M, Huang J, Miller M B, Heidecorn K, 2015. Measuring and modeling mercury in the atmosphere: A critical review. Atmospheric Chemistry and Physics, 15: 5697-5713.

Gustin M S, Huang J Y, Miller M B, Peterson C, Jaffe D A, Ambrose J, Finley B D, Lyman S N, Call K, Talbot R, Feddersen D, Mao H T, Lindberg S E, 2013. Do we understand what the mercury speciation instruments are actually measuring? Results of RAMIX. Environmental Science & Technology, 47: 7295-7306.

Gustin M S, Weiss-Penzias P S, Peterson C, 2012. Investigating sources of gaseous oxidized mercury in dry deposition at three sites across Florida, USA. Atmospheric Chemistry and Physics, 12: 9201-9219.

Han Y J, Holsen T A, Hopke P K, Yi S M, 2005. Comparison between back-trajectory based modeling and Lagrangian backward dispersion modeling for locating sources of reactive gaseous mercury. Environmental Science & Technology, 39: 3887.

Han Y J, Holsen T M, Hopke P K, 2007. Estimation of source locations of total gaseous mercury measured in New York State using trajectory-based models. Atmospheric Environment, 41: 6033-6047.

Han Y J, Holsen T M, Lai S O, Hopke P K, Yi S M, Liu W, Pagano J, Falanga L, Milligan M, Andolina C, 2004. Atmospheric gaseous mercury concentrations in New York State: Relationships with meteorological data and other pollutants. Atmospheric Environment, 38: 6431-6446.

Hartman J S, Weisberg P J, Pillai R, Ericksen J A, Kuiken T, Lindberg S E, Zhang H, Rytuba J J, Gustin M S, 2009. Application of a rule-based model to estimate mercury exchange for three background biomes in the Continental United States. Environmental Science & Technology, 43: 4989-4994.

Hedgecock I M, Pirrone N, 2004. Chasing quicksilver: Modeling the atmospheric lifetime of $Hg^0_{(g)}$ in the marine boundary layer at various latitudes. Environmental Science & Technology, 38: 69-76.

Henry R C, Lewis C W, Hopke P K, Williamson H J, 1984. Review of receptor model fundamentals. Atmospheric Environment, 18: 1507-1515.

Holmes C D, Jacob D J, Corbitt E S, Mao J, Yang X, Talbot R, Slemr F, 2010. Global atmospheric model for mercury including oxidation by bromine atoms. Atmospheric Chemistry and Physics, 10: 12037-12057.

Hopke P K, 2003. Recent developments in receptor modeling. Journal of Chemometrics, 17: 255-265.

Horowitz H M, Jacob D J, Zhang Y X, Dibble T S, Slemr F, Amos H M, Schmidt J A, Corbitt E S, Marais E A, Sunderland E M, 2017. A new mechanism for atmospheric mercury redox chemistry: implications for the global mercury budget. Atmospheric Chemistry and Physics, 17: 6353-6371.

Howard D, Nelson P F, Edwards G C, Morrison A L, Fisher J A, Ward J, Harnwell J, Van Der Schoot M, Atkinson B, Chambers S D, Griffiths A D, Werczynski S, Williams A G, 2017. Atmospheric mercury in the Southern Hemisphere tropics: seasonal and diurnal variations and influence of inter-hemispheric transport. Atmospheric Chemistry and Physics, 17: 11623-11636.

Huang J, Kang S C, Guo J M, Zhang Q G, Cong Z Y, Sillanpaa M, Zhang G S, Sun S W, Tripathee L, 2016. Atmospheric particulate mercury in Lhasa city, Tibetan Plateau. Atmospheric Environment, 142: 433-441.

Huang J Y, Gustin M S, 2015. Uncertainties of gaseous oxidized mercury measurements using KCl-coated denuders, Cation-exchange membranes, and Nylon membranes: Humidity influences. Environmental Science & Technology, 49: 6102-6108.

Huang J Y, Miller M B, Weiss-Penzias P, Gustin M S, 2013. Comparison of gaseous oxidized Hg measured by KCl-coated denuders, and Nylon and cation exchange Membranes. Environmental Science & Technology, 47: 7307-7316.

Iverfeldt A, 1991. Occurrence and turnover of atmospheric mercury over the Nordic countries. Water Air and Soil Pollution, 56: 251-265.

Keeler G, Glinsorn G, Pirrone N, 1995. Particulate Mercury in the atmosphere: Its significance, transport, transformation and sources. Water Air and Soil Pollution, 80: 159-168.

Keeler G J, Landis M S, Norris G A, Christianson E M, Dvonch J T, 2006. Sources of mercury wet deposition in Eastern Ohio, USA. Environmental Science & Technology, 40: 5874-5881.

Kellerhals M, Beauchamp S, Belzer W, Blanchard P, Froude F, Harvey B, Mcdonald K, Pilote M, Poissant L, Puckett K, Schroeder B, Steffen A, Tordon R, 2003. Temporal and spatial variability of total gaseous mercury in Canada: Results from the Canadian Atmospheric Mercury Measurement Network (CAMNet). Atmospheric Environment, 37: 1003-1011.

Kim K H, Kim M Y, Kim J, Lee G, 2003. Effects of changes in environmental conditions on atmospheric mercury exchange: Comparative analysis from a rice paddy field during the two spring periods of 2001 and 2002. Journal of Geophysical Research-Atmospheres, 108(D19), 4607, doi: 10.1029/2003JD003375.

Kim K H, Lindberg S E, Meyers T P, 1995. Micrometeorological measurements of mercury-vapor fluxes over background forest soils in Eastern Tennessee. Atmospheric Environment, 29: 267-282.

Kim K H, Mishra V K, Hong S, 2006. The rapid and continuous monitoring of gaseous elemental mercury (GEM) behavior in ambient air. Atmospheric Environment, 40: 3281-3293.

Kim S H, Han Y J, Holsen T M, Yi S M, 2009. Characteristics of atmospheric speciated mercury concentrations (TGM, Hg(II) and Hg(P)) in Seoul, Korea. Atmospheric Environment, 43: 3267-3274.

Kock H H, Bieber E, Ebinghaus R, Spain T G, Thees B, 2005. Comparison of long-term trends and seasonal variations of atmospheric mercury concentrations at the two European coastal monitoring stations Mace Head, Ireland, and Zingst, Germany. Atmospheric Environment, 39: 7549-7556.

Kuiken T, Zhang H, Gustin M, Lindberg S, 2008. Mercury emission from terrestrial background surfaces in the eastern USA. Part I: Air/surface exchange of mercury within a southeastern deciduous forest (Tennessee) over one year. Applied Geochemistry, 23: 345-355.

Lamborg C H, Fitzgerald W F, O'donnell J, Torgersen T, 2002. A non-steady-state compartmental model of global-scale mercury biogeochemistry with interhemispheric atmospheric gradients. Geochimica Et Cosmochimica Acta, 66: 1105-1118.

Lan X, Talbot R, Castro M, Perry K, Luke W, 2012. Seasonal and diurnal variations of atmospheric mercury across the US determined from AMNet monitoring data. Atmospheric Chemistry and Physics, 12: 10569-10582.

Landis M S, Stevens R K, Schaedlich F, Prestbo E M, 2002. Development and characterization of an annular denuder methodology for the measurement of divalent inorganic reactive gaseous mercury in ambient air. Environmental Science & Technology, 36: 3000-3009.

Laurier F, Mason R, 2007. Mercury concentration and speciation in the coastal and open ocean boundary layer. Journal of Geophysical Research-Atmospheres, 112, D06302, doi: 10.1029/2006JD007320.

Laurier F J G, Mason R P, Whalin L, Kato S, 2003. Reactive gaseous mercury formation in the North Pacific Ocean's marine boundary layer: A potential role of halogen chemistry. Journal of Geophysical Research-Atmospheres, 108, doi: 10.1029/2003JD003625.

Lee D S, Nemitz E, Fowler D, Kingdon R D, 2001. Modelling atmospheric mercury transport and deposition across Europe and the UK. Atmospheric Environment, 35: 5455-5466.

Lee J H, Hopke P K, 2006. Apportioning sources of $PM_{2.5}$ in St. Louis, MO using speciation trends network data. Atmospheric Environment, 40: S360-S377.

Lin C J, Pan L, Streets D G, Shetty S K, Jang C, Feng X, Chu H W, Ho T C, 2010. Estimating mercury emission outflow from East Asia using CMAQ-Hg. Atmospheric Chemistry and Physics, 10: 1853-1864.

Lin C J, Pongprueksa P, Lindberg S E, Pehkonen S O, Byun D, Jang C, 2006. Scientific uncertainties in atmospheric mercury models I: Model science evaluation. Atmospheric Environment, 40: 2911-2928.

Lindberg S, Bullock R, Ebinghaus R, Engstrom D, Feng X B, Fitzgerald W, Pirrone N, Prestbo E, Seigneur C, 2007. A synthesis of progress and uncertainties in attributing the sources of mercury in deposition. Ambio, 36: 19-32.

Lindberg S E, Brooks S, Lin C J, Scott K J, Landis M S, Stevens R K, Goodsite M, Richter A, 2002a. Dynamic oxidation of gaseous mercury in the Arctic troposphere at polar sunrise. Environmental Science & Technology, 36: 1245-1256.

Lindberg S E, Dong W J, Meyers T, 2002b. Transpiration of gaseous elemental mercury through vegetation in a subtropical wetland in Florida. Atmospheric Environment, 36: 5207-5219.

Lindberg S E, Stratton W J, 1998. Atmospheric mercury speciation: Concentrations and behavior of reactive gaseous mercury in ambient air. Environmental Science & Technology, 32: 49-57.

Lindberg S E, Stratton W J, Pai P, Allan M A, 2000. Measurements and modeling of a water soluble gas-phase mercury species in ambient air. Fuel Processing Technology, 65: 143-156.

Liu B, Keeler G J, Dvonch J T, Barres J A, Lynam M M, Marsik F J, Morgan J T, 2007. Temporal variability of mercury speciation in urban air. Atmospheric Environment, 41: 1911-1923.

Liu B, Keeler G J, Dvonch J T, Barres J A, Lynam M M, Marsik F J, Morgan J T, 2010. Urban-rural differences in atmospheric mercury speciation. Atmospheric Environment, 44: 2013-2023.

Liu M, Chen L G, Xie D H, Sun J R, He Q S, Cai L M, Gao Z Q, Zhang Y Q, 2016. Monsoon-driven transport of atmospheric mercury to the South China Sea from the Chinese mainland and Southeast Asia-Observation of gaseous elemental mercury at a background station in South China. Environmental Science and Pollution Research, 23: 21631-21640.

Liu S L, Nadim F, Perkins C, Carley R J, Hoag G E, Lin Y H, Chen L T, 2002. Atmospheric mercury monitoring survey in Beijing, China. Chemosphere, 48: 97-107.

Liu W, Hopke P K, Han Y J, Yi S M, Holsen T M, Cybart S, Kozlowski K, Milligan M, 2003. Application of receptor modeling to atmospheric constituents at Potsdam and Stockton, NY. Atmospheric Environment, 37: 4997-5007.

Lu J Y, Schroeder W H, Berg T, Munthe J, Schneeberger D, Schaedlich F, 1998. A device for sampling and determination of total particulate mercury in ambient air. Analytical Chemistry, 70: 2403-2408.

Lyman S N, Gustin M S, 2009. Determinants of atmospheric mercury concentrations in Reno, Nevada, USA. Science of the Total Environment, 408: 431-438.

Lyman S N, Jaffe D A, 2012. Formation and fate of oxidized mercury in the upper troposphere and lower stratosphere. Nature Geoscience, 5: 114-117.

Lyman S N, Jaffe D A, Gustin M S, 2010. Release of mercury halides from KCl denuders in the presence of ozone. Atmospheric Chemistry and Physics, 10: 8197-8204.

Lynam M M, Keeler G J, 2005. Automated speciated mercury measurements in Michigan. Environmental Science & Technology, 39: 9253-9262.

Lynam M M, Keeler G J, 2002. Comparison of methods for particulate phase mercury analysis: Sampling and analysis. Analytical and Bioanalytical Chemistry, 374: 1009-1014.

Lynam M M, Keeler G J, 2006. Source-receptor relationships for atmospheric mercury in urban Detroit, Michigan. Atmospheric Environment, 40: 3144-3155.

Malcolm E G, Keeler G J, 2007. Evidence for a sampling artifact for particulate-phase mercury in the marine atmosphere. Atmospheric Environment, 41: 3352-3359.

Malcolm E G, Keeler G J, Landis M S, 2003. The effects of the coastal environment on the atmospheric mercury cycle. Journal of Geophysical Research-Atmospheres, 108(D12), 4357, doi: 10.1029/2002JD003084.

Mao H, Talbot R W, Sigler J M, Sive B C, Hegarty J D, 2008. Seasonal and diurnal variations of Hg degrees over New England. Atmospheric Chemistry and Physics, 8: 1403-1421.

Mason R P, Sheu G R, 2002. Role of the ocean in the global mercury cycle. Global Biogeochemical Cycles, 16(4), 1093, doi: 10.1029/2001GB001440.

Mcclure C D, Jaffe D A, Edgerton E S, 2014. Evaluation of the KCl denuder method for gaseous oxidized mercury using $HgBr_2$ at an In-Service AMNet Site. Environmental Science & Technology, 48: 11437-11444.

Muller D, Wip D, Warneke T, Holmes C D, Dastoor A, Notholt J, 2012. Sources of atmospheric mercury in the tropics: Continuous observations at a coastal site in Suriname. Atmospheric Chemistry and Physics, 12: 7391-7397.

Munthe J, Wangberg I, Iverfeldt A, Lindqvist O, Stromberg D, Sommar J, Gardfeldt K, Petersen G, Ebinghaus R, Prestbo E, Larjava K, Siemens V, 2003. Distribution of atmospheric mercury species in Northern Europe: Final results from the MOE project. Atmospheric Environment, 37: S9-S20.

Munthe J, Wangberg I, Pirrone N, Iverfeldt A, Ferrara R, Ebinghaus R, Feng X, Gardfeldt K, Keeler G, Lanzillotta E, Lindberg S E, Lu J, Mamane Y, Prestbo E, Schmolke S, Schroeder W H, Sommar J, Sprovieri F, Stevens R K, Stratton W, Tuncel G, Urba A, 2001. Intercomparison of methods for sampling and analysis of atmospheric mercury species. Atmospheric Environment, 35: 3007-3017.

Murphy D M, Hudson P K, Thomson D S, Sheridan P J, Wilson J C, 2006. Observations of mercury-containing aerosols. Environmental Science & Technology, 40: 3163-3167.

Nguyen D L, Kim J Y, Shim S G, Zhang X S, 2011. Ground and shipboard measurements of atmospheric gaseous elemental mercury over the Yellow Sea region during 2007-2008. Atmospheric Environment, 45: 253-260.

Nguyen H T, Kim K H, Kim M Y, Hong S M, Youn Y H, Shon Z H, Lee J S, 2007. Monitoring of atmospheric mercury at a global atmospheric watch (GAW) site on An-Myun Island, Korea. Water Air and Soil Pollution, 185: 149-164.

Nguyen H T, Kim M Y, Kim K H, 2010. The influence of long-range transport on atmospheric mercury on Jeju Island, Korea. Science of the Total Environment, 408: 1295-1307.

Obrist D, Agnan Y, Jiskra M, Olson C L, Colegrove D P, Hueber J, Moore C W, Sonke J E, Helmig D, 2017. Tundra uptake of atmospheric elemental mercury drives Arctic mercury pollution. Nature, 547: 201-204.

Obrist D, Tas E, Peleg M, Matveev V, Fain X, Asaf D, Luria M, 2011. Bromine-induced oxidation of mercury in the mid-latitude atmosphere. Nature Geoscience, 4: 22-26.

Osawa T, Ueno T, Fu F F, 2007. Sequential variation of atmospheric mercury in Tokai-mura, seaside area of eastern central Japan. Journal of Geophysical Research-Atmospheres, 112, D19107, doi: 10.1029/2007JD008538.

Paatero P, Tapper U, 1994. Positive matrix factorization: A nonnegative factor model with optimal utilization of error-estimates of data values. Environmetrics, 5: 111-126.

Pacyna E G, Pacyna J M, Fudala J, Strzelecka-Jastrzab E, Hlawiczka S, Panasiuk D, 2006a. Mercury emissions to the atmosphere from anthropogenic sources in Europe in 2000 and their scenarios until 2020. Science of the Total Environment, 370: 147-156.

Pacyna E G, Pacyna J M, Steenhuisen F, Wilson S, 2006b. Global anthropogenic mercury emission inventory for 2000. Atmospheric Environment, 40: 4048-4063.

Pacyna E G, Pacyna J M, Sundseth K, Munthe J, Kindbom K, Wilson S, Steenhuisen F, Maxson P, 2010. Global emission of mercury to the atmosphere from anthropogenic sources in 2005 and projections to 2020. Atmospheric Environment, 44: 2487-2499.

Pan L, Carmichael G R, Adhikary B, Tang Y H, Streets D, Woo J H, Friedli H R, Radke L F, 2008. A regional analysis of the fate and transport of mercury in East Asia and an assessment of major uncertainties. Atmospheric Environment, 42: 1144-1159.

Pan L, Woo J H, Carmichael G R, Tang Y H, Friedli H R, Radke L F, 2006. Regional distribution and emissions of mercury in east Asia: A modeling analysis of Asian Pacific Regional Aerosol Characterization Experiment (ACE-Asia) observations. Journal of Geophysical Research-Atmospheres, 111.

Petersen G, Bloxam R, Wong S, Munthe J, Kruger O, Schmolke S R, Kumar A V, 2001. A comprehensive Eulerian modelling framework for airborne mercury species: Model development and applications in Europe. Atmospheric Environment, 35: 3063-3074.

Pirrone N, Cinnirella S, Feng X, Finkelman R B, Friedli H R, Leaner J, Mason R, Mukherjee A B, Stracher G B, Streets D G, Telmer K, 2010. Global mercury emissions to the atmosphere from anthropogenic and natural sources. Atmospheric Chemistry and Physics, 10: 5951-5964.

Poissant L, Pilote M, Beauvais C, Constant P, Zhang H H, 2005. A year of continuous measurements of three atmospheric mercury species (GEM, RGM and Hg-p) in southern Quebec, Canada. Atmospheric Environment, 39: 1275-1287.

Poissant L, Pilote M, Constant P, Beauvais C, Zhang H H, Xu X H, 2004. Mercury gas exchanges over selected bare soil and flooded sites in the bay St. Francois wetlands (Quebec, Canada). Atmospheric Environment, 38: 4205-4214.

Polissar A V, Hopke P K, Harris J M, 2001. Source regions for atmospheric aerosol measured at Barrow, Alaska. Environmental Science & Technology, 35: 4214-4226.

Rutter A P, Schauer J J, 2007. The impact of aerosol composition on the particle to gas partitioning of reactive mercury. Environmental Science & Technology, 41: 3934-3939.

Rutter A P, Snyder D C, Stone E A, Schauer J J, Gonzalez-Abraham R, Molina L T, Marquez C, Cardenas B, De Foy B, 2009. *In situ* measurements of speciated atmospheric mercury and the identification of source regions in the Mexico City Metropolitan Area. Atmospheric Chemistry and Physics, 9: 207-220.

Ryaboshapko A, Bullock O R, Christensen J, Cohen M, Dastoor A, Ilyin I, Petersen G, Syrakov D, Artz R S, Davignon D, Draxler R R, Munthe J, 2007. Intercomparison study of atmospheric mercury models: 1. Comparison of models with short-term measurements. Science of the Total Environment, 376: 228-240.

Schroeder W H, Anlauf K G, Barrie L A, Lu J Y, Steffen A, Schneeberger D R, Berg T, 1998. Arctic springtime depletion of mercury. Nature, 394: 331-332.

Schroeder W H, Munthe J, 1998. Atmospheric mercury: An overview. Atmospheric Environment, 32: 809-822.

Seigneur C, Karamchandani P, Lohman K, Vijayaraghavan K, Shia R L, 2001. Multiscale modeling of the atmospheric fate and transport of mercury. Journal of Geophysical Research-Atmospheres, 106: 27795-27809.

Seigneur C, Lohman K, 2008. Effect of bromine chemistry on the atmospheric mercury cycle. Journal of Geophysical Research-Atmospheres, 113.

Seigneur C, Vijayaraghavan K, Lohman K, Karamchandani P, Scott C, 2004. Global source attribution for mercury deposition in the United States. Environmental Science & Technology, 38: 555-569.

Selin N E, Jacob D J, Park R J, Yantosca R M, Strode S, Jaegle L, Jaffe D, 2007. Chemical cycling and deposition of atmospheric mercury: Global constraints from observations. Journal of Geophysical Research-Atmospheres, 112.

Shah V, Jaegle L, Gratz L E, Ambrose J L, Jaffe D A, Selin N E, Song S, Campos T L, Flocke F M, Reeves M, Stechman D, Stell M, Festa J, Stutz J, Weinheimer A J, Knapp D J, Montzka D D, Tyndall G S, Apel E C, Hornbrook R S, Hills A J, Riemer D D, Blake N J, Cantrell C A, Mauldin R L, 2016. Origin of oxidized mercury in the summertime free troposphere over the southeastern US. Atmospheric Chemistry and Physics, 16: 1511-1530.

Shetty S K, Lin C J, Streets D G, Jang C, 2008. Model estimate of mercury emission from natural sources in East Asia. Atmospheric Environment, 42: 8674-8685.

Sheu G R, Lin N H, Wang J L, Lee C T, Yang C F O, Wang S H, 2010. Temporal distribution and potential sources of atmospheric mercury measured at a high-elevation background station in Taiwan. Atmospheric Environment, 44: 2393-2400.

Sheu G R, Mason R P, 2001. An examination of methods for the measurements of reactive gaseous mercury in the atmosphere. Environmental Science & Technology, 35: 1209-1216.

Skov H, Christensen J H, Goodsite M E, Heidam N Z, Jensen B, Wahlin P, Geernaert G, 2004. Fate of elemental mercury in the arctic during atmospheric mercury depletion episodes and the load of atmospheric mercury to the arctic. Environmental Science & Technology, 38: 2373-2382.

Slemr F, Angot H, Dommergue A, Magand O, Barret M, Weigelt A, Ebinghaus R, Brunke E G, Pfaffhuber K A, Edwards G, Howard D, Powell J, Keywood M, Wang F, 2015. Comparison of mercury concentrations measured at several sites in the Southern Hemisphere. Atmospheric Chemistry and Physics, 15: 3125-3133.

Slemr F, Brunke E G, Ebinghaus R, Kuss J, 2011. Worldwide trend of atmospheric mercury since 1995.

Atmospheric Chemistry and Physics, 11: 4779-4787.

Slemr F, Brunke E G, Ebinghaus R, Temme C, Munthe J, Wangberg I, Schroeder W, Steffen A, Berg T, 2003. Worldwide trend of atmospheric mercury since 1977. Geophysical Research Letters, 30(10), 1516, doi: 10.1029/2003GL016954.

Slemr F, Ebinghaus R, Brenninkmeijer C a M, Hermann M, Kock H H, Martinsson B G, Schuck T, Sprung D, Van Velthoven P, Zahn A, Ziereis H. Gaseous mercury distribution in the upper troposphere and lower stratosphere observed, 2009 onboard the CARIBIC passenger aircraft. Atmospheric Chemistry and Physics, 9: 1957-1969.

Slemr F, Scheel H E, 1998. Trends in atmospheric mercury concentrations at the summit of the Wank mountain, southern Germany. Atmospheric Environment, 32: 845-853.

Slemr F, Weigelt A, Ebinghaus R, Kock H H, Bödewadt J, Brenninkmeijer C a M, Rauthe-Schröch A, Weber S, Hermann M, Becker J, Zahn A, Martinsson B, 2016. Atmospheric mercury measurements onboard the CARIBIC passenger aircraft. Atmospheric Measurement Techniques, 9: 2291-2302.

Soerensen A L, Sunderland E M, Holmes C D, Jacob D J, Yantosca R M, Skov H, Christensen J H, Strode S A, Mason R P, 2010. An improved global model for air-sea exchange of mercury: High concentrations over the North Atlantic. Environmental Science & Technology, 44: 8574-8580.

Sommar J, Andersson M E, Jacobi H W, 2010. Circumpolar measurements of speciated mercury, ozone and carbon monoxide in the boundary layer of the Arctic Ocean. Atmospheric Chemistry and Physics, 10: 5031-5045.

Song Y, Dai W, Shao M, Liu Y, Lu S H, Kuster W, Goldan P, 2008. Comparison of receptor models for source apportionment of volatile organic compounds in Beijing, China. Environmental Pollution, 156: 174-183.

Sprovieri F, Pirrone N, Bencardino M, D'amore F, Carbone F, Cinnirella S, Mannarino V, Landis M, Ebinghaus R, Weigelt A, Brunke E G, Labuschagne C, Martin L, Munthe J, Wangberg I, Artaxo P, Morais F, Barbosa H D J, Brito J, Cairns W, Barbante C, Dieguez M D, Garcia P E, Dommergue A, Angot H, Magand O, Skov H, Horvat M, Kotnik J, Read K A, Neves L M, Gawlik B M, Sena F, Mashyanov N, Obolkin V, Wip D, Bin Feng X, Zhang H, Fu X W, Ramachandran R, Cossa D, Knoery J, Marusczak N, Nerentorp M, Norstrom C, 2016. Atmospheric mercury concentrations observed at ground-based monitoring sites globally distributed in the framework of the GMOS network. Atmospheric Chemistry and Physics, 16: 11915-11935.

Sprovieri F, Pirrone N, Hedgecock I M, Landis M S, Stevens R K, 2002. Intensive atmospheric mercury measurements at Terra Nova Bay in Antarctica during november and december 2000. Journal of Geophysical Research-Atmospheres, 107.

Steen A O, Berg T, Dastoor A P, Durnford D A, Engelsen O, Hole L R, Pfaffhuber K A, 2011. Natural and anthropogenic atmospheric mercury in the European Arctic: A fractionation study. Atmospheric Chemistry and Physics, 11: 6273-6284.

Steffen A, Douglas T, Amyot M, Ariya P, Aspmo K, Berg T, Bottenheim J, Brooks S, Cobbett F, Dastoor A, Dommergue A, Ebinghaus R, Ferrari C, Gardfeldt K, Goodsite M E, Lean D, Poulain A J, Scherz C, Skov H, Sommar J, Temme C, 2008. A synthesis of atmospheric mercury depletion event chemistry in the atmosphere and snow. Atmospheric Chemistry and Physics, 8: 1445-1482.

Steffen A, Schroeder W, Macdonald R, Poissant L, Konoplev A, 2005. Mercury in the Arctic atmosphere: An analysis of eight years of measurements of GEM at Alert (Canada) and a comparison with observations at Amderma (Russia) and Kuujjuarapik (Canada). Science of the

Total Environment, 342: 185-198.

Stein A F, Draxler R R, Rolph G D, Stunder B J B, Cohen M D, Ngan F, 2015. NOAA's hysplit atmospheric transport and dispersion modeling system. Bulletin of the American Meteorological Society, 96: 2059-2077.

Stohl A, Forster C, Frank A, Seibert P, Wotawa G, 2005. Technical note: The Lagrangian particle dispersion model FLEXPART version 6.2. Atmospheric Chemistry and Physics, 5: 2461-2474.

Streets D G, Devane M K, Lu Z F, Bond T C, Sunderland E M, Jacob D J, 2011. All-time releases of mercury to the atmosphere from human activities. Environmental Science & Technology, 45: 10485-10491.

Streets D G, Zhang Q, Wu Y, 2009. Projections of global mercury emissions in 2050. Environmental Science & Technology, 43: 2983-2988.

Strode S A, Jaegle L, Jaffe D A, Swartzendruber P C, Selin N E, Holmes C, Yantosca R M, 2008. Trans-Pacific transport of mercury. Journal of Geophysical Research-Atmospheres, 113, D15305, doi: 10.1029/2007JD009428.

Strode S A, Jaegle L, Selin N E, Jacob D J, Park R J, Yantosca R M, Mason R P, Slemr F, 2007. Air-sea exchange in the global mercury cycle. Global Biogeochemical Cycles, 21, GB1017, doi: 10.1029/2006GB002766.

Swartzendruber P C, Jaffe D A, Finley B, 2009. Development and first results of an aircraft-based, high time resolution technique for gaseous elemental and reactive (oxidized) gaseous mercury. Environmental Science & Technology, 43: 7484-7489.

Swartzendruber P C, Jaffe D A, Prestbo E M, Weiss-Penzias P, Selin N E, Park R, Jacob D J, Strode S, Jaegle L, 2006. Observations of reactive gaseous mercury in the free troposphere at the Mount Bachelor Observatory. Journal of Geophysical Research-Atmospheres, 111: D24301.

Talbot R, Mao H, Scheuer E, Dibb J, Avery M, 2007. Total depletion of Hg degrees in the upper troposphere-lower stratosphere. Geophysical Research Letters, 34, L23804, doi: 10.1029/2007GL 031366.

Tan H, He J L, Liang L, Lazoff S, Sommer J, Xiao Z F, Lindqvist O, 2000. Atmospheric mercury deposition in Guizhou, China. Science of the Total Environment, 259: 223-230.

Tang S L, Feng X B, Qiu H R, Yin G X, Yang Z C, 1998. Mercury speciation and emissions from coal combustion in Guiyang, southwest China. Environmental Research, 2007, 105: 175-182.

Tekran: Model 2357A Principles of Operation. 1998. Tekran Inc. Toronto, Canada.

Temme C, Blanchard P, Steffen A, Banic C, Beauchamp S, Poissant L, Tordon R, Wiens B, 2007. Trend, seasonal and multivariate analysis study of total gaseous mercury data from the Canadian atmospheric mercury measurement network (CAMNet). Atmospheric Environment, 41: 5423-5441.

Temme C, Einax J W, Ebinghaus R, Schroeder W H, 2003. Measurements of atmospheric mercury species at a coastal site in the Antarctic and over the south Atlantic Ocean during polar summer. Environmental Science & Technology, 37: 22-31.

Thurston G D, Spengler J D, 1985. A quantitative assessment of source contributions to inhalable particulate matter pollution in Metropolitan Boston. Atmospheric Environment, 19: 9-25.

Timonen H, Ambrose J L, Jaffe D A, 2013. Oxidation of elemental Hg in anthropogenic and marine air masses. Atmos. Chem. Phys., 13: 2827-2836.

Travnikov O, 2005. Contribution of the intercontinental atmospheric transport to mercury pollution in the Northern Hemisphere. Atmospheric Environment, 39: 7541-7548.

Urba A, Kvietkus K, Sakalys J, Xiao Z, Lindqvist O, 1995. A new sensitive and portable mercury-vapor analyzer Gardis-1A. Water Air and Soil Pollution, 80: 1305-1309.

Valente R J, Shea C, Humes K L, Tanner R L, 2007. Atmospheric mercury in the Great Smoky Mountains compared to regional and global levels. Atmospheric Environment, 41: 1861-1873.

Van Donkelaar A, Martin R V, Brauer M, Kahn R, Levy R, Verduzco C, Villeneuve P J, 2010. Global estimates of ambient fine farticulate matter concentrations from satellite-based aerosol optical depth: Development and application. Environmental Health Perspectives, 118: 847-855.

Wang F, Saiz-Lopez A, Mahajan A S, Martin J C G, Armstrong D, Lemes M, Hay T, Prados-Roman C, 2014a. Enhanced production of oxidised mercury over the tropical Pacific Ocean: A key missing oxidation pathway. Atmospheric Chemistry and Physics, 14: 1323-1335.

Wang L, Wang S X, Zhang L, Wang Y X, Zhang Y X, Nielsen C, Mcelroy M B, Hao J M, 2014b. Source apportionment of atmospheric mercury pollution in China using the GEOS-Chem model. Environmental Pollution, 190: 166-175.

Wang X, Bao Z D, Lin C J, Yuan W, Feng X B, 2016a. Assessment of global mercury deposition through litterfall. Environmental Science & Technology, 50: 8548-8557.

Wang X, Lin C J, Feng X, 2014c. Sensitivity analysis of an updated bidirectional air-surface exchange model for elemental mercury vapor. Atmospheric Chemistry and Physics, 14: 6273-6287.

Wang X, Lin C J, Yuan W, Sommar J, Zhu W, Feng X B, 2016b. Emission-dominated gas exchange of elemental mercury vapor over natural surfaces in China. Atmospheric Chemistry and Physics, 16: 11125-11143.

Wang X, Zhang H, Lin C J, Fu X W, Zhang Y P, Feng X B, 2015. Transboundary transport and deposition of Hg emission from springtime biomass burning in the indo-China Peninsula. Journal of Geophysical Research-Atmospheres, 120: 9758-9771.

Wang Y G, Huang J Y, Hopke P K, Rattigan O V, Chalupa D C, Utell M J, Holsen T M, 2013. Effect of the shutdown of a large coal-fired power plant on ambient mercury species. Chemosphere, 92: 360-367.

Wang Z W, Chen Z S, Duan N, Zhang X S, 2007. Gaseous elemental mercury concentration in atmosphere at urban and remote sites in China. Journal of Environmental Sciences-China, 19: 176-180.

Wangberg I, Munthe J, Pirrone N, Iverfeldt A, Bahlman E, Costa P, Ebinghaus R, Feng X, Ferrara R, Gardfeldt K, Kock H, Lanzillotta E, Mamane Y, Mas F, Melamed E, Osnat Y, Prestbo E, Sommar J, Schmolke S, Spain G, Sprovieri F, Tuncel G, 2001. Atmospheric mercury distribution in Northern Europe and in the Mediterranean region. Atmospheric Environment, 35: 3019-3025.

Watson J G, Chen L W A, Chow J C, Doraiswamy P, Lowenthal D H, 2008. Source apportionment: Findings from the US supersites program. Journal of the Air & Waste Management Association, 58: 265-288.

Weiss-Penzias P, Amos H M, Selin N E, Gustin M S, Jaffe D A, Obrist D, Sheu G R, Giang A, 2015. Use of a global model to understand speciated atmospheric mercury observations at five high-elevation sites. Atmospheric Chemistry and Physics, 15: 2225.

Weiss-Penzias P, Gustin M S, Lyman S N, 2009. Observations of speciated atmospheric mercury at three sites in Nevada: Evidence for a free tropospheric source of reactive gaseous mercury. Journal of Geophysical Research-Atmospheres, 114, D14302, doi: 10.1029/2008JD011607.

Weiss-Penzias P, Jaffe D A, Mcclintick A, Prestbo E M, Landis M S, 2003. Gaseous elemental mercury in the marine boundary layer: Evidence for rapid removal in anthropogenic pollution.

Environmental Science & Technology, 37: 3755-3763.

Weiss-Penzias P S, Gustin M S, Lyman S N, 2011. Sources of gaseous oxidized mercury and mercury dry deposition at two southeastern US sites. Atmospheric Environment, 45: 4569-4579.

Wilson S J, Steenhuisen F, Pacyna J M, Pacyna E G, 2006. Mapping the spatial distribution of global anthropogenic mercury atmospheric emission inventories. Atmospheric Environment, 40: 4621-4632.

Wu Q R, Wang S X, Li G L, Liang S, Lin C J, Wang Y F, Cai S Y, Liu K Y, Hao J M, 2016. Temporal trend and spatial distribution of speciated atmospheric mercury emissions in China during 1978-2014. Environmental Science & Technology, 50: 13428-13435.

Wu Y, Wang S X, Streets D G, Hao J M, Chan M, Jiang J K, 2006. Trends in anthropogenic mercury emissions in China from 1995 to 2003. Environmental Science & Technology, 40: 5312-5318.

Xiao Z, Sommar J, Wei S, Lindqvist O, 1997. Sampling and determination of gas phase divalent mercury in the air using a KCl coated denuder. Fresenius Journal of Analytical Chemistry, 358: 386-391.

Xiao Z F, Munthe J, Schroeder W H, Lindqvist O, 1991. Vertical fluxes of volatile mercury over forest soil and lake surfaces in Sweden. Tellus Series B-Chemical and Physical Meteorology, 43: 267-279.

Xin M, Gustin M S, 2007. Gaseous elemental mercury exchange with low mercury containing soils: Investigation of controlling factors. Applied Geochemistry, 22: 1451-1466.

Xu L L, Chen J S, Niu Z C, Yin L Q, Chen Y T, 2013. Characterization of mercury in atmospheric particulate matter in the southeast coastal cities of China. Atmospheric Pollution Research, 4: 454-461.

Xu L L, Chen J S, Yang L M, Niu Z C, Tong L, Yin L Q, Chen Y T, 2015. Characteristics and sources of atmospheric mercury speciation in a coastal city, Xiamen, China. Chemosphere, 119: 530-539.

Xu X, Akhtar U S, 2010. Identification of potential regional sources of atmospheric total gaseous mercury in Windsor, Ontario, Canada using hybrid receptor modeling. Atmospheric Chemistry and Physics, 10: 7073-7083.

Yang Y K, Chen H, Wang D Y, 2009. Spatial and temporal distribution of gaseous elemental mercury in Chongqing, China. Environmental Monitoring and Assessment, 156: 479-489.

Yu B, Wang X, Lin C J, Fu X W, Zhang H, Shang L H, Feng X B, 2015. Characteristics and potential sources of atmospheric mercury at a subtropical near-coastal site in East China. Journal of Geophysical Research-Atmospheres, 120: 8563-8574.

Zhang H, Fu X, Lin C J, Shang L, Zhang Y, Feng X, Lin C, 2016a. Monsoon-facilitated characteristics and transport of atmospheric mercury at a high-altitude background site in southwestern China. Atmospheric Chemistry and Physics, 16: 13131-13148.

Zhang H, Fu X W, Lin C-J, Wang X, Feng X B, 2015a. Observation and analysis of speciated atmospheric mercury in Shangri-La, Tibetan Plateau, China. Atmospheric Chemistry and Physics, 15: 653-665.

Zhang H H, Poissant L, Xu X H, Pilote M, 2005. Explorative and innovative dynamic flux bag method development and testing for mercury air-vegetation gas exchange fluxes. Atmospheric Environment, 39: 7481-7493.

Zhang L, Blanchard P, Gay D A, Prestbo E M, Risch M R, Johnson D, Narayan J, Zsolway R, Holsen T M, Miller E K, Castro M S, Graydon J A, St Louis V L, Dalziel J, 2012. Estimation of speciated and total mercury dry deposition at monitoring locations in eastern and central North America.

Atmospheric Chemistry and Physics, 12: 4327-4340.

Zhang L, Vet R, Wiebe A, Mihele C, Sukloff B, Chan E, Moran M D, Iqbal S, 2008. Characterization of the size-segregated water-soluble inorganic ions at eight Canadian rural sites. Atmospheric Chemistry and Physics, 8: 7133-7151.

Zhang L, Wang S X, Wang L, Hao J M, 2013. Atmospheric mercury concentration and chemical speciation at a rural site in Beijing, China: Implications of mercury emission sources. Atmospheric Chemistry and Physics, 13: 10505-10516.

Zhang L, Wang S X, Wang L, Wu Y, Duan L, Wu Q R, Wang F Y, Yang M, Yang H, Hao J M, Liu X, 2015b. Updated emission inventories for speciated atmospheric mercury from anthropogenic sources in China. Environmental Science & Technology, 49: 3185-3194.

Zhang L M, Gong S L, Padro J, Barrie L, 2001. A size-segregated particle dry deposition scheme for an atmospheric aerosol module. Atmospheric Environment, 35: 549-560.

Zhang L M, Wright L P, Blanchard P, 2009. A review of current knowledge concerning dry deposition of atmospheric mercury. Atmospheric Environment, 43: 5853-5864.

Zhang Y Q, Liu R H, Wang Y, Cui X Q, Qi J H, 2015c. Change characteristic of atmospheric particulate mercury during dust weather of spring in Qingdao, China. Atmospheric Environment, 102: 376-383.

Zhang Y X, Jacob D J, Horowitz H M, Chen L, Amos H M, Krabbenhoft D P, Slemr F, St Louis V L, Sunderland E M, 2016b. Observed decrease in atmospheric mercury explained by global decline in anthropogenic emissions. Proceedings of the National Academy of Sciences of the United States of America, 113: 526-531.

Zhu J, Wang T, Talbot R, Mao H, Hall C B, Yang X, Fu C, Zhuang B, Li S, Han Y, Huang X, 2012. Characteristics of atmospheric total gaseous mercury (TGM) observed in urban Nanjing, China. Atmospheric Chemistry and Physics, 12: 12103-12118.

Zhu J, Wang T, Talbot R, Mao H, Yang X, Fu C, Sun J, Zhuang B, Li S, Han Y, Xie M, 2014. Characteristics of atmospheric mercury deposition and size-fractionated particulate mercury in urban Nanjing, China. Atmospheric Chemistry and Physics, 14: 2233-2244.

Zhu J S, Wang D Y, Ma M, 2013. Mercury release flux and its influencing factors at the air-water interface in paddy field in Chongqing, China. Chinese Science Bulletin, 58: 266-274.

Zielonka U, Hlawiczka S, Fudala J, Wangberg I, Munthe J, 2005. Seasonal mercury concentrations measured in rural air in Southern Poland: Contribution from local and regional coal combustion. Atmospheric Environment, 39: 7580-7586.

第 6 章 高山地区汞的生物地球化学循环

本章导读

- 简要介绍高山地区生态环境的特性和汞的生物地球化学循环研究关注的问题及其研究特点；择要介绍了高山地区汞的特殊环境行为，包括大气汞形态和特征，大气汞沉降和海拔梯度分布及效应。
- 以冰芯和湖芯为主，概述高山地区汞的历史记录；选取高山地区汞循环的几个关键生物化学过程，论述冰川消融对汞的释放和迁移的影响、冻土的汞库效应和向大气释汞以及高山森林汞的生物地球化学过程。
- 阐述高山汞循环的复杂性和重要性及其未来变化趋势的不确定性，并提出未来研究的要点和方向。

6.1 高山地区生态环境与汞的生物地球化学循环

6.1.1 高山地区生态环境

山地是具有一定海拔高度和坡度的陆面单元，以垂向的突出性和水平的延伸性为特点（王明业，1988）。山地区域气候和水分条件随海拔梯度发生显著变化，是自然地理要素密集交叉、相互作用强烈和反馈多变的区域。山地地区尤其是高山地区的生态环境系统是陆地生态系统中极其重要、敏感且复杂的生态系统之一。

全球范围内，亚洲内陆以青藏高原为主体的高海拔山体和区域极大改变了亚洲的地理格局与气候格局，发源了长江、黄河和恒河等亚洲大河，被称为"世界屋脊"、"亚洲水塔"，是地球"第三极"；欧洲的阿尔卑斯山脉是中欧温带大陆性温润气候和南欧亚热带气候的分界线，发源了多瑙河和莱茵河等欧洲大河；落基山脉被称为北美洲的"脊梁"，包括密西西比河在内的几乎所有北美河流都发源于此；南美洲的安第斯山脉是全球陆地上最长的山脉，是世界上流量最大、流域面积最广的大河——亚马孙河的源区。

高山地区的海拔更高，人类活动更为稀少。随着海拔升高，气温变低，高山地区发育了冰川、积雪和冻土等特有环境介质；植被类型呈现从落叶林到针叶林再到高山草甸的变化分布；鱼类等动物的生理特征也与平原地区存在显著不同。这些都造成高山地区生态环境系统中物质和能量的循环表现为独特性、多样性和复杂性。在高山生态环境系统开展元素分布和迁移转化及其生态环境效应研究有助于进一步拓展和深化生物地球化学领域的理论和方法，揭示全球气候变化和人类活动进程对区域生态环境的影响。

6.1.2 高山地区汞的生物地球化学循环

汞是常温下唯一可以气态存在的重金属元素，可随大气环流进行区域和全球传输。一般而言，高山地区远离人类活动聚集区，受到人为汞排放的直接影响较小，大气汞的长距离传输是汞进入高山地区的主要途径。

高山地区汞的生物地球化学循环研究关注长距离传输汞到达山区后的沉降、分布、迁移转化过程和生态环境效应等。汞在各类环境介质中的本底含量和形态分布是开展高山地区汞生物地球化学循环研究的基础。同时，由于高山地区的偏远属性，汞的含量和历史记录常被用于理解人为汞排放对区域环境影响和历史变化的重要方面；高山地区气候环境因子随海拔梯度变化大且剧烈，汞的迁移转化随温度、气压和降水量等梯度变化的变异性更强；高山地区各类环境单元和要素集中耦合，因此在开展研究时应注重汞的生物地球化学循环过程的完整性；在以上基础上认识和评价汞污染的局地生态环境效应及其潜在的远程影响，可以为保护人居生态环境和人体健康提供科学建议和决策支撑。

6.2 高山地区汞的重要环境行为

6.2.1 大气汞形态和特征

高山地区远离人类源区，且海拔较高，是开展区域大气汞含量背景监测和自由大气中汞行为研究的理想地区。全球高山地区代表性的观测站点有：亚洲的青藏高原地区，包括瓦里关、纳木错、珠穆朗玛峰和香格里拉等观测站，海拔均在 3000 m 以上，主要受亚洲季风和西风气团影响；北美的学士山观测站（Mount Bachelor Observatory）和风暴峰天文台（Storm Peak Laboratory），主要受太平洋海洋性气团影响；欧洲的比利牛斯山 Pic du Midi 天文台和阿尔卑斯山少女峰（Mt. Jungfraujoch）观测站，受欧洲大陆排放影响较大。南美地区高海拔山区大气汞的长期连续观测还未见公开数据报道，在玻利维亚的 Chacaltaya 山已开展了连续的在线大气汞观测。此外，一些区域内相对海拔较高的高山站也是监测大气汞本底和长距离传输过程的重要站点。

高山地区大气汞形态与平原地区有显著差异。一般而言，大气总汞中95%以上是气态单质汞，活性气态汞和颗粒态汞的比例在一些平原地区甚至低于1%。因此，以全球背景区大气单质汞的平均浓度（1.74±0.71）ng/m^3计算（参见5.2.2节），平原地区活性气态汞和颗粒态汞的浓度一般仅为几个pg/m^3，但在一些高山地区，大气活性气态汞平均浓度较高甚至出现数百个pg/m^3的异常值。例如美国学士山观测站和风暴峰天文台的大气活性气态汞平均浓度分别为43 pg/m^3和（20±21）pg/m^3，是美国低海拔背景区活性气态汞浓度的10~20倍（Fain et al.，2009；Swartzendruber et al.，2006），最高值甚至达到600 pg/m^3。在学士山观测站2008年春季短短两个月的观测时间段内，出现了8次活性气态汞高值事件，最高值为50~137 pg/m^3，这期间颗粒态汞的含量也有微弱升高。活性气态汞与臭氧、气溶胶微粒和一氧化碳等污染物未显示出显著相关，说明这些异常高值的活性气态汞并非主要来源于局地对流和长距离传输的人为汞污染气团。分析表明，相对湿度与活性气态汞变化最为密切，高值活性气态汞全部对应于相对湿度40%以下时段。研究还提出了学士山观测站大气活性气态汞的3种主要来源模式（Timonen et al.，2013）：一是自由对流层顶部和平流层富含活性气态汞大气气团的入侵，主要是从高纬度向中纬度入侵；二是陆源长距离传输汞，主要是亚洲排放的大气单质汞在自由对流层长距离传输过程中被氧化为活性气态汞，长距离传输至美国西部的高海拔地区；三是海洋源长距离传输汞，是来自太平洋边界层的大气单质汞在相对湿度极低的条件下被源自大洋的卤族氧化物氧化为活性气态汞，继而随大气环流爬升输送至美国西部的高海拔地区。欧洲的Pic du Midi天文台观测显示（Fu et al.，2016），大气各形态汞均明显高于全球背景值，其中大气单质汞主要受到法国南部低海拔地区汞排放的影响，大气颗粒态汞和活性气态汞主要受到长距离大气传输的影响，其中颗粒态汞主要来自于北美和北大西洋地区的自由对流层，而活性气态汞则主要来自于低纬度大西洋的自由对流层下部和高纬度地区的自由对流层上部地区，这些地区较高的大气氧化物浓度有利于大气单质汞发生氧化。

大气汞在对流层和平流层中的垂直分布一般表现为：气态单质汞随海拔升高而降低，在平流层保持较低的稳定值；而活性气态汞则正好相反，其在对流层顶部和平流层中较高（图6-1）。全球航测资料和对活性气态汞的垂直剖面理论计算有一定吻合。比如飞机航测对得克萨斯州自由对流层（海拔7000 m）活性气态汞的监测发现其平均浓度达到了（266±38）pg/m^3，约是背景区边界层的50~100倍（Gratz et al.，2015）。研究发现，赤道辐合带空气的强烈对流能导致海洋排放的卤族元素传输到平流层，进而发生大规模的大气单质汞氧化现象，形成的活性气态汞进而通过长距离的传输输送到美国中部（Gratz et al.，2015）。高山地区的汞观测和研究为理解汞在垂直梯度上的分布和行为以及对流层自由大气中汞的转化和传输提供了不可多得的信息。

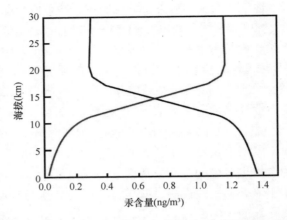

图 6-1 模型计算显示的大气气态单质汞和氧化态汞（包括活性气态汞和颗粒态汞）的垂直分布模式（黑线表示气态单质汞，红线表示氧化态汞）

以上这些观测和分析说明高山地区更接近平流层，且处在自由大气之中，易受到高层大气入侵或长距离传输影响。此外，高海拔地区特殊大气物理和化学组分组合条件下存在气态单质汞的氧化，包括长距离传输过程中的氧化和原位氧化。

由于高山地区气温较低，气溶胶含量一般较低且空间差异较大。不同区域的对比发现，近海区域的气态氧化汞高。而地处大陆深处的青藏高原则未表现出这一特征，如在纳木错（de Foy et al., 2016）、瓦里关（Fu et al., 2012）和贡嘎山（Fu et al., 2008）的研究都无一例外地表现为颗粒态汞高于活性气态汞，这可能显示了 Br 和 Cl 等海洋源卤族元素对气态单质汞氧化的重要性和对区域大气汞形态分布的重要影响。

从机理角度而言，在研究高山地区大气汞形态分布时，大气汞的气粒分配是一个重要指标，是大气汞化学反应中的非均相反应的第一步。大气活性汞气粒分配是指 Hg(Ⅱ)按不同比例以活性气态汞和颗粒态汞两种形式存在的状态及分配过程。Hg(Ⅱ)在气相和在干燥气溶胶颗粒物之间的吸附分配系数，主要受颗粒物的化学构成及环境温度的影响（Rutter and Schauer, 2007a; 2007b）。颗粒物化学构成方面，在以硝酸盐和氯化盐为主的颗粒物中，Hg(Ⅱ)具有更高的气-固分配系数（如硝酸盐：$100 \sim 1000 \ m^3/\mu g$；氯化盐：$10 \sim 100 \ m^3/\mu g$；有机质/硫酸盐：约 $1 \ m^3/\mu g$），从而使得 Hg(Ⅱ)更容易吸附在以此为主要成分的颗粒物表面。而以在硫酸铵或有机质为主的气溶胶中，其分配系数则比前者要低一个数量级以上，导致其中Hg(Ⅱ)更易于以气相存在。另一方面，随着环境温度升高，颗粒物上附着的 Hg(Ⅱ) 更倾向于分配到气相中。

根据观测数据，Hg(Ⅱ)气粒分配系数可以利用活性气态汞、颗粒态汞、$PM_{2.5}$颗粒物以及温度建立经验公式。

$$K = (PBM/PM_{2.5})/RGM \tag{6-1}$$

式中，K 为气粒分配系数，PBM 为颗粒态汞浓度（pg/m^3），RGM 为活性气态汞浓度（pg/m^3），$PM_{2.5}$ 为细颗粒物浓度（$\mu g/m^3$）。该经验公式给出了气态 Hg(Ⅱ)和固态 Hg(Ⅱ)之间的均衡关系。Hg(Ⅱ)的两种主要形式，即活性气态汞和颗粒态汞具有相似的挥发性，因而单一的均衡常数可以适用。式（6-1）中利用 $PM_{2.5}$ 浓度进行标准化处理引入了另一个假设前提，即气溶胶颗粒对 Hg(Ⅱ)的吸附量与气溶胶质量浓度成比例增长，在设定气溶胶体积和表面积比例一定的前提下，气溶胶的吸附能力可用其质量浓度来表征。

基于多个站的监测数据进行拟合（Amos et al., 2012），得到气粒分配系数 K 与温度 T 的关系如下：

$$\log_{10}(K^{-1}) = (10 \pm 1) - (2500 \pm 300)/T \tag{6-2}$$

式中，K 为气粒分配系数，T 为温度（K）。该拟合结果介于前人理论计算结果和实测结果之间。需要指出的是，目前对环境中相对湿度、光照等条件对 Hg(Ⅱ)气粒分配产生的影响尚不明确。

6.2.2 大气汞沉降

汞的干湿沉降是大气汞经过长距离传输进入陆表生态系统的主要途径，相较于气态单质汞，活性气态汞和颗粒态汞更易沉降或被降水清除。高山地区地形垂直多变，气温、降水类型和降雨量等随着海拔和坡向会发生变化，导致大气汞的干湿沉降量发生变化甚至有显著差异。

关于降水类型对大气汞的清除和沉降的影响研究较少，主要工作和进展基本来自美国汞沉降观测网（Mercury Deposition Network，MDN）的采样和数据分析。Shanley 等（2015）在波多黎各卢基约（Luquillo）山对降水开展以周为时间尺度的收集，对 2005 年 4 月至 2007 年 3 月连续两年的采样分析意外发现，这一地区大气中氧化态汞的含量小于 10 pg/m^3，接近全球背景值，是非常洁净的地区。但降雨中总汞的体积加权平均浓度为 9.8 ng/L，夏季最高，汞的降水量效应（即随降水增加汞含量减小的趋势）非常微弱；汞的年湿沉降通量高达 27.9 $\mu g/m^3$，是北美地区最高的地区之一。进一步分析表明，降水汞含量与降水最大高度的长期变化趋势一致 [图 6-2（a）]，月尺度降水汞含量与 NOAA NEXRAD 雷达站测试的降水云回波顶高度（体积加权平均）呈现正相关 [图 6-2（b）]。这说明高空雨滴在大气混合层以上对流云顶部能清除较多的氧化态汞，造成地面降水汞含量高，进而产生高汞湿沉降量。

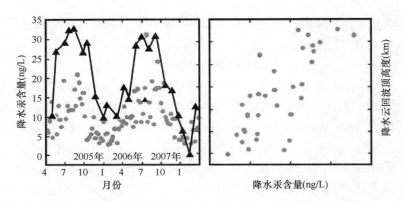

图 6-2 （a）周尺度降水汞浓度与降水云回波顶高度（当月最高值）变化趋势；
（b）降水汞的浓度的月混合值与降水云回波顶高度（体积加权平均）关系
［引用并修改自（Shanley et al., 2015）］

研究者进而开展降水特征对降水汞浓度和沉降影响的研究。Holmes 等（2016）对 MDN 东部 7 个站点的 817 次单场降水事件中汞含量和降水特征做了分析，结果显示，强对流风暴等降水中汞浓度较弱对流降水和层云降水高 50%。结合雷达和卫星数据分析表明，强对流天气系统可达对流层上部，这一层位的氧化态汞含量较高，因而降水中的汞含量较高。降水天气特征，特别是风暴频次和总降水量是造成不同站点汞沉降差异的主要原因。研究者在分析的基础上系统提出了降水类型对汞沉降影响的机理假设（图 6-3）。

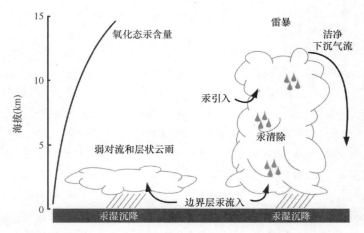

图 6-3 不同降水类型对降水中汞含量影响原理示意图
［引用并修改自（Holmes et al., 2016）］

Kaulfus 等（2017）进一步对 MDN 更多站点的 525 次单场降雨事件的汞浓度进行了分析，并利用雷达观测将降水类型进行了细致划分。结果显示，降水中的汞含量与降水量呈幂函数关系，而降水量则受降水类型控制（图 6-4）。在同一降水量水平上，汞沉降按降水类型不同从高到低依次为超级单体雷暴（supercell thunderstorms）、紊乱雷暴（disorganized thunderstorms）、准线形对流系统（quasi-linear convective systems QLCS）、温带气旋（extratropical cyclones）、小雨（light rain）和登陆热带气旋（land-falling tropical cyclones）。前三种对流形态的降水比后三种非对流形态降水能提高汞沉降至少 1.6 倍。在剔除降水量和地形等因素后，发现汞的湿沉降在高海拔站点较高，特别是在夏季，由于对流形降雨较多，湿沉降汞量更高。这说明高山地区多发的对流降水可能会诱发更高的汞湿沉降通量。

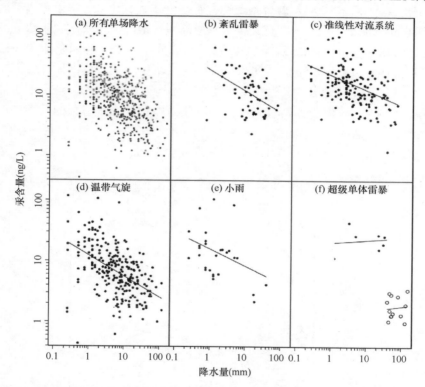

图 6-4　不同降水类型中汞含量与降水量关系图［引用并修改自（Kaulfus et al.，2017）］
(a) 单场降水事件散点图：黄色代表超级单体雷暴，红色代表紊乱雷暴，绿色代表准线形对流系统，蓝色代表温带气旋，黑色代表小雨，灰色代表登陆热带气旋；(b~f) 不同降水类型汞浓度

此外，需要注意的是，高山地区随海拔升高降雨类型多变，雨雪比例变化较大，高海拔地区的降水多以降雪形式发生。有观测显示降雪对汞的清除效率比降

水低（Lombard et al.，2011；Sigler et al.，2009）。迎风坡和背风坡降雨量和降雨类型的不同也会导致大气干湿沉降有显著差异。山地地区强对流降雨较多，部分山区降水量随海拔升高而增多，并在某一个高度出现最大降水量带。因此高山地区降水中的汞含量和沉降量会因山区独特的降水类型和分布趋势不同，表现出变化的降水汞浓度和沉降通量。

6.2.3 海拔梯度分布及效应

海拔梯度分布是指汞在某一区域某一种介质中分布时呈现的随海拔梯度变化，汞形态和含量也随着海拔梯度变化呈现规律性变化，一般被称为海拔效应。

在高海拔地区，冰川和积雪分布广泛，常被用于反映大气干湿沉降信息。Susong 等（1999）于 1999 年就曾报道称在瓦萨奇和提顿山脉，年汞积累量随海拔提升 1000 m 增加了 1~1.7 倍。有关青藏高原高海拔地区冰川雪冰中汞浓度的海拔梯度变化的报道较多：Huang 等（2012b）在高原不同地区的四条冰川（5200~6600 m）采样分析发现，虽然不同冰川表层雪汞浓度的浓度范围差异较大，但对各条冰川而言，汞浓度均呈现随海拔梯度增高的趋势（图 6-5），这主要可能是由于海拔较高的地区气温更低，有利于大气氧化态中颗粒态汞组分的升高并增强其向雪冰的沉降；然而，随后在青藏高原北部的老虎沟 12 号冰川（4400~5100 m）和小冬克玛底冰川（5400~5700 m）的多次采样中却并未发现表层雪中汞浓度随

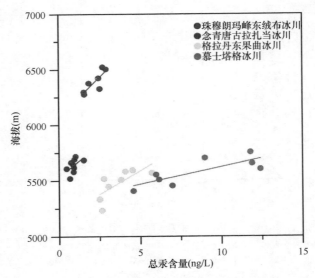

图 6-5 青藏高原四条典型冰川表层雪中汞浓度随海拔变化趋势
［引用并修改自（Huang et al.，2012b）］

海拔梯度增高的整体规律（Paudyal et al.，2017；Sun et al.，2018a）。同时，在冰川附加冰带中意外发现了极高的汞含量（Huang et al.，2014），且高原北部的老虎沟12号冰川和南部的珠穆朗玛峰东绒布冰川中附加冰中的汞含量均呈现出随海拔梯度升高增高的现象（Huang et al.，2014；Sun et al.，2018a）。由于冰川附加冰是由于冰川雪淋融作用和融水再冻结而形成的，雪冰中汞的淋融流失随海拔升高气温降低而逐渐减弱，因此保存了更多的汞。

在北美内华达山脉，积雪的总雪柱中总汞和溶解态汞浓度随海拔上升而增高（Pearson et al.，2015），研究者给出的解释是海拔越高，积雪深度和密度越大，而85%左右的光致还原作用都发生在积雪上层10 cm范围内，由于高海拔地区积雪的形态不利于阳光穿透积雪，还原作用减弱，促使积雪中保留了更多汞。

综上可见，高海拔冰川雪冰中汞的海拔梯度变化呈现出较强的空间变异性。汞在雪冰中的浓度受大气干湿沉降量、雪冰类型、局地气象条件等因素影响。单纯的浓度和海拔高度间的变化还仅仅是揭示现象，对其控制机理的阐明还需要获取更多因子进行综合分析。

土壤中的汞分布也有类似的海拔效应，尤以林地土壤中的研究最为丰富且系统。对北美洲14个典型森林林地的土壤汞分析表明（Obrist et al.，2011），土壤中的汞含量与纬度、降水量和有机质含量关系最为密切，海拔高度与土壤汞含量的关系并不一致，但仍是潜在的影响因素。具体到更小的空间尺度，Stankwitz等（2012）在美国Green山的Camels峰沿540~1160 m海拔以每60 m高度变化间距采集表层土壤（有机层），发现土壤中汞累积量从0.99 mg/m^{2}增加到7.65 mg/m^2；Townsen等（2014）在美国纽约州Catskill山的研究同样发现土壤中汞含量从海拔200 m到1200 m增加了4.4倍。来自Whiteface山的研究表明，表层土壤有机层，特别是高分解有机层中汞的含量与海拔的关系呈显著正相关（Blackwell and Driscoll，2015a）。甲基汞含量在中高纬度的针叶林区含量最高（0.39 ng/g±0.07 ng/g），远高于低海拔地区的落叶林（0.17 ng/g±0.02 ng/g），但甲基汞比率随海拔升高而降低。高山林地一般呈现显著的垂直分布规律，从低到高依次为落叶林、针叶林和高山灌木及草甸，降水量、降水类型及土壤的理化性质（特别是有机质含量）也随着发生变化。相应地，大气汞沉降输入量及土壤中汞的积累量也发生变化，这些因素的集合塑造了高山林地土壤汞含量的海拔变化趋势。Gerson等（2017）基于系统研究，给出了林地系统总汞和甲基汞主要输入因子的通量和土壤中的汞积累量，细致描绘了不同海拔区域土壤中汞积累量及其可能来源。

生物体中的汞含量随海拔变化的趋势也有研究和报道。Blais等（2006）发现法国比利牛斯山11个高海拔湖泊中的鱼类肌肉的总汞含量与海拔变化显著正相关，而与鱼龄、营养级和生长速率无关，指出高海拔地区食物网的汞沉降量和富集效率可能较高。但美国纽约州 44 个湖泊的浮游动物、小龙虾和鱼类的研究表明（Yu et al.，2011），生物体中的汞含量并未显示出与海拔的相关性，而浮游动物和鱼类的甲基汞富集效率与海拔有一定正相关。

综上所述，高山地区地形多变，海拔梯度大，环境介质中汞的形态、含量和富集特征等随海拔梯度的变化成为研究高山地区汞生物地球化学循环的关注点。普遍认为，高海拔地区大气汞沉降随海拔升高、气温降低而增强，是长距离传输大气汞的"汇"，海拔越高，其"汇"的表现和作用越强，汞随海拔梯度的分布具备正向增强的趋势。但实际的调查研究则显示了汞海拔梯度分布的不均一性，大范围空间内汞的分布与海拔并非简单的正相关，同一地区不同时间段针对相同介质得到的汞海拔梯度分布也不尽相同，甚至完全相反，这说明高山地区汞分布的差异性和复杂性。

海拔梯度效应是环境介质中的汞在垂直空间分布上的一种表象，其实质是海拔梯度上的基本自然因子，特别是气温、降水、光照等因素的规律变化直接或间接影响环境介质属性和汞行为的综合效应。汞在高山某一环境介质中的含量可简化为汞的收入和支出间的差值，即

$$[Hg] = f[Hg]_{收入} - f[Hg]_{支出}$$

式中，[Hg]为汞含量，$f[Hg]_{收入}$或$f[Hg]_{支出}$是对汞含量起到增加或减少的效应的总和。对于汞随海拔梯度增加的分布趋势，可以有以下三种情况：①$f[Hg]_{收入}$随海拔升高增大，而$f[Hg]_{支出}$随海拔升高减小；②$f[Hg]_{收入}$随海拔升高增大，但$f[Hg]_{支出}$随海拔升高不变或变化微弱；③$f[Hg]_{收入}$随海拔升高不变或变化微弱，但$f[Hg]_{支出}$随海拔升高而增大。汞的海拔分布规律研究应在注重综合评价汞的输入和输出因子的基础上进行，深入辨析高山地区汞的空间分布规律和主控因子。

6.3 高山地区汞的历史记录

6.3.1 湖泊沉积物

湖泊沉积物是记录湖泊及其流域气候环境信息的有效载体，具有连续性好、分辨率高、信息量丰富和地理覆盖广等优势。高山地区湖泊分布广泛，且沉积物样品相对冰芯等较易获取，定年技术和方法成熟，因而利用湖泊沉积物重建和认识汞的历史变化研究较多。全球范围内高山地区的湖泊沉积物记录反映了近几百年来人为排放汞增加对环境的扰动，但不同区域记录的历史变化有一定差异。

欧洲西班牙境内 Pre-Pyrenees 山 Montcortes 湖沉积物的汞记录显示（Corella et al.，2017），汞的沉积通量自 1550 年起显著上升，与西班牙当时的矿业开发兴起

时段相对应，小冰期期间的冷湿气候环境有利于汞沉积通量的增加，降水量变化和矿产开发是工业革命前期汞沉积通量变化的主要影响因素；工业革命后，汞的沉积通量开始上升，并于 20 世纪 40 年代达到最高，与当地汞矿产量的峰值期一致，近几十年来汞的沉积通量开始下降，反映了欧洲人为汞排放下降的趋势。Guedron 等（2016）利用法国阿尔卑斯山 Luitel 湖沉积物恢复了工业革命后汞沉积通量的变化趋势，并以汞同位素判断了汞的来源，汞沉积通量自 1860 年的 45 μg/(m²·a)到 1915 年左右的第一次世界大战期间持续上升，到第二次世界大战末期达到 250 μg/(m²·a)的峰值，随后持续下降到 90 μg/(m²·a)。研究者指出大气远距离传输沉降是湖泊中汞的主要来源，认为虽然汞同位素无法给出沉积物中汞的具体人为来源，但能够用于反映工业活动和城市生活源相对贡献的变化趋势：汞的非质量分馏和沉积通量变化特征可能反映了 1860~1910 年期间工业汞排放以及 1950~1980 年期间现代工业和城市排放增加的趋势；汞的质量分馏能作为判断人为燃烧源（表现为 $\delta^{202}Hg$ 偏负）和化工活动来源（表现为 $\delta^{202}Hg$ 偏正）的依据。

来自北美东部 Whiteface 山的两个高山湖泊（Clear 湖和 West Pine 湖）的汞历史记录也显示（Sarkar et al.，2015），汞的沉积通量为 2~50 μg/(m²·a)，与周边地区湖泊记录一致。汞沉积通量自 19 世纪 90 年代开始上升，20 世纪 50~70 年达到峰值期，之后两个湖泊的趋势出现分异，其中 Clear 湖汞记录自 1965 年后缓慢下降，但 West Pine 湖则保持持续上升，可能与两个湖泊流域内汞的输入来源和贡献差异有关。北美西部落基山 9 个高山湖泊的沉积物记录显示（Mast et al.，2010），多数湖泊沉积物中汞含量自 1900 年上升，1980 年后达到峰值，1875 年前汞的沉积通量为 5.7~42 μg/(m²·a)，1985 年后的沉积通量为 17.7~141 μg/(m²·a)。

亚洲中部青藏高原湖泊众多，为利用湖芯记录认识区域大气汞沉降和空间变化提供了可能。Wang 等（2010）利用青藏高原东北部青海湖浅湖芯揭示了过去一百年汞的变化趋势，1970 年后汞含量和沉积通量开始急剧上升，这与中国改革开放后经济的持续增长吻合。Yang 等（2010b）在高原北、中、南部三个区域各选取三个湖泊研究了沉积物中汞的历史变化，与欧美地区不同，高原湖泊沉积物的汞沉积通量自 20 世纪 70 年代后才开始显著上升，20 世纪 90 年代后上升速度更快，这种趋势与亚洲特别是中国和印度的经济增长同步。作者认为高原地区湖泊中的汞主要来自周边地区大气汞的长距离传输沉积，反映了亚洲地区人为汞排放的进程和变化。随后，Kang 等（2016）利用喜马拉雅南北坡 8 个湖芯再次证实了这一"亚洲趋势"，并进一步指出高原内陆与喜马拉雅山南坡高海拔湖泊中汞的总体历史变化趋势较为一致，与亚洲特别是南亚地区人为汞生产和排放的历史进程基本一致。这些研究表明大空间尺度上"阵列式"湖泊沉积物研究可以揭示长距离传输汞的源区和历史变化趋势。

在研究湖泊沉积物汞的历史序列时，不同时期汞的沉积通量比值，即现代（如1950年之后或2000年之后）与工业革命前（例如1860年前）的比值（Modern/Pre-industrial）常用于评价人为汞排放对汞的环境本底的影响程度。普遍认为，由于人为汞排放的增加，使得全球范围内大气汞沉降通量水平较工业革命前增加了3倍左右（UNEP，2013）。研究显示，高海拔地区部分湖泊沉积物中现代与工业革命前汞沉积通量的比值高于通常认为的3倍幅度（Drevnick et al.，2010；Yang et al.，2010b）。例如，内华达山Lake Tahoe中现代与工业革命前汞沉积通量的比值为7.5~10，研究者认为这一地区海拔较高，夜间山风促使高层自由大气输入，造成气态氧化汞和氧化物均较高，增强了大气汞向湖泊系统沉降的量。同样，青藏高原海拔高于3000 m湖泊汞沉降通量的现代与工业革命前比值也远远大于3（Yang et al.，2010b），也再次印证了高海拔地区湖泊中汞沉积通量增幅更显著这一观点。但来自加拿大中西部班夫国家公园9个高山湖泊的沉积物汞记录显示（Phillips et al.，2011），虽然在过去150年内汞沉积通量也显示了上升的趋势，但其20世纪后期与1800~1850年时段的比值仅为1.8。另外，非洲乌干达Rwenzori Mountain高山湖泊Lake Bujuku（4000 m）和Lake Kitandara（2700 m）的记录显示（Yang et al.，2010a），其现代与工业革命前汞沉积通量的比值为3，并未显示异常。研究者认为可能是由于赤道附近大气边界层较高，不利于富含气态氧化汞的高层自由大气入侵；再者，赤道附近大气中的氧化物（如臭氧等）也较中纬度地区低。

总之，高海拔地区湖泊中汞的沉积通量及其历史变化趋势和幅度虽然存在区域差异，但无一例外显示了其现代本底较工业革命前的增加，也间接证实了汞可以长距离传输进入高山地区并对环境产生扰动。湖泊沉积物中的汞主要来源于大气干湿沉降直接输入和流域内物质输入，湖内汞的沉降和沉积后过程也会影响沉积物中汞的记录，这些都应在阐释湖泊沉积物汞历史记录中予以考虑。例如，一般认为湖泊沉积物可忠实记录大气汞沉降历史记录，但也有研究指出（Yang，2015），对于径流补给较多的流域，近期受气候变化影响，湖泊沉积物记录了大气沉降和流域输入的叠加信号，不能忠实反映北美等地区20世纪80年代后大气汞排放下降的趋势（图6-6）。

6.3.2 冰芯

冰川的形态改变和其自身保存的大气沉降是气候环境变化的敏感指标。冰川的物质来自大气干湿沉降，是高山地区最能直接反映大气环境变化的信息载体。山地冰川相对于极地冰川具有较高的积累量，其位置也更为接近人类活动区。这些特点使得山地冰川成为重建古气候环境变迁以及人类和地球相互作用的优良介质。

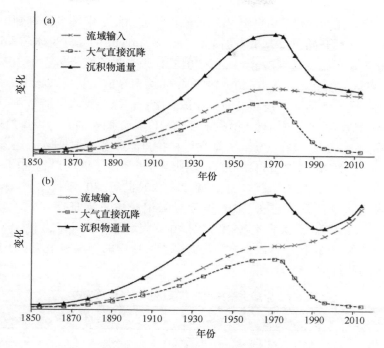

图 6-6 湖泊沉积物记录的污染物通量变化示意图

1970 年前统一为上升趋势，之后分为两种模态：(a) 流域输入缓慢下降而大气直接沉降快速下降；(b) 流域输入上升而大气直接输入下降 [引用并修改自 (Yang, 2015)]

山地冰芯中关于汞历史记录的研究报道很少，仅见于北半球欧洲阿尔卑斯 Col du Dome 冰芯、北美地区美国怀俄明州 Upper Freemont Glacier (UFG) 冰芯和加拿大 Mount Logan 冰芯、亚洲青藏高原中部格拉丹东冰芯和西伯利亚地区阿尔泰山 Belukha 冰芯 (表 6-1)。欧洲 Col du Dome 冰芯中总汞和甲基汞含量分别为 0.86~5.96 ng/L 和 0.25~3.96 ng/L，总汞含量自 20 世纪初不断增加，到 1965 年达到峰值，其后开始下降。作者认为这主要是由于阿尔卑斯山周边国家采取严格的汞排放控制措施使得大气汞含量明显下降 (Jitaru et al., 2003)。Schuster 等 (2002) 于 2002 年研究发表的北美 UFG 冰芯汞历史记录颇具影响力，为认识过去几百年人为和自然汞排放对大气环境的影响提供了重要参考 (UNEP, 2013)。Schuster 将工业革命前冰芯中的总汞含量作为自然背景值，以此为标准初步量化了 20 世纪工业化迅速发展时代汞的人为源与自然源的贡献，这一方法也成为解释冰芯汞记录环境意义的通用方法之一。但需要着重指出的是，Chellman 等 (2017) 对 UFG 冰芯的年代序列提出质疑，通过多定年指标及与 UFG 周边树轮序列进行对比，修正了 UFG 冰芯的定年错误，调整后的 UFG 汞记录时间序列显示，总汞含量从 19 世纪末期开始升高,在 20 世纪 20 年代出现阶段性峰值,此后继续上升直到 1975 年左右达到最高,其后开始下降。来自加拿大 Mount Logan 的高分辨冰芯汞记录揭示了过去 500 年汞污染变化(Beal et al., 2015),汞含量和沉降量在 1850~1900 年"淘金热"期间小幅上升,在 1940~1975 年间达到峰值,其后开始下降,这与北美地区实施《空气清洁法》控制人为汞排放有关;但冰芯汞沉降通量在 1993 年后又出现上升趋势,可能是由于 20 世纪末亚洲地区持续增长的人为汞排放通过长距离传输沉降到这一地区。在亚洲,青藏高原格拉丹东冰芯汞记录显示(Kang et al., 2016),总汞含量和沉降通量自 1500 年起围绕背景值上下波动,到 20 世纪 40 年代第二次世界大战后开始急剧上升并持续至 20 世纪 80 年代。阿尔泰山地区 Belukha 冰川冰芯恢复的过去 320 年汞沉降记录也再次证明了这一变化趋势(Eyrikh et al., 2017),总汞含量在 20 世纪 40 年代后急剧上升,20 世纪 80 年代小幅下降,但进入 90 年代后再次上升,这归因于区域内的燃煤与手工和小规模采金等人为汞使用和排放。

表 6-1 全球主要高山冰芯汞历史记录信息简表

地点	海拔(m)	时间范围	总汞含量*(ng/L)	沉降通量	参考文献
欧洲 Col du Dome	4250	1855~1990 年	0.25~3.96	—	(Jitaru et al., 2003)
美国 UFG	4100	1719~1993 年	1.2~35(8.9)	—	(Schuster et al., 2002)
加拿大 Mt. Logan	5340	1410~1998 年	<0.01~3.5(0.14)	0.01~1.2(0.12)	(Beal et al., 2015)
青藏高原格拉丹东果曲冰川	5750	1477~1982 年	<0.20~9.80(0.84)	<0.01~2.5(0.28)	(Kang et al., 2016)
阿尔泰山 Belukha	3895	1680~2001 年	0.07~8.90(0.89)	0.04~1.9(0.47)	(Eckley et al., 2018)

*括号内为平均值。

从以上总结可见,北半球高山地区的冰芯汞记录一定程度上反映了全球和区域自然和人为汞排放的历史变迁。总体而言(图 6-7),冰芯中汞的含量在工业革命前保持自然背景的稳定状态,进入 20 世纪后都出现显著增长,在 20 世纪中后期出现峰值阶段,但此后北美和亚洲的冰芯汞记录趋势略有不同,欧美在 20 世纪 70 年代实施了大气汞排放控制政策,大气汞含量和沉降量开始下降,这在冰芯记录中有所反映。而亚洲地区冰芯汞记录在 20 世纪后期仍保持增长,反映了区域人为汞生产和排放的持续增长。北美 Mount Logan 冰芯的汞记录在 20 世纪末的增长则体现了人为汞长距离传输所产生的跨区域影响(Beal et al., 2015)。

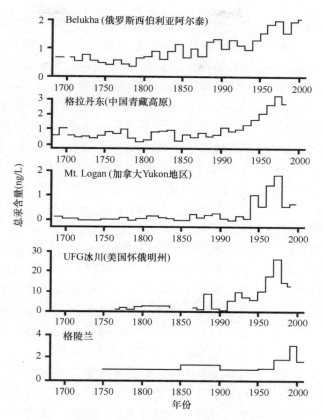

图 6-7　世界范围内主要高山冰芯汞历史记录
［引用并修改自（Eyrikh et al.，2017）］

6.3.3　其他环境介质

高山地区环境介质丰富，研究者还利用泥炭、树轮和特有植物等建立了汞的历史序列。

泥炭是植物遗体在沼泽中经过泥碳化作用形成的一种松散富含水分的有机物聚集物，基本不受地表径流等外界干扰，其中的养分几乎全部来自于大气干湿沉降。汞通过长距迁移并通过大气沉降进入泥炭，因此泥炭汞记录可以反映大气汞污染的历史变化。Enrico 等（2016）在法国 Pyrenees 山区泥炭地开展现代过程研究，发现泥炭藓和新近形成的表层泥炭中的汞同位素信号与大气湿沉降存在显著差异，这表明泥炭中汞的主要来源并非来自大气降水。研究者进一步通过控制实验量化了表层泥炭中汞的来源，其中 79%来自气态单质汞的干沉降，这一结果为准确认识和解释泥炭汞历史记录的环境意义提供了基础。随后其又利用泥炭柱重建了过去数千年的大气气态单质汞变化历史,发现气态单质汞含量在 1970 年左右达到最高值 3.9 ng/m^3 并开始下降,是全新世背景值 0.27 ng/m^3 的 15 倍左右(Enrico et al.,2017)。我国小兴安岭的泥炭汞记录也显示,汞含量和沉积通量在 20 世纪 70 年代达到最高,分别是工业革命前背景值的 3.6 倍和 13.1 倍(Tang et al.,2012)。泥炭中汞含量和通量的工业革命前后比值比前述的湖泊沉积物和冰芯记录的比值都大,这可能与其主要反映大气气态汞含量和沉降有关。再者,这也说明了工业革命后人类活动排汞对大气和地表环境汞背景的巨大改变(Enrico et al.,2017)。泥炭中的汞同位素还指示了人为源汞和自然源汞的历史变化。青藏高原东北部的红原泥炭中汞及其同位素的时间序列显示,泥炭中汞含量在工业革命之后逐渐增加,表明人类向大气的排汞逐渐增加。汞同位素 δ^{202}Hg 呈现出变轻的趋势,指示了人为源汞的逐渐增加(侍文芳等,2011)。

树木可以通过叶片表面吸收大气干沉降的水溶性和颗粒态汞,通过树木气孔直接吸收气态单质汞,通过根部吸收土壤中的可溶态汞(Rea et al.,2002)。这些进入树木的汞最终有少部分积累在年轮木质部,为利用树轮重建大气汞历史记录提供了理论基础。Wright 等(2014)采集了美国内华达州和加利福尼亚州山区的树木年轮,并重建了汞历史变化,发现树轮中的汞含量记录可以反映局地甚至区域的大气汞污染变化历史,例如 Bald 山区的树轮记录清晰显示了汞自 1900 年以来的升高,以及近期随着局地汞矿开发的兴衰而出现的小幅波动。

一些地表植被如苔藓和地衣等也能敏感指示大气汞含量变化。垫状点地梅是分布于青藏高原及周边高寒区域的一种特殊类型的植物,其当年生长枝节在每年生长期过后并不会直接凋落,而是木质化后仍保留在原有植株上,并与新生长的枝节呈先后排列,每年植株会生长一层新的枝节并与原有死亡枝节有明显区分。利用垫状点地梅的这一特性,研究者结合大气汞被动采样建立了垫状点地梅叶片汞含量与大气汞含量间的转换关系,恢复了过去十年来高原内陆高海拔山区大气汞变化趋势,发现自 2010 年起大气汞呈下降趋势(Tong et al.,2016),这与我国和世界其他地区近期的大气直接观测研究结果一致(Zhang et al.,2016),也表明高山地区对全球人为汞排放下降的敏感响应。

6.3.4 高山多环境介质汞历史记录的集成研究

从 6.3.1～6.3.3 节的论述可以看出,高山地区不同环境介质的汞时间序列能够刻画人为汞排放对区域和全球环境本底造成的影响。当前,利用高山地区多环境介质记录集成认识和评价汞污染的历史变化成为趋势和必然,这是因为高山地区各类环境介质丰富,且一般远离人为汞排放点源的直接干扰,记录的汞污染水平一般具有区域代表性,为集成多环境介质认识区域汞污染变化趋势提供了基本条件；高山地区各类环境介质往往随海拔变化集中出现在同一地区，这些环境介质间天然发生联系或记录的气候环境信号具有一致性，便于进行协同对比研究。比如对 UFG 冰芯汞年代序列进行修订的研究中，除了对冰芯样品进行定年的多指标测试分析外，还利用冰芯与邻近树轮记录的区域温度变化进行了年份关联（Chellman et al., 2017），树轮序列准确而高分辨率的定年成为约束和修正冰芯序列的有力辅助。另外，高山地区不同环境介质的自身变化和相互作用强烈，也为开展多环境介质汞记录集成研究提出了必然要求。例如，近年来山地冰川的积累区正经历强烈消融和减薄，部分冰川积累区物质损失甚至消耗了近几十年的净积累（Zhang et al., 2015b）。冰芯近几十年来的汞记录遭到破坏甚至消失（Kang et al., 2016），而开展冰芯与湖泊汞记录的集成研究，则可能为理解近期气候强烈变暖背景下汞污染的变化趋势提供新的可能。

不同环境介质的历史记录各有特点：湖泊分布广泛，便于开展多区域多支湖芯记录的比对研究，但湖芯记录一般时间分辨率较低，且有研究指出，近期受气候变化影响，部分湖泊沉积物记录了大气沉降和流域输入的叠加信号，不能真实反映北美等地区 20 世纪 80 年代后大气汞排放下降的趋势（Yang, 2015）；山地冰川冰芯记录分辨率高，但汞在雪冰及雪气界面迁移转化活跃，冰芯汞记录的可靠性和环境指示意义仍存疑问（Durnford and Dastoor, 2011）；泥炭能反映气态单质汞含量和沉降变化信息。在高山地区综合多环境介质记录研究汞污染历史具有较大潜力和广阔前景。首先，可相互弥补不同环境介质记录时间尺度和分辨率不足，交叉比对提高年代序列准确性。其次，有利于汞污染历史过程和演化的立体解析，各类环境介质记录提供了汞污染的多维历史印记，集成研究可以为认识关键时段的汞污染过程提供全景视角。因此，未来利用高山地区多环境介质记录集成研究汞污染历史，识别人为和自然等因素的影响可能成为全球汞污染新研究中新的增长点。

6.4 高山地区汞的关键生物地球化学过程

6.4.1 山地冰川消融和汞的迁移释放

冰川雪冰中的汞主要来源于大气汞干湿沉降，气态单质汞沉降后可能会暂时储存在雪冰中（Durnford and Dastoor, 2011），雪冰空隙中发生的扩散与通风使部分气态单质汞重释入大气；活性汞既有可能发生氧化作用而与一些颗粒物质相结合暂时储存，也可能被还原转化为气态单质汞而再次进入大气；较之于气态单质汞的不稳定性和活性汞的复杂性，颗粒态汞惰性较强，在雪冰消融前会暂存在雪冰中（Durnford et al.，2012）。高山地区冰川雪冰中的汞多以颗粒态形式存在，且比例较高。例如，对中国西部9条冰川采集的14个雪坑样品分析表明（Zhang et al.，2012），雪坑中总汞与颗粒态汞具有显著的相关性，以颗粒态汞为主。一些山地冰川雪冰中颗粒态汞占总汞的比例可高达90%（Huang et al.，2012a）。

融雪初期，大量离子脉冲式释放进入到融水，其中就包括Hg^{2+}以及汞与其他离子的络合物（Durnford and Dastoor，2011）。有研究指出融雪初期近90%的表层雪中汞会以脉冲式的特征释放进入融水（Dommergue et al.，2003）。如前所述，山地冰川颗粒态汞含量较高，在消融过程中以气态单质汞释放进入大气的比例较低，绝大部分雪冰中储存的汞可随消融进入融水。颗粒态汞可能在消融初期被滞留在积雪内，随后随着消融增强而得到释放（Dommergue et al.，2003）。如研究发现冰川雪坑中总汞浓度的峰值出现在污化层和附加冰层（Huang et al.，2012a；Zhang et al.，2012），这有可能是雪冰消融中汞随融水下渗而造成的富集。随着消融加剧与融水增多，原先保留的颗粒态汞释放进入径流，加上径流对河床的侵蚀作用，消融中后期冰川径流中颗粒态汞比重可能会逐渐提高（Fain et al.，2007；Mitchell et al.，2008）。例如，对青藏高原扎当冰川融水径流的研究表明，上下游河水中的汞都以颗粒态汞为主且与悬浮颗粒物具有较好的相关性（Sun et al.，2017b）。

山地冰川径流具有显著的日变化与季节变化特征，消融季径流量较大，而在冬季径流量下降甚至断流（Immerzeel et al.，2010；Li et al.，2014）。一般而言，径流中汞的传输与流量存在显著的相关性，随流量增大而增大（Brigham et al.，2009；Schuster et al.，2011）。对青藏高原中南部扎当冰川及其补给的曲嘎切河的连续加密观测和采样显示，河水中总汞浓度与流量具有较一致的变化趋势，受其影响明显（图6-8）（Sun et al.，2017b）。总体而言，冰川径流汞传输过程中存在以下变化趋势：一方面，消融季输送量大，非消融季则较少；另一方面，受气温日变化的影响，冰川消融量具有日变化特征，使得径流汞的浓度也存在显著的日变化特征。

青藏高原东缘贡嘎山冰川区的径流水体采样分析表明（梅露等，2016），从地表雪样经冰川河水到壤中流总汞浓度逐渐降低，这主要是水体传输过程中大量的颗粒态汞发生了沉淀作用，富集于土壤中。对青藏高原枪勇冰川流域的冰川径流和冰前湖进行采样和分析发现（Sun et al.，2016），进入冰前湖的径流中总汞浓度（2.5 ng/L）高于流出的径流（2.0 ng/L），说明湖泊在冰川径流汞传输过程中可能扮演"汇"的角色。估算显示，过去40年间中国西部有近500 km^3 冰川消融，融水向下游直接输出的汞约为2.5 t（Zhang et al.，2012）。

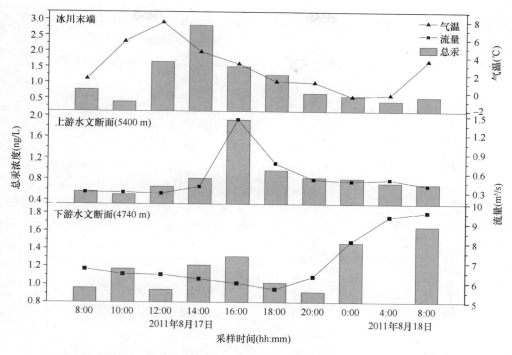

图 6-8　青藏高原扎当冰川流域径流汞传输过程［引自（Sun et al.，2017b）］

基于冰川物质平衡、冰川融水径流量以及总汞含量，还可以估算山地冰川及其流域的年释汞总量和通量。以青藏高原扎当冰川为例，冰川的年输出汞量以及冰川补给径流曲嘎切的年总汞输出量分别为 8.76 g/a 和 157.85 g/a，远小于北极及亚北极地区冰川作用区河流汞输出量，这主要与曲嘎切流域面积较小有关。来自雪冰消融直接释放的汞量较小，但冰川融水径流在向下游运移过程中通过水动力条件对流域侵蚀释放的汞量较大。需要指出的是，曲嘎切流域径流汞输出通量 $[2.74\ \mu g/(m^2\cdot a)]$ 高于北极地区，表明山地冰川对区域汞输出和循环的影响更为突出。

冰川可以作为大气沉降污染物的临时储库，沉降到消融区以及季节性积雪区的部分大气污染物包括汞等会在当年气温回升时随雪冰消融而释放。在气候变暖背景下，冰川消融区扩大，前期沉降并保存在雪冰中的历史污染物也会快速释放（图 6-9）。相对于极地冰川，山地冰川对气候更为敏感，在最近几十年处于快速萎缩中（Zemp et al.，2015；Zhang et al.，2015a）。山地冰川径流是一些地区动植物和人类赖以生存的重要水资源，其丰水期、被补给区水资源利用的旺盛期与汞通量输送高峰期具有高度同期性。同时汞在高山地区湖泊、湿地和农田草场等生态系统中甲基化和积累可能更为显著。例如 Sun 等（2018b）指出冰川径流中的汞以颗粒态为主，生物可利用性和环境风险较低，但在流经下游湿地时甲基化率提升，增强了对下游补给生态系统中动植物的汞污染风险。Zhang 等（2014）指出青藏高原高海拔河流与湖泊生态系统中鱼体内汞富集显著，表明高山生态系统对汞输入的敏感性。

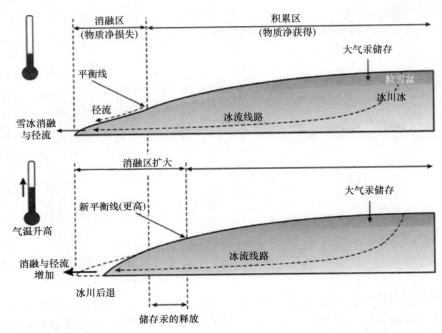

图 6-9 气候变暖背景下冰川消融释汞过程简图［引用并修改自（Stern et al.，2012）］

综上所述，山地冰川所处地区气候和环境因子的季节与日变化显著，自身分布海拔跨度大，汞在其消融过程中的行为和迁移转化较为复杂且可能较极地有所差异；另外，山地冰川对气候变化更加敏感，响应更为剧烈（Diaz et al.，2003），且距离人类居住区较近，冰川融水对人类生产生活的影响和潜在风险较大。气候变化加剧背景下，部分地区山地冰川消融导致融水和径流增加，造成的直接和间接汞释放量将增加，这一高山地区汞的生物地球化学过程及其对下游生态系统的影响正成为新的研究热点。

6.4.2 冻土的汞库效应

冻土是在 0℃或 0℃以下，含有冰的各种岩石或土，是由矿物颗粒、冰、未冻水、气体和有机质等组成。全球多年冻土主要分布在北半球的极地地区以及北美和亚洲的高山地区，南半球的安第斯山和南极大陆也有少量多年冻土分布。据统计，北半球的高山多年冻土分布面积可达 230 余万平方千米，其中 68%分布在青藏高原和喜马拉雅山。

覆盖于多年冻土之上，夏季融化而冬季冻结的土层称为冻土活动层，它具有夏季单向融化、冬季双向冻结的特征。这一特点使得冻土区土壤的物理化学性质和微生物活动在周期性的冻融作用下发生强烈变化，对存储于活动层中的汞的分布和行为具有重要影响。对北极地区覆盖阿拉斯加内陆和北坡冻土区 13 个 1~2.5 m 土壤柱的 588 个样品的分析表明（Schuster et al., 2018），土壤总汞含量和总汞与总碳比值中值分别为（43±30）ng/g 和（1.6±0.9）μg Hg/g C。利用这一比值关系和冻土土壤碳分布数据，估算得到了北半球冻土区土壤中汞的储量为（1656±962）Gg，其中（793±461）Gg 储存在冻土中，（863±501）Gg 分布在活动层中。这一量值是全世界冻土区之外土壤、海洋和大气中汞储量总和的两倍，凸显了冻土区土壤作为汞库的巨大效应。

我国青藏高原是全世界高山多年冻土最为发育的地区。青藏高原东北部疏勒河流域典型冻土发育地区的调查表明（Sun et al., 2017a），23 个土壤垂直剖面 218 个土壤样品中总汞含量为 6.3~29.1 ng/g，大部分土壤剖面中总汞在近地表层位最高，随深度逐渐降低，这主要与土壤中的有机碳的分布趋势相关。土壤总汞含量与海拔高度呈现负相关趋势，这主要是因为高海拔地区土壤以粉质土为主要组分，有机质含量偏低。初步估算认为，冻土退化可能使土壤中约 62.4%的汞"活化"并参与到流域地表汞的生物地球化学循环。青藏高原中部 4700 m 海拔典型多年冻土区的地气界面汞通量观测显示（Ci et al., 2016），冻土区表层土在暖季总体表现为大气汞的源，汞释放通量具有显著的季节变化，日变化表现为白天以释放为主而夜间以沉降为主，紫外辐射强度是地表汞释放强度的主控因子。在观测期间，还发现干旱土壤的汞释放量在降水事件后的几个小时内增高。对这一地区土壤剖面气态单质汞的变化趋势观测表明（Ci et al., 2018），土壤不同层位中的气态单质汞与土壤温度呈显著的指数正相关关系（图 6-10），其中表层 5 cm 土壤中气态单质汞含量对土壤温度上升的响应最为敏感，值得注意的是，这一指数关系在 0℃以上表现最为剧烈。暖季土壤表层中气态单质汞含量甚至高于冻土区环境大气气态单质汞含量，表明冻土土壤是显著的大气汞源，这与此前的地气交换测试结论一致。基于实测得到的土壤气态单质的温度敏感性关系，预估认为在 RCP8.5 排放情景下，21 世纪末青藏高原冻土区向大气的汞释放可能增加 54.9%。气候变暖导致土壤升温与冻土退化，可能增强冻土土壤向大气的汞释放，其主要的驱动机制可能包括（Ci et al., 2018）：①地表土壤温度升高会导致土壤中特别是表层土壤中气态单质汞的含量增加；②升温可能促使土壤从冻结状态转换为非冻结状态，加深活动层厚度并"释放"部分"冻结"汞，同时还能增强活动层中的微生物活动，有利于气态单质汞的形成和向大气释放;③土壤温度上升可能改变土壤微生物群落并促使可溶性有机碳分解,进而导致土壤中可还原性汞转化为气态单质汞并释放进大气;④土壤中气态单质汞含量的增高意味着其与大气中汞含量的梯度差异变大,可能增强土壤汞的释放通量。

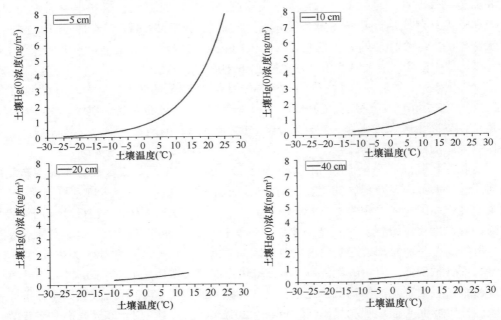

图 6-10 青藏高原冻土活动层不同层位土壤中气态单质汞 Hg(0) 与土壤温度的关系
[引用并修改自(Ci et al., 2018)]

需要注意的是,基于当前对冻土所含的全部汞储量值而言,冻土的汞库效应巨大,气候变化导致冻土退化,可能致使其中的一部分汞在活动层发生迁移转化并有机会参与到区域和全球地表系统的汞循环,但并非冻土中所有的汞都会被清除而进入地表环境。

6.4.3 高山森林汞的生物地球化学过程

森林占全球陆地总面积的 31%(Keenan et al., 2015)。森林生态系统是陆地上最复杂的生态系统,在区域和全球汞的循环中扮演着重要的角色(Fleck et al., 1999)。森林汞循环中几个关键环节和参数包括大气降水与穿透雨降水的汞输入、凋落物汞输入、土壤汞的排放与地表径流汞的输出等。通过调查和量化这些因子可以估算森林生态系统汞的输入和输出通量,确定森林汞的源汇关系。王训等(2017)对森林生态系统汞的重要生物地球化学过程作了归纳总结,主要有四个方面：①植物叶片-大气界面的汞交换；②植物-土壤系统的汞交换；③土壤-大气汞交换；④地表径流-土壤-地下渗流系统汞交换。

当前对森林生态系统在全球大气汞的循环中角色的认识还不清楚（Agnan et al.，2016）。但近期的研究越发重视和深入揭示森林生态系统汞汇的作用和机制。森林生态系统的汞汇作用主要是指森林生态系统直接接受大气降雨带来的汞，或通过叶片吸收与吸附大气中的汞并以穿冠水或者凋落物沉降等方式将大气汞转运累积在森林土壤中的作用过程，这其中后者的汞输入即干沉降汞的通量至关重要。森林生态系统汞的干沉降常用叶片凋落物汞沉降通量加上穿冠水与空旷降雨汞沉降的差异值予以替代（St Louis et al.，2001）。从全球尺度上看，最新的研究表明，全球森林凋落物汞的沉降总量为（1180±710）Mg/a（Wang et al.，2016），是大气汞沉降到森林生态系统的主要方式，可占总沉降的70%～85%（Wang et al.，2016）。一般而言，综合输入与输出汞通量，大气向森林生态系统输入的汞绝大部分能被保留在森林土壤中（王训等，2017），这凸显了森林对大气汞的吸收固定作用。

高山地区一般没有或远离汞的人为源排放，可以吸收和固定长距离传输大气汞，通常被认为是区域和全球人为排放汞重要的汇。Fu等（2010）对中国西南地区雷公山气态单质汞进行了长期观测，并分析了空旷降雨、穿冠水和森林凋落物汞通量，发现虽然降雨和穿冠水中的总汞和甲基汞通量与欧美地区相当，但凋落物汞通量却较高，说明汞的干沉降在高山地区的重要作用，同时指出雷公山地区较高的气态单质汞含量可能促使叶片吸收更多大气汞而沉降进入森林土壤。来自哀牢山的研究也显示（Zhou et al.，2013），凋落物汞沉降通量可达总汞沉降量的92.8%，极大提升了森林地区汞的沉降总量。对青藏高原东部近25个森林站点的凋落物的汞同位素组成分析表明（Wang et al.，2017），在海拔小于3600 m的区域，森林凋落物中的$\Delta^{199}Hg$趋于0，指示了局地人为源的影响；而在海拔3700～4300 m的区域，森林凋落物中的$\Delta^{199}Hg$则明显偏负，指示了长距离传输大气汞的影响。

山地地形和气候条件复杂，森林生态系统的组成分布变化大，导致对汞在森林土壤中的累积过程和效应缺乏系统认识。越来越多的研究指出，相比低海拔地区，高山地区大气汞的沉降通量有增加的趋势。报道显示，山地森林土壤（Gerson et al.，2017；Zhang et al.，2013）、植物和野生动物体内都出现汞富集效应（Townsend et al.，2014）。山地森林系统可能是大气汞重要的吸收汇集区（Blackwell and Driscoll，2015b；Zhang et al.，2013）。在研究和阐释高山地区汞的分布累积和源汇效应时，对森林吸收固定大气汞在其中的作为和重要性的认识也逐步提升。此前，研究者倾向强调海拔上升，降雨量增加，温度降低利于大气汞沉降，这些因素叠加导致高海拔地区汞的湿沉降增加（Blackwell and Driscoll，2015b；Stankwitz et al.，2012），更多强调的是高海拔地区的湿沉降增加效应。但近期基于雷公山和青藏高原东南地区高山森林生态系统的汞同位素研究表明，随海拔升高变化的降雨与温度可以通过控制凋落物的生物量间接影响土壤汞的累积（Wang et al., 2017; Zhang et al., 2013），且以凋落物为主的汞的干沉降才是山地森林土壤汞的主要来源和分布主控因素（Zheng et al., 2016）。这也说明高山地区森林系统可以增强大气汞在高山地区的沉降和储存。

总之，森林生态系统的存在增加了高山地表与大气汞相互作用的复杂性。考虑到森林生态系统对大气汞吸收和固定的重要性，高山地区可能成为长距离传输大气汞的汇，使得高山地区作为整体在全球和区域汞的生物地球化学循环中的角色和意义更为突出。随着全球变暖，高山地区降水量和降水类型可能发生改变，造成森林生态系统吸收和富集大气汞发生变化。此外，累积在土壤中的汞还可随地表径流迁移到下游的水生生态系统，经甲基化作用生成毒性更强的甲基汞。人为活动特别是对山区森林的砍伐利用也会改变山区水文和生物地球化学过程，进而影响山区汞的固定和输出。例如 Eckley 等（2018）通过对比实验证明，相对于未砍伐的山区林地系统，砍伐后的山区林地地表径流量增加了 32%，颗粒态汞输出量没有显著变化，但溶解态汞输出量却增加了 80% 以上，这一结果表明在强调高山森林的汞汇效应的同时，也要注意这一系统储存的汞可能随外因变化时而转变为汞源。需要特别指出的是，我国是一个多山地的国家，90% 以上的森林均为山地森林，因而研究我国山地森林汞的累积与源汇关系意义显得尤为重大。

6.5 高山汞循环的复杂性及其变化趋势

6.5.1 高山地区汞循环的复杂性和重要性

较海洋和陆地平原地区，高山地区的汞循环表现出独特性和复杂性。高山在一定的经纬度平面范围内具有最大的表面积，是各类陆表环境要素分布最全和最密集的区域，各种要素在较小空间跨度的集合和不连续分布是高山地区的汞循环复杂性的根本原因。高山各环境要素相互作用强烈，使得汞在高山地区的迁移转化速度和强度趋向增加。关于高山地区的汞分布和循环过程往往有"出人意料"的发现。比如关于青藏高原水生生态系统汞富集的调查研究显示（Zhang et al., 2014），在高原水体等各类环境介质汞含量非常低的背景下，野生鱼体的汞含量相对较高，具有显著的富集作用，这与中国其他地区高汞污染环境中鱼体汞含量较低的结果形成鲜明对照（Liu et al., 2012）。造成这一现象的原因是高原特有野生鱼类在高寒寡营养环境中具有极低的生长率、较长的寿命和较低的甲基汞去除率，此外，高原地区水生生态系统食物链甲基汞富集效率高。这些原因都与高山的自然环境和高山生物本身特有的生理现象相关。此外，高山地区汞的分布和行为规律还表现为很强的异质性，如 6.2.3 节关于冰川雪冰中汞的海拔梯度分布所述，不同地区观测到的汞分布规律不同，甚至针对同一地区同一冰川，几次不同时段的采样分析得到的结果不同甚至完全相反（Sun et al., 2018a）。这与高山地区多变的气候环境因子引发的雪冰环境变化有关。

高山地区还具有很强的"汞库效应"。高山地区广泛发育的冰川、冻土和森林等都被认为是长距离传输大气汞的汇。目前还少有将高山地区作为独立单元对其中各类生态系统汞储量进行估算，但如前所述，一些高山地区自身人为汞排放强度很低，其中特有的环境介质具有"捕获和固定"长距离传输汞的潜力，在区域和全球汞循环中可能充当汇的作用。再者，研究高山地区特有环境介质中汞的分布和行为也对解释汞在自然环境中的本底和变化具有重要科学贡献。例如高山地区的大气汞观测是认识自由大气中汞形态和转化的重要手段，虽然飞机航测能提供高层自由大气汞的形态和含量等信息，但其长期性和连续性欠佳。将高山定位观测实验和飞机航测数据相结合可以更好认识大气汞的垂直分布及其在自由大气中的长距离传输，同时还有助于区域和全球大气汞传输模型的验证和提高（UNEP, 2013）。

6.5.2 高山地区汞循环的变化趋势

气候变化和人类活动是地表环境变化和物质能量循环的两大驱动因子。就气候变化而言，高海拔地区被认为有更强的增温效应（Pepin et al., 2015），是对全球变暖背景响应更敏感的地区。青藏高原当前和未来的气候变化以变暖变湿润为主要特征（陈德亮等，2015）；阿尔卑斯山区除了变暖以外，降水的季节性也将发生变化，表现为夏季降水减少而冬季增多（Gobiet et al., 2014）。温度和降水等气候因子的改变会深刻影响高山地区环境要素的状态和分布格局。变暖背景下山地冰川近期正以前所未有速度退缩（Zemp et al., 2015），冻土区活动层加厚和冻土退化（Yang et al., 2010c），可能导致其中历史时期沉降和储存的汞（Legacy mercury）重新活化和释放进入高山其他环境中，这些是高山地区汞的源效应可能增强的表现；但同时暖湿可能会造成高山植被生态系统生物量增长，也可能促使其吸收和储存汞的汇效应增强。就人为活动而言，旨在控制和削减全球人为汞排放和含汞产品使用的《水俣公约》已全面生效，全球包括高山地区已有观测研究显示由于人为汞排放的降低，大气汞含量出现下降（Tong et al., 2016；Zhang et al., 2016），这一因素会减少大气中的汞储量，间接降低高山地区吸收汞的量。未来全球变化会对全球汞循环在排放、传输、沉降、转化和储存等各个环节都产生影响（Krabbenhoft and Sunderland, 2013），而高山地区环境要素和变化的复杂性会增加高山地区的汞循环变化的不确定性。

6.5.3 高山地区汞循环研究的未来趋势

当前高山地区的汞循环研究存在一些困难和问题：高山地区的汞观测和调查资料少，长周期系统性的数据极为匮乏，这与高山地区自身的偏远性和自然条件较为恶劣有关；高山地区汞的分布和行为研究还多以离散的描述性报道为主，对控制汞分布和迁移转化的机理阐述不足；目前还缺乏将高山地区作为整体考虑其汞循环规律和变化的研究。在这方面，北极地区的汞污染和循环是环境研究的热点，已有系统的综述和评估报告对其现状、影响、变化趋势和相关政策建议等进行全面阐释（AMAP，2011；Stern et al.，2012）。

对汞在高山地区的生物地球化学循环研究，应特别关注影响高山地区汞输入输出和环境作用的关键行为与过程机制，包括高山地区的大气汞干湿沉降过程，活跃环境要素如冰川、冻土和草甸等变化对汞的活化释放作用，环境中汞的生物可利用性及环境风险等；应加强在多环境要素耦合的典型高山区开展系统的汞循环过程和机理研究，汞同位素技术的发展和应用可提升对这些问题的精细解释，应尝试对不同山区汞循环开展对比和综合研究，探寻内在控制规律，整体认识高山地区的汞输入输出对区域和全球汞循环的影响。

总之，鉴于高山地区汞循环的复杂性和重要性及其未来变化趋势的不确定性，未来的研究应在充分考虑气候变化和人类活动的双重影响前提下，关注高山地区生态环境快速变化背景下，汞的行为特征和迁移转化及局地和远程环境效应，将区域甚至全球高山地区作为整体研究汞的生物地球化学过程，揭示高山地区在区域和全球汞循环中的源汇角色。

参 考 文 献

陈德亮, 徐柏青, 姚檀栋, 郭正堂, 崔鹏, 陈发虎, 张人禾, 张宪洲, 张镱锂, 樊杰, 侯增谦, 张天华, 2015. 青藏高原环境变化科学评估: 过去、现在与未来. 科学通报, 60: 3025-3035.

梅露, 王训, 冯新斌, 2016. 青藏高原贡嘎山冰川区水体中 Hg 的空间分布及其源汇特征. 环境化学, 35: 1549-1556.

侍文芳, 冯新斌, 张干, 明荔莉, 尹润生, 赵志琦, 王静, 2011. 150 年以来红原雨养型泥炭中高分辨率的汞同位素沉积记录. 科学通报, 56: 583-588.

王明业, 1988. 中国的山地. 成都: 四川科学技术出版社.

王训, 袁巍, 冯新斌, 2017. 森林生态系统汞的生物地球化学过程. 化学进展, 29: 970-980.

Agnan Y, Le Dantec T, Moore C W, Edwards G C, Obrist D, 2016. New constraints on terrestrial surface atmosphere fluxes of gaseous elemental mercury using a global database. Environmental Science & Technology, 50: 507-524.

AMAP, 2011. Arctic Pollution 2011. Arctic Monitoring and Assessment Programme (AMAP), Oslo. vi + 38.

Amos H M, Jacob D J, Holmes C D, Fisher J A, Wang Q, Yantosca R M, Corbitt E S, Galarneau E, Rutter A P, Gustin M S, Steffen A, Schauer J J, Graydon J A, St Louis V L, Talbot R W, Edgerton E S, Zhang Y, Sunderland E M, 2012. Gas-particle partitioning of atmospheric Hg(II) and its effect on global mercury deposition. Atmospheric Chemistry and Physics, 12: 591-603.

Beal S A, Osterberg E C, Zdanowicz C M, Fisher D A, 2015. Ice core perspective on mercury pollution during the past 600 years. Environmental Science & Technology, 49: 7641-7647.

Blackwell B D, Driscoll C T, 2015a. Deposition of mercury in forests along a montane elevation gradient. Environmental Science & Technology, 49: 5363-5370.

Blackwell B D, Driscoll C T, 2015b. Using foliar and forest floor mercury concentrations to assess spatial patterns of mercury deposition. Environmental Pollution, 202: 126-134.

Blais J M, Charpentie S, Pick F, Kimpe L E, Amand A S, Regnault-Roger C, 2006. Mercury, polybrominated diphenyl ether, organochlorine pesticide, and polychlorinated biphenyl concentrations in fish from lakes along an elevation transect in the French Pyrenees. Ecotoxicology and Environmental Safety, 63: 91-99.

Brigham M E, Wentz D A, Aiken G R, Krabbenhoft D P, 2009. Mercury cycling in stream ecosystems. 1. Water column chemistry and transport. Environmental Science & Technology, 43: 2720-2725.

Chellman N, Mcconnell J R, Arienzo M, Pederson G T, Aarons S M, Csank A, 2017. Reassessment of the upper fremont glacier ice-core chronologies by synchronizing of ice-core-water isotopes to a nearby tree-ring chronology. Environmental Science & Technology, 51: 4230-4238.

Ci Z J, Peng F, Xue X, Zhang X S, 2018. Temperature sensitivity of gaseous elemental mercury in the active layer of the Qinghai-Tibet Plateau permafrost. Environmental Pollution, 238: 508-515.

Ci Z J, Peng F, Xue X A, Zhang X S, 2016. Air-surface exchange of gaseous mercury over permafrost soil: An investigation at a high-altitude (4700 m a.s.l.) and remote site in the central Qinghai-Tibet Plateau. Atmospheric Chemistry and Physics, 16: 14741-14754.

Corella J P, Valero-Garces B L, Wang F, Martinez-Cortizas A, Cuevas C A, Saiz-Lopez A, 2017. 700 years reconstruction of mercury and lead atmospheric deposition in the Pyrenees (NE Spain). Atmospheric Environment, 155: 97-107.

De Foy B, Tong Y D, Yin X F, Zhang W, Kang S C, Zhang Q G, Zhang G S, Wang X J, Schauer J J, 2016. First field-based atmospheric observation of the reduction of reactive mercury driven by sunlight. Atmospheric Environment, 134: 27-39.

Diaz H F, Grosjean M, Graumlich L, 2003. Climate variability and change in high elevation regions: Past, present and future. Climatic Change, 59: 1-4.

Dommergue A, Ferrari C P, Gauchard P A, Boutron C F, Poissant L, Pilote M, Jitaru P, Adams F C, 2003. The fate of mercury species in a sub-Arctic snowpack during snowmelt. Geophysical Research Letters, 30: 1621.

Drevnick P E, Shinneman A L C, Lamborg C H, Engstrom D R, Bothner M H, Oris J T, 2010. Mercury flux to sediments of Lake Tahoe, California-Nevada. Water Air and Soil Pollution, 210: 399-407.

Durnford D, Dastoor A, 2011. The behavior of mercury in the cryosphere: A review of what we know from observations. Journal of Geophysical Research-Atmospheres, 116: 30.

Durnford D, Dastoor A, Ryzhkov A, Poissant L, Pilote M, Figueras-Nieto D, 2012. How relevant is the deposition of mercury onto snowpacks? Part 2: A modeling study. Atmospheric Chemistry and Physics, 12: 9251-9274.

Eckley C S, Eagles-Smith C, Tate M T, Kowalski B, Danehy R, Johnson S L, Krabbenhoft D P, 2018. Stream mercury export in response to contemporary timber harvesting methods (Pacific Coastal Mountains, Oregon, USA). Environmental Science & Technology, 52: 1971-1980.

Enrico M, Le Roux G, Heimburger L-E, Van Beek P, Souhaut M, Chmeleff J, Sonke J E, 2017. Holocene atmospheric mercury levels reconstructed from peat bog mercury stable isotopes. Environmental Science & Technology, 51: 5899-5906.

Enrico M, Le Roux G, Marusczak N, Heimbuerger L-E, Claustres A, Fu X, Sun R, Sonke J E, 2016. Atmospheric mercury transfer to peat bogs dominated by gaseous elemental mercury dry deposition. Environmental Science & Technology, 50: 2405-2412.

Eyrikh S, Eichler A, Tobler L, Malygina N, Papina T, Schwikowski M, 2017. A 320 year ice-core record of atmospheric Hg pollution in the Altai, Central Asia. Environmental Science & Technology, 51: 11597-11606.

Fain X, Grangeon S, Bahlmann E, Fritsche J, Obrist D, Dommergue A, Ferrari C P, Cairns W, Ebinghaus R, Barbante C, Cescon P, Boutron C, 2007. Diurnal production of gaseous mercury in the alpine snowpack before snowmelt. Journal of Geophysical Research-Atmospheres, 112: D21311.

Fain X, Obrist D, Hallar A G, Mccubbin I, Rahn T, 2009. High levels of reactive gaseous mercury observed at a high elevation research laboratory in the Rocky Mountains. Atmospheric Chemistry and Physics, 9: 8049-8060.

Fleck J A, Grigal D F, Nater E A, 1999. Mercury uptake by trees: An observational experiment. Water Air and Soil Pollution, 115: 513-523.

Fu X W, Feng X, Dong Z Q, Yin R S, Wang J X, Yang Z R, Zhang H, 2010. Atmospheric gaseous elemental mercury (GEM) concentrations and mercury depositions at a high-altitude mountain peak in south China. Atmospheric Chemistry and Physics, 10: 2425-2437.

Fu X W, Feng X, Liang P, Deliger, Zhang H, Ji J, Liu P, 2012. Temporal trend and sources of speciated atmospheric mercury at Waliguan GAW station, Northwestern China. Atmospheric Chemistry and Physics, 12: 1951-1964.

Fu X W, Feng X B, Zhu W Z, Zheng W, Wang S F, Lu J Y, 2008. Total particulate and reactive gaseous mercury in ambient air on the eastern slope of the Mt. Gongga area, China. Applied Geochemistry, 23: 408-418.

Fu X W, Marusczak N, Heimburger L E, Sauvage B, Gheusi F, Prestbo E M, Sonke J E, 2016. Atmospheric mercury speciation dynamics at the high-altitude Pic du Midi Observatory, southern France. Atmospheric Chemistry and Physics, 16: 5623-5639.

Gerson J R, Driscoll C T, Demers J D, Sauer A K, Blackwell B D, Montesdeoca M R, Shanley J B, Ross D S, 2017. Deposition of mercury in forests across a montane elevation gradient: Elevational and seasonal patterns in methylmercury inputs and production. Journal of Geophysical Research-Biogeosciences, 122: 1922-1939.

Gobiet A, Kotlarski S, Beniston M, Heinrich G, Rajczak J, Stoffel M, 2014. 21st Century climate change in the European Alps: A review. Science of the Total Environment, 493: 1138-1151.

Gratz L E, Ambrose J L, Jaffe D A, Shah V, Jaegle L, Stutz J, Festa J, Spolaor M, Tsai C, Selin N E, Song S, Zhou X, Weinheimer A J, Knapp D J, Montzka D D, Flocke F M, Campos T L, Apel E, Hornbrook R, Blake N J, Hall S, Tyndall G S, Reeves M, Stechman D, Stell M, 2015. Oxidation of mercury by bromine in the subtropical Pacific free troposphere. Geophysical Research Letters, 42: 10494-10502.

Guedron S, Amouroux D, Sabatier P, Desplanquee C, Develle A-L, Barre J, Feng C, Guiter F, Arnaud F, Reyss J L, Charlet L, 2016. A hundred year record of industrial and urban development in French Alps combining Hg accumulation rates and isotope composition in sediment archives from Lake Luitel. Chemical Geology, 431: 10-19.

Holmes C D, Krishnamurthy N P, Caffrey J M, Landing W M, Edgerton E S, Knapp K R, Nair U S, 2016. Thunderstorms increase mercury wet deposition. Environmental Science & Technology, 50: 9343-9350.

Huang J, Kang S, Guo J, Zhang Q, Xu J, Jenkins M G, Zhang G, Wang K, 2012a. Seasonal variations, speciation and possible sources of mercury in the snowpack of Zhadang glacier, Mt. Nyainqentanglha, southern Tibetan Plateau. Science of the Total Environment, 429: 223-230.

Huang J, Kang S C, Guo J M, Sillanpaa M, Zhang Q G, Qin X, Du W T, Tripathee L, 2014. Mercury distribution and variation on a high-elevation mountain glacier on the northern boundary of the Tibetan Plateau. Atmospheric Environment, 96: 27-36.

Huang J, Kang S C, Zhang Q G, Jenkins M G, Guo J M, Zhang G S, Wang K, 2012b. Spatial distribution and magnification processes of mercury in snow from high-elevation glaciers in the Tibetan Plateau. Atmospheric Environment, 46: 140-146.

Immerzeel W W, Van Beek L P H, Bierkens M F P, 2010. Climate change will affect the Asian water towers. Science, 328: 1382-1385.

Jitaru P, Infante H G, Ferrari C P, Dommergue A, Boutron C F, Adams F C, 2003. Present century record of mercury species pollution in high altitude alpine snow and ice. Journal De Physique Iv, 107: 683-686.

Kang S C, Huang J, Wang F Y, Zhang Q G, Zhang Y L, Li C L, Wang L, Chen P F, Sharma C M, Li Q, Sillanpaa M, Hou J Z, Xu B Q, Guo J M, 2016. Atmospheric mercury depositional chronology reconstructed from lake sediments and ice core in the Himalayas and Tibetan Plateau. Environmental Science & Technology, 50: 2859-2869.

Kaulfus A S, Nair U, Holmes C D, Landing W M, 2017. Mercury wet scavenging and deposition differences by precipitation type. Environmental Science & Technology, 51: 2628-2634.

Keenan R J, Reams G A, Achard F, De Freitas J V, Grainger A, Lindquist E, 2015. Dynamics of global forest area: Results from the FAO Global Forest Resources Assessment 2015. Forest Ecology and Management, 352: 9-20.

Krabbenhoft D P, Sunderland E M, 2013. Global change and mercury. Science, 341: 1457-1458.

Li B, Yu Z, Liang Z, Acharya K, 2014. Hydrologic response of a high altitude glacierized basin in the central Tibetan Plateau. Global and Planetary Change, 118: 69-84.

Liu B, Yan H Y, Wang C P, Li Q H, Guedron S, Spangenberg J E, Feng X B, Dominik J, 2012. Insights into low fish mercury bioaccumulation in a mercury-contaminated reservoir, Guizhou, China. Environmental Pollution, 160: 109-117.

Lombard M A S, Bryce J G, Mao H, Talbot R, 2011. Mercury deposition in Southern New Hampshire, 2006-2009. Atmospheric Chemistry and Physics, 11: 7657-7668.

Mast M A, Manthorne D J, Roth D A, 2010. Historical deposition of mercury and selected trace elements to high-elevation National Parks in the Western US inferred from lake-sediment cores. Atmospheric Environment, 44: 2577-2586.

Mitchell C P J, Branfireun B A, Kolka R K, 2008. Total mercury and methylmercury dynamics in upland-peatland watersheds during snowmelt. Biogeochemistry, 90: 225-241.

Obrist D, Johnson D W, Lindberg S E, Luo Y, Hararuk O, Bracho R, Battles J J, Dail D B, Edmonds R

L, Monson R K, Ollinger S V, Pallardy S G, Pregitzer K S, Todd D E, 2011. Mercury distribution across 14 US forests. Part I: Spatial patterns of concentrations in biomass, litter, and soils. Environmental Science & Technology, 45: 3974-3981.

Paudyal R, Kang S C, Huang J, Tripathee L, Zhang Q G, Li X F, Guo J M, Sun S W, He X B, Sillanpaa M, 2017. Insights into mercury deposition and spatiotemporal variation in the glacier and melt water from the central Tibetan Plateau. Science of the Total Environment, 599: 2046-2053.

Pearson C, Schumer R, Trustman B D, Rittger K, Johnson D W, Obrist D S, 2015. Nutrient and mercury deposition and storage in an alpine snowpack of the Sierra Nevada, USA. Biogeoscience, 12: 3665-3680.

Pepin N, Bradley R S, Diaz H F, Baraer M, Caceres E B, Forsythe N, Fowler H, Greenwood G, Hashmi M Z, Liu X D, Miller J R, Ning L, Ohmura A, Palazzi E, Rangwala I, Schoner W, Severskiy I, Shahgedanova M, Wang M B, Williamson S N, Yang D Q, Mt Res Initiative E D W W G, 2015. Elevation-dependent warming in mountain regions of the world. Nature Climate Change, 5: 424-430.

Phillips V J A, St Louis V L, Cooke C A, Vinebrooke R D, Hobbs W O, 2011. Increased mercury loadings to Western Canadian Alpine Lakes over the past 150 years. Environmental Science & Technology, 45: 2042-2047.

Rea A W, Lindberg S E, Scherbatskoy T, Keeler G J, 2002. Mercury accumulation in foliage over time in two northern mixed-hardwood forests. Water Air and Soil Pollution, 133: 49-67.

Rutter A P, Schauer J J, 2007a. The effect of temperature on the gas-particle partitioning of reactive mercury in atmospheric aerosols. Atmospheric Environment, 41: 8647-8657.

Rutter A P, Schauer J J, 2007b. The impact of aerosol composition on the particle to gas partitioning of reactive mercury. Environmental Science & Technology, 41: 3934-3939.

Sarkar S, Ahmed T, Swami K, Judd C D, Bari A, Dutkiewicz V A, Husain L, 2015. History of atmospheric deposition of trace elements in lake sediments, similar to 1880 to 2007. Journal of Geophysical Research-Atmospheres, 120: 5658-5669.

Schuster P F, Krabbenhoft D P, Naftz D L, Cecil L D, Olson M L, Dewild J F, Susong D D, Green J R, Abbott M L, 2002. Atmospheric mercury deposition during the last 270 years: A glacial ice core record of natural and anthropogenic sources. Environmental Science & Technology, 36: 2303-2310.

Schuster P F, Schaefer K M, Aiken G R, Antweiler R C, Dewild J F, Gryziec J D, Gusmeroli A, Hugelius G, Jafarov E, Krabbenhoft D P, Liu L, Herman-Mercer N, Mu C C, Roth D A, Schaefer T, Striegl R G, Wickland K P, Zhang T J, 2018. Permafrost stores a globally significant amount of mercury. Geophysical Research Letters, 45: 1463-1471.

Schuster P F, Striegl R G, Aiken G R, Krabbenhoft D P, Dewild J F, Butler K, Kamark B, Dornblaser M, 2011. Mercury export from the Yukon River basin and potential response to a changing climate. Environmental Science & Technology, 45: 9262-9267.

Shanley J B, Engle M A, Scholl M, Krabbenhoft D P, Brunette R, Olson M L, Conroy M E, 2015. High mercury wet deposition at a "Clean Air" site in Puerto Rico. Environmental Science & Technology, 49: 12474-12482.

Sigler J M, Mao H, Sive B C, Talbot R, 2009. Oceanic influence on atmospheric mercury at coastal and inland sites: A springtime nor, easter in New England. Atmospheric Chemistry and Physics, 9: 4023-4030.

St Louis V L, Rudd J W M, Kelly C A, Hall B D, Rolfhus K R, Scott K J, Lindberg S E, Dong W, 2001.

Importance of the forest canopy to fluxes of methyl mercury and total mercury to boreal ecosystems. Environmental Science & Technology, 35: 3089-3098.

Stankwitz C, Kaste J M, Friedland A J, 2012. Threshold increases in soil lead and mercury from tropospheric deposition across an elevational gradient. Environmental Science & Technology, 46: 8061-8068.

Stern G A, Macdonald R W, Outridge P M, Wilson S, Chetelat J, Cole A, Hintelmann H, Loseto L L, Steffen A, Wang F Y, Zdanowicz C, 2012. How does climate change influence arctic mercury? Science of the Total Environment, 414: 22-42.

Sun S, Kang S, Huang J, Li C, Guo J, Zhang Q, Sun X, Tripathee L, 2016. Distribution and transportation of mercury from glacier to lake in the Qiangyong Glacier Basin, southern Tibetan Plateau, China. Journal of Environmental Sciences, 44: 213-223.

Sun S W, Kang S C, Guo J M, Zhang Q G, Paudyal R, Sun X J, Qin D H, 2018a. Insights into mercury in glacier snow and its incorporation into meltwater runoff based on observations in the southern Tibetan Plateau. Journal of Environmental Sciences, 68: 130-142.

Sun S W, Kang S C, Huang J, Chen S Y, Zhang Q G, Guo J M, Liu W J, Neupane B, Qin D H, 2017a. Distribution and variation of mercury in frozen soils of a high-altitude permafrost region on the northeastern margin of the Tibetan Plateau. Environmental Science and Pollution Research, 24: 15078-15088.

Sun X, Zhang Q, Kang S, Guo J, Li X, Yu Z, Zhang G, Qu D, Huang J, Cong Z, Wu G, 2018b. Mercury speciation and distribution in a glacierized mountain environment and their relevance to environmental risks in the inland Tibetan Plateau. Science of the Total Environment, 631-632: 270-278.

Sun X J, Wang K, Kang S C, Guo J M, Zhang G S, Huang J, Cong Z Y, Sun S W, Zhang Q G, 2017b. The role of melting alpine glaciers in mercury export and transport: An intensive sampling campaign in the Qugaqie Basin, inland Tibetan Plateau. Environmental Pollution, 220: 936-945.

Susong D D, Abbott M, Krabbenhoft D P, 1999. Reconnaissance of mercury concentrations in snow from Teton and Wasatch ranges to assess the atmospheric deposition of mercury from an urban area. Eos, Transactions, American Geophysical Union, 80: H12b-06.

Swartzendruber P C, Jaffe D A, Prestbo E M, Weiss-Penzias P, Selin N E, Park R, Jacob D J, Strode S, Jaegle L, 2006. Observations of reactive gaseous mercury in the free troposphere at the Mount Bachelor Observatory. Journal of Geophysical Research-Atmospheres, 111: 12.

Tang S, Huang Z, Liu J, Yang Z, Lin Q, 2012. Atmospheric mercury deposition recorded in an ombrotrophic peat core from Xiaoxing'an Mountain, Northeast China. Environmental Research, 118: 145-148.

Timonen H, Ambrose J L, Jaffe D A, 2013. Oxidation of elemental Hg in anthropogenic and marine airmasses. Atmospheric Chemistry and Physics, 13: 2827-2836.

Tong Y, Yin X, Lin H, Buduo, Danzeng, Wang H, Deng C, Chen L, Li J, Zhang W, Schauer J J, Kang S, Zhang G, Bu X, Wang X, Zhang Q, 2016. Recent decline of atmospheric mercury recorded by androsace tapete on the Tibetan Plateau. Environmental Science & Technology, 50: 13224-13231.

Townsend J M, Driscoll C T, Rimmer C C, Mcfarland K P, 2014. Avian, salamander, and forest floor mercury concentrations increase with elevation in a terrestrial ecosystem. Environmental Toxicology and Chemistry, 33: 208-215.

UNEP, 2013. Global Mercury Asessment 2013: Sources, Emissions, Releases and Environmental

Transport Geneva, Switzerland.

Wang X, Bao Z D, Lin C J, Yuan W, Feng X B, 2016. Assessment of global mercury deposition through litterfall. Environmental Science & Technology, 50: 8548-8557.

Wang X, Luo J, Yin R S, Yuan W, Lin C J, Sommar J, Feng X B, Wang H M, Lin C, 2017. Using mercury isotopes to understand mercury accumulation in the montane forest floor of the Eastern Tibetan Plateau. Environmental Science & Technology, 51: 801-809.

Wang X P, Yang H D, Gong P, Zhao X, Wu G J, Turner S, Yao T D, 2010. One century sedimentary records of polycyclic aromatic hydrocarbons, mercury and trace elements in the Qinghai Lake, Tibetan Plateau. Environmental Pollution, 158: 3065-3070.

Wright G, Woodward C, Peri L, Weisberg P J, Gustin M S, 2014. Application of tree rings dendrochemistry for detecting historical trends in air Hg concentrations across multiple scales. Biogeochemistry, 120: 149-162.

Yang H, Engstrom D R, Rose N L, 2010a. Recent changes in atmospheric mercury deposition recorded in the sediments of remote equatorial lakes in the Rwenzori Mountains, Uganda. Environmental Science & Technology, 44: 6570-6575.

Yang H D, 2015. Lake sediments may not faithfully record decline of atmospheric pollutant deposition. Environmental Science & Technology, 49: 12607-12608.

Yang H D, Battarbee R W, Turner S D, Rose N L, Derwent R G, Wu G J, Yang R Q, 2010b. Historical reconstruction of mercury pollution across the Tibetan Plateau using lake sediments. Environmental Science & Technology, 44: 2918-2924.

Yang M X, Nelson F E, Shiklomanov N I, Guo D L, Wan G N, 2010c. Permafrost degradation and its environmental effects on the Tibetan Plateau: A review of recent research. Earth-Science Reviews, 103: 31-44.

Yu X, Driscoll C T, Montesdeoca M, Evers D, Duron M, Williams K, Schoch N, Kamman N C, 2011. Spatial patterns of mercury in biota of Adirondack, New York lakes. Ecotoxicology, 20: 1543-1554.

Zemp M, Frey H, Gartner-Roer I, Nussbaumer S U, Hoelzle M, Paul F, Haeberli W, Denzinger F, Ahlstrom A P, Anderson B, Bajracharya S, Baroni C, Braun L N, Caceres B E, Casassa G, Cobos G, Davila L R, Granados H D, Demuth M N, Espizua L, Fischer A, Fujita K, Gadek B, Ghazanfar A, Hagen J O, Holmlund P, Karimi N, Li Z Q, Pelto M, Pitte P, Popovnin V V, Portocarrero C A, Prinz R, Sangewar C V, Severskiy I, Sigurosson O, Soruco A, Usubaliev R, Vincent C, Correspondents W N, 2015. Historically unprecedented global glacier decline in the early 21st century. Journal of Glaciology, 61: 745-762.

Zhang H, Yin R-S, Feng X-B, Sommar J, Anderson C W N, Sapkota A, Fu X-W, Larssen T, 2013. Atmospheric mercury inputs in montane soils increase with elevation: Evidence from mercury isotope signatures. Scientific Reports, 3: 3322.

Zhang Q, Kang S, Gabrielli P, Loewen M, Schwikowski M, 2015a. Vanishing high mountain glacial archives: Challenges and perspectives. Environmental Science & Technology, 49: 9499-9500.

Zhang Q G, Huang J, Wang F Y, Mark L W, Xu J Z, Armstrong D, Li C L, Zhang Y L, Kang S C, 2012. Mercury distribution and deposition in glacier snow over Western China. Environmental Science & Technology, 46: 5404-5413.

Zhang Q G, Kang S C, Gabrielli P, Loewen M, Schwikowski M, 2015b. Vanishing high mountain glacial archives: Challenges and perspectives. Environmental Science & Technology, 49: 9499-9500.

Zhang Q G, Pan K, Kang S C, Zhu A J, Wang W X, 2014. Mercury in wild fish from high-altitude aquatic ecosystems in the Tibetan Plateau. Environmental Science & Technology, 48: 5220-5228.

Zhang Y X, Jacob D J, Horowitz H M, Chen L, Amos H M, Krabbenhoft D P, Slemr F, St Louis V L, Sunderland E M, 2016. Observed decrease in atmospheric mercury explained by global decline in anthropogenic emissions. Proceedings of the National Academy of Sciences of the United States of America, 113: 526-531.

Zheng W, Obrist D, Weis D, Bergquist B A, 2016. Mercury isotope compositions across North American forests. Global Biogeochemical Cycles, 30: 1475-1492.

Zhou J, Feng X, Liu H, Zhang H, Fu X, Bao Z, Wang X, Zhang Y, 2013. Examination of total mercury inputs by precipitation and litterfall in a remote upland forest of Southwestern China. Atmospheric Environment, 81: 364-372.

第 7 章 极地地区汞生物地球化学循环

本章导读
- 由于汞的长距离传输以及极地生态环境的脆弱性,极地地区汞生物地球化学循环引起极大关注。
- 极地地区是重要的汞汇与汞的储存库。
- 总结极地地区环境介质中汞的分布与影响因素,以及汞的关键生物地球化学过程。
- 气候变化下极地地区汞生物地球化学循环应引起未来研究的重视。

极地地区包括北极和南极,因气候环境恶劣、人烟稀少被认为是全球受人类活动影响最小的区域之一。作为地球极端环境的典型代表,极地地区的汞循环研究具有重要科学意义(Barkay and Poulain, 2007; Braune et al., 2015; Chetelat et al., 2015; Gamberg et al., 2015)。汞(Hg)是常温下唯一以液态存在的重金属,具有较高的饱和蒸气压,能够以气态形式存在于大气中,可随大气进行长距离传输与沉降,从而造成偏远地区的汞污染。虽然极地地区汞本地污染源较少,但由于全球大气输运作用,低纬度排放的大量汞可以被传输到极地地区,通过食物链传递后威胁到生态系统和人体健康。实际上,极地地区的某些人群(如因纽特人、法罗群岛居民)以海洋哺乳动物和海洋鱼类为主要食物来源,甲基汞暴露风险较高并已引起广泛关注。因此,极地地区汞的分布、迁移与转化是全球汞污染的一个研究热点。

7.1 极地地区汞的主要来源

7.1.1 极地汞的主要来源与收支

环境中汞的污染源主要包括人为源和自然源,其中自然源主要包括火山喷发、海洋和陆地释放等,人为源包括燃煤燃烧、金属冶炼等。极地地区是地球上最为洁净的区域之一,人为汞排放源较少,大气单质汞的长距离传输被认为是极地区域汞的最重要来源（Durnford et al.，2010）。极地汞的传输路径受全球大气环流趋势影响。北极污染物传输主要受太平洋-北美信号（Pacific North America teleconnection）和北大西洋涛动（North Altantic oscillation）影响（Steffen et al.，2005）。此外，中纬度大气阻塞（mid-latitude atmospheric blocking）也在污染物传输到北极过程中起着重要作用（Iversen and Joranger，1985）。全球模型估算结果表明北极地区绝大部分的大气汞污染来自于北半球地区的汞排放，其中来自于亚洲地区的大气长距离传输对北极大气汞的污染贡献达到 43%，其余的污染分别来自于俄罗斯（27%）、北美（16%）和欧洲（14%）（Durnford et al.，2010）。南极洲虽然与世隔绝，但其不仅受气候变化的影响，也从低纬度地区输送大气污染物。由于南极大陆几乎完全被冰雪覆盖，对许多寿命较长的大气污染物来说，是一个巨大的储库（Eisele et al.，2008），估算表明南极平原每年大约能固定 60 t 汞（Brooks et al.，2008a）。陆地、海洋和大气界面汞的再释放过程也是极地地区汞的一个主要来源。由于表层海水溶解气态汞（dissolved gaseous mercury，DGM）处于过饱和状态，海水每年也会向大气释放大量工业革命前到现在累积的大量汞。根据南极西部区域的监测结果表明，南大洋每年可以向大气释放 30 t 左右的汞（Mastromonaco et al.，2017）。此外，随着极地的开发利用，科研活动和旅游业发展等活动也成为极地包括汞等污染物的一个潜在来源。生物载体也是极地环境中一个重要的汞来源。在南极詹姆斯·罗斯岛的北部一海豹尸体下采集的土壤样本中，汞含量达到了背景水平的 5 倍。甲基汞占总汞含量的3%，而在背景样品中，甲基汞含量不足 0.1%。释放出的汞向较低的土壤层缓慢地垂直迁移。并且汞和甲基汞的浓度都随着尸体的腐烂而增加（Zverina et al.，2017）。

7.1.2 极地大气汞"亏损"事件对汞输入贡献

1995 年研究者在加拿大阿勒特（Alert）地区首次发现了春季（4～5月）对流层大气中气态总汞（total gaseous mercury，TGM）浓度突然急剧下降至很低的现象（Schroeder et al.，1998），后经多年观测发现该现象在北极地区每年都有发生（Steffen et al.，2008），且广泛存在（Berg et al.，2003；Henrik et al.，2004；Lindberg et al.，2001；Poissant and Pilote，2003）。除在北极区域外，2000 年以来在南极对流层大气中也发现了类似的大气汞"亏损"事件（atmospheric mercury depletion event，AMDE）现象（Ebinghaus et al.，2002）。大气汞"亏损"事件的发生表明极地大气中存在着汞的快速去除机制。大气汞"亏损"事件发生时大气中气态单质汞浓度急剧下降而活性气态汞浓度急剧上升，表明气态单质汞转化为活性气态汞（Steffen et al.，2005）。同时，大气中的活性气态汞（gaseous oxidized mercury，GOM）浓度急剧升高，说明气态单质汞（gaseous elemental mercury，GEM）被快速氧化（Braune et al.，2015）。有研究发现大气汞"亏损"事件与臭氧层空洞事件具有很好的相关性（Schroeder et al.，1998），进而推测 AMDE 可能与臭氧层空洞事件机制类似，即溴氧化物可以将气态元素汞氧化成活性气态汞，进而更容易从大气交换到冰、雪表面（Lindberg et al.，2002；Schroeder et al.，1998）（图 7-1）。AMDE 一般从极地日出开始一直持续至融雪结束，汞沉积在极地冰雪表面，主要受阳光、温度及来自海盐的卤素等因素的影响。虽然北极地区汞的排放源有限，但由于 AMDE 可使积雪中颗粒态汞的浓度高于 1000 ng/L，远高于低纬度地区的冰雪中的浓度（Douglas et al.，2017）。有研究估算表明 AMDE 每年可使高达 100 t 的汞沉降到北极地区（Dastoor et al.，2015；Douglas et al.，2012；Skov et al.，2004），是极地地区汞的重要来源。然而近期也有研究表明超过 80% AMDE 沉降的活性气态汞经光化学还原为气态单质汞重新释放到大气中（Ferrari et al.，2008），AMDE 对极地汞的净输入贡献要低于之前的估算。AMDE 来源的汞会在融雪季节汇入地表径流中（McNamara et al.，1998），雪冰中的汞也可被吸附在植被或土壤表面（Grigal，2003），通过微生物作用沉积在土壤、湖泊（Semkin et al.，2005），排放到海洋中（Dommergue et al.，2003）等，进而影响到极地整体汞的循环。

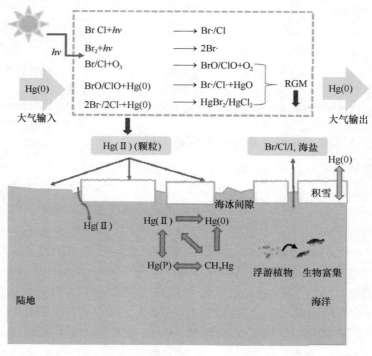

图 7-1　极地汞循环图［根据 Steffen 等（2008）图重新绘制］

7.2 极地地区环境介质中汞的分布与影响因素

在自然环境中,汞主要以三种形态存在,即零价汞[Hg(0)]、甲基汞(MeHg)和二价汞[Hg(Ⅱ)](Schroeder et al.,1998;Steffen et al.,2008)。气态单质汞是极地大气中汞的主要形态,同时也存在于土壤或者雪的空隙中(O'Driscoll et al.,2005)。二价汞主要存在于大气颗粒物、雪、冰及水环境中(O'Driscoll et al.,2008)。甲基汞是通过食物链富集的主要汞形态,这也导致虽然极地地区汞整体污染水平较低,但高营养等级的海洋哺乳动物和人体汞浓度有时候会远高于其他地区(Braune et al.,2005;Su et al.,2011)。极地地区环境介质中不同形态汞的水平见表7-1所示。

7.2.1 极地地区大气汞及其影响因素

大气中汞主要包括气态单质汞(GEM)、活性气态汞(RGM)和颗粒态汞[particle-bound Hg,Hg(P)]。气态单质汞在大气中的停留时间可以长达一年(Schroeder and Munthe,1998),这使得汞可以通过长距离大气传输在全球分布。而RGM和Hg(P)由于停留时间较短,一般排放后很快就会沉降下来。由于北半球汞人为排放源多于南半球,北半球大气总汞浓度要高于南半球(Steffen et al.,2008),北半球大气总汞的背景值为 1.5～1.7 ng/m^3,南半球大气总汞的背景值为 1.1～1.3 ng/m^3。总体而言,北极大气汞监测开始时间较早、站点较多、时间序列较为完整,而南极相关数据相对匮乏,缺乏长期的大气监测数据(Dommergue et al.,2010)。如表7-1所示,北极的大气GEM浓度(1.1～1.8 ng/m^3)明显高于南极(0.54～1.2 ng/m^3),也反映了北半球工业发达、人为排放汞源较多。极地大气汞的主要来源为自然汞排放和人为汞排放。极地区域汞分布特征受采样站位海拔、气象条件及大气传输扩散条件、汞的氧化还原化学、区域自然源/人为源的位置和强度等因素的影响(Nguyen et al.,2009;Sprovieri and Pirrone,2000)。北极大气汞具有明显的季节变化趋势(Costa and Liss,1999),春季最低,夏季和冬季相对秋季较高。模型模拟结果表明冬季由于气象因素中纬度地区向北极输入了更多的汞,这可能导致了北极地区冬季相较于秋季更高的大气汞水平(Dastoor and Larocque,2004)。春季大气汞浓度较低主要是由于春季 AMDE 导致大气中汞的消除(Berg et al.,2003;Henrik et al.,2004;Lindberg et al.,2001;Poissant and Pilote,2003)。由于夏季北极地区空气流通较弱,中纬度地区向极地区域汞的输送较少(Dastoor and Larocque,2004),夏季北极地区大气汞浓度的升高不是由中纬度地区汞的输送引起的。冰、雪表面和海洋汞释放的增加可能是夏季北极地区大气中汞浓度升高的

表 7-1 极地区环境介质中不同形态汞的浓度水平

编号	采样区域	检测时间	介质	形态及浓度	参考文献
1	北极：加拿大阿勒特	1995~2000 年；2002 年	大气	GEM 均值：1.21~1.62 ng/m^3	(Kellerhals et al., 2003)
2	北极：北部站	2011~2014 年	大气	GEM 均值：1.11~1.57 ng/m^3	(Angot et al., 2016)
3	北极：挪威安岛	2011~2015 年	大气	GEM 均值：1.50~1.61 ng/m^3	(Angot et al., 2016)
4	北极：加拿大库居阿尔阿皮克	1999~2000 年	大气	GEM 均值：1.80 ng/m^3，范围：0.30~8.60 ng/m^3	(Steffen et al., 2005)
5	北极：俄罗斯阿德马	2001~2002 年	大气	GEM 均值：1.70 ng/m^3，范围：0.01~4.60 ng/m^3	(Steffen et al., 2005)
6	北极：美国阿拉斯加州巴罗	1998~2000 年	大气	GEM 均值：1.79 ng/m^3；RGM 均值：0.025 ng/m^3	(Lindberg, 2001)
7	北极：美国阿拉斯加州巴罗	2000 年	积雪	TGM 范围：1000~70000 ng/m^3	(Lindberg, 2001)
8	北极：挪威新奥尔松	1996~1997 年	大气	GEM 均值：1.43 ng/m^3，范围：0.63~3.55 ng/m^3；RGM 均值：0.125 ng/m^3；Hg(P)均值：0.0027 ng/m^3，范围：0.001~0.2 ng/m^3	(Berg et al., 2001)
9	北极：挪威新奥尔松	1996~1997 年	积雪	TGM 均值：14200 ng/m^3，范围：3800~31000 ng/m^3	(Berg et al., 2001)
10	北极：挪威新奥尔松	2011~2015 年	大气	Hg(0) 范围：1.47~1.51 ng/m^3	(Angot et al., 2016)
11	北极：芬兰帕拉斯	1996~1997 年	大气	GEM 均值：1.26 ng/m^3，范围：0.15~1.80 ng/m^3；Hg^P 均值：0.0014 ng/m^3，范围：0.0003~0.0058 ng/m^3	(Berg et al., 2001)
12	北极：芬兰帕拉斯	1996~1997 年	积雪	TGM 均值：15500 ng/m^3，范围：3900~51700 ng/m^3	(Berg et al., 2001)
13	北极：阿拉斯加	1993~1994 年	积雪	TGM 均值：4210 ng/m^3，范围：980~7500 ng/m^3	(SnyderConn et al., 1997)

第 7 章 极地地区汞生物地球化学循环

续表

编号	采样区域	检测时间	介质	形态及浓度	参考文献
14	北极：北冰洋	1997~1998 年	积雪	TGM 均值：7800~34000 ng/m^3	(Lu et al., 2001)
15	北极：加拿大群岛哈得孙湾	1997~1998 年	积雪	TGM 范围：25000~160000 ng/m^3	(Lu et al., 2001)
16	北极：加拿大图克托亚图克	1997~1998 年	积雪	TGM 均值：2200 ng/m^3	(Lu et al., 2001)
17	北极：温尼伯湖南岸	1996 年	积雪	TGM 均值：1800 ng/m^3	(Lu et al., 2001)
18	北极：阿拉斯加西部 Kasegaluk Lagoon 1	1996 年	积雪	TGM 均值：100000 ng/m^3	(Garbarino et al., 2002)
19	北极：阿拉斯加西部 Kasegaluk Lagoon 2	1996 年	积雪	TGM 均值：214000 ng/m^3	(Garbarino et al., 2002)
20	北极：Admiralty 湾（巴罗站位东部）	1996 年	积雪	TGM 均值：10000 ng/m^3	(Garbarino et al., 2002)
21	北极：Elson Lagoon（巴罗站位东部）	1996 年	积雪	TGM 均值：11000 ng/m^3	(Garbarino et al., 2002)
22	北极：迪斯湾	1996 年	积雪	TGM 均值：23000 ng/m^3	(Garbarino et al., 2002)
23	北极：巴芬湾	2008 年	海水	总汞（total Hg，THg）均值：419.23 ng/m^3	(Zdanowicz et al., 2013)
24	北极：加拿大北极群岛	2005 年	海冰	THg 均值：585.72 ng/m^3	(Kirk et al., 2008)
25	北极：波弗特海	—	海冰	THg 范围：130.38~12235.99 ng/m^3	(Beattie et al., 2014)
26	北极：加拿大北极群岛	—	海冰	THg 范围：100.30~802.36 ng/m^3	(Chaulk et al., 2011)
27	北极：楚科奇海	2008 年；2010 年	底栖沉积物	THg 均值：31 ng/g，范围：5~55 ng/g	(Fox et al., 2014)
28	北极：挪威南部泰勒马克郡	2013 年	湖泊水体	THg 范围：1000~3000 ng/m^3；MeHg 范围：10~60 ng/m^3	(Okelsrud et al., 2016)
29	北极：波弗特海	2004~2011 年	北极熊	THg 均值：3500 ng/m^3，范围：600~13300 ng/m^3	(McKinney et al., 2017)
30	北极：利奥波德王子岛	1975~2014 年	海鸟蛋	THg 范围：160~770 ng/m^3	(Braune et al., 2016)
31	北极：挪威	2014~2015 年	土壤	THg 均值：111 ng/g，范围：41~254 ng/g	(Halbach et al., 2017)

续表

编号	采样区域	检测时间	介质	形态及浓度	参考文献
32	北极：加拿大北极和格陵兰岛	1877~2007年	象牙海鸥羽毛	MeHg 范围：90~4110 ng/g	(Bond et al., 2015)
33	北极：西格陵兰岛中部	1500年	人头发	THg 均值：3100 ng/g	(Hansen JC, 1989)
34	北极：西格陵兰岛中部	1975年	人头发	THg 均值：9800 ng/g	(Hansen JC, 1989)
35	北极：挪威北部	1971~1972年	人牙齿	THg 均值：3230 ng/g	(Tvinnereim et al., 2000)
36	南极：麦克默多站	2003年	大气	GEM 均值：1.20 ng/m^3，范围：<0.01~11.16 ng/m^3，RGM 均值：0.116 ng/m^3，范围：0.029~0.275 ng/m^3，Hg(P) 均值：0.049 ng/m^3，范围：0.005~0.182 ng/m^3	(Brooks et al., 2008b)
37	南极：南极研究站	2003年	大气	GEM 均值：0.539 ng/m^3，RGM 均值：0.344 ng/m^3	(Brooks et al., 2008a)
38	南极：南极研究站	2003年	雪	TGM 均值：198000 ng/m^3	(Brooks et al., 2008a)
39	南极：诺伊迈尔站	2000~2001年	大气	GEM 均值：1.06 ng/m^3	(Ebinghaus et al., 2002)
40	南极：特拉诺瓦湾	2000~2001年	大气	GEM 均值：0.90 ng/m^3，范围：0.29~2.30 ng/m^3，RGM 均值：0.116 ng/m^3，范围：0.011~0.334 ng/m^3	(Sprovieri et al., 2002)
41	南极：特罗尔站	2011~2015年	大气	GEM 均值：0.90~0.98 ng/m^3	(Angot et al., 2016)
42	南极：多姆C站	2012~2013年；2015年	大气	GEM 均值：0.76~1.06 ng/m^3	(Angot et al., 2016)
43	南极：迪蒙迪维尔站	2012~2015年	大气	GEM 均值：0.85~0.91 ng/m^3	(Angot et al., 2016)
44	南极西部海域威德尔海	2013年	海水	DGM 均值：9 ng/m^3，范围：0.1~44 ng/m^3，GEM/TGM 均值：1.2 ng/m^3，范围：0.8~1.7 ng/m^3	(Mastromonaco et al., 2017)

续表

编号	采样区域	检测时间	介质	形态及浓度	参考文献
45	南极西部海域威德尔海	2013 年	海水	DGM 均值: 12 ng/m³, 范围: 0~42 ng/m³, GEM/TGM 均值: 1.1 ng/m³, 范围: 0.2~1.9 ng/m³	(Mastromonaco et al., 2017)
46	南极西部海域别林斯高晋海, 阿蒙森海, 罗斯海	2010~2011 年	海水	DGM 均值: 7 ng/m³, 范围: 0.2~72 ng/m³, GEM/TGM 均值: 1.4 ng/m³, 范围: 0.3~2.0 ng/m³	(Mastromonaco et al., 2017)
47	南极西部海域威德尔海	2013 年	海冰	DGM 均值: 46 ng/m³, 范围: 4~84 ng/m³, GEM/TGM 均值: 1.0 ng/m³, 范围: 0.1~1.8 ng/m³	(Mastromonaco et al., 2017)
48	南极西部海域别林斯高晋海, 阿蒙森海, 罗斯海	2010~2011 年	海冰	DGM 均值: 20 ng/m³, 范围: 0.1~72 ng/m³, GEM/TGM 均值: 1.2 ng/m³, 范围: 0.1~2.0 ng/m³	(Mastromonaco et al., 2017)
49	南极大陆: 泰勒谷	—	植物	THg 均值: 105 ng/g, 范围: 99~110 ng/g (苔藓)	(Matsumoto, 1983)
50	南极大陆: 麦克默多干河谷	—	植物	THg 均值: 190 ng/g, 范围: 12~710 ng/g (藻类)	(Matsumoto, 1983)
51	南极大陆: 麦克默多冰架	—	植物	THg 均值: 46 ng/g	(Camacho et al., 2015)
52	南极大陆: 格兰纳特港	—	植物	THg 均值: 78 ng/g, 范围: 27~134 ng/g (苔藓), THg 均值: 207 ng/g, 范围: 112~318 ng/g (地衣)	(Bargagli et al., 2005)
53	南极大陆: 默里山	—	植物	THg 均值: 35 ng/g, 范围: 22~55 ng/g (藻类), THg 均值: 310 ng/g, 范围: 129~422 ng/g (地衣)	(Bargagli et al., 2005)

续表

编号	采样区域	检测时间	介质	形态及浓度	参考文献
54	南极大陆：南森冰原（海岸波利尼亚）	—	植物	THg 均值：366 ng/g,范围：184～570 ng/g（苔藓）THg 均值：715 ng/g,范围：310～1661 ng/g（地衣）THg 均值：158 ng/g,范围：111～358 ng/g（藻类）	(Bargagli et al., 2005)
55	南极大陆：Edmonson 点	—	植物	THg 均值：82 ng/g,范围：42～150 ng/g（苔藓）THg 均值：195 ng/g,范围：87～321 ng/g（地衣）THg 均值：60 ng/g,范围：43～85 ng/g（藻类）	(Bargagli et al., 2005)
56	南极：格雷厄姆地	—	植物	THg 均值：68 ng/g,范围：50～80 ng/g（地衣）	(Bargagli et al., 2005)
57	南极：乔治王群岛	—	植物	THg 均值：43 ng/g,范围：26～61 ng/g（地衣）	(Bargagli et al., 2005)
58	南极：南设德兰群岛、欺骗岛	—	植物	THg 均值：213 ng/g,范围：190～253 ng/g（地衣）	(Bargagli et al., 2005)
59	南极：拜尔斯半岛	—	植物	THg 均值：56 ng/g（藻类）	(Camacho et al., 2015)
60	南极：希望湾（邻近埃斯佩兰萨站）	—	植物	THg 均值：399 ng/g（藻类）	(Camacho et al., 2015)
61	南极：乔治王群岛	—	植物	THg 范围：23～40 ng/g（苔藓）THg 均值：36 ng/g（地衣）	(Dos Santos et al., 2006)
62	南极：乔治王群岛	—	植物	THg 范围：89～213 ng/g（地衣）	(Bubach et al., 2016)
63	南极：乔治王群岛	—	植物	THg 均值：260 ng/g（苔藓）THg 均值：180 ng/g（地衣）	(Wojtun et al., 2013)
64	南极：詹姆斯·罗斯岛	—	植物	THg 均值：1590 ng/g,范围：72～2730 ng/g（地衣）	(Zverina et al., 2014)
65	南极：金钟湾，乔治王岛	—	海洋沉积物	THg 范围：53～210 ng/g	(Favaro, 2004)

续表

编号	采样区域	检测时间	介质	形态及浓度	参考文献
66	南极：金钟湾，乔治王岛	—	土壤和海岸沉积物	THg 范围：16.3～29.9 ng/g	(Dos Santos et al., 2006)
67	南极：乔治王岛	—	鸟粪沉积	THg 均值：54.8 ng/g，范围：35.5～100 ng/g	(Liu et al., 2005)
68	南极：威德尔海架	—	海洋沉积物	THg 均值：22.4 ng/g，范围：14～44 ng/g	(Niemisto and Perttila, 1995)
69	南极：维多利亚北部	—	土壤	THg 均值：34 ng/g，范围：7～96 ng/g	(Bargagli et al., 1993)
70	南极：特拉诺瓦湾	—	海洋沉积物	THg 均值：12 ng/g，范围：6～27 ng/g	(Bargagli et al., 1998)
71	南极：维多利亚北部	—	表层土壤	THg 均值：31 ng/g，范围：12～86 ng/g	(Bargagli et al., 2005)
72	南极：维多利亚北部	—	海洋沉积物（<2000 μm）	THg 均值：41.86 ng/g，范围：11～85 ng/g	(Riva et al., 2004)
73	南极：麦克默多站	—	河口沉积物	THg 均值：74 ng/g，范围：63～87 ng/g	(Negri et al., 2006)
74	南极：金钟湾，乔治王岛	—	海洋沉积物	THg 范围：53～210 ng/g	(Favaro, 2004)
75	南极洲罗斯海的麦克默多湾	2011年	鱼类	THg 范围：21.9～71.3 ng/g	(Wintle et al., 2015)
76	南极	—	南极雪海燕血液	THg 均值：2200～3600ng/g	(Tartu et al., 2014)
77	南极乔治王岛企鹅区	2010年	企鹅	THg 均值：22～197 ng/g（血液）297～742 ng/g（羽毛）	(Polito et al., 2016)
78	南极乔治王岛企鹅区	2010年	鱼虾类	THg 均值：2 ng/g（磷虾）；8 ng/g（浅层鱼类）；31 ng/g（中层鱼类）；41 ng/g（底栖鱼）	(Polito et al., 2016)
79	南极洲阿德利	2010～2012年	鱼类	THg 均值：65～355 ng/g	(Goutte et al., 2015)

主要原因（Douglas et al., 2012; Poulain et al., 2004b; Soerensen et al., 2016; Yu et al., 2014）。模型模拟结果（Dastoor and Durnford, 2014）也表明夏季北极区域大气汞有两个明显的高峰：第一个峰值主要是由积雪融化后汞的再挥发过程导致的；第二个峰值主要是由海洋向大气释放汞引起的。

大气汞长期监测结果表明全球大气汞的含量从20世纪70年代末至80年代呈明显增加的趋势，然后至1996年呈递减趋势，并保持基本恒定（Slemr et al., 2003）。目前在北极地区已有一些较长时间的大气汞观测数据，最长的监测数据在加拿大阿勒特（Alert），始于1995年，然而报道的整体变化趋势尚有一定差异。在加拿大阿勒特1995~2002年的观测（Costa and Liss, 1999）结果表明，春季AMDE发生期间大气GEM年际变化呈波动趋势，并未呈现明显变化趋势，而夏季有明显降低趋势。而另有研究表明，1995~2005年间阿勒特夏季大气GEM并未呈现明显变化趋势（Temme et al., 2007）。此外，通过对1974~2000年加拿大地区大气总颗粒态汞的观测（Li et al., 2009）显示，夏季和秋季总颗粒态汞含量分别以3.0%±0.8%和3.1%±0.9%的速度下降，与全球人为排汞的降低速率（每年3.3%）相当。冬季和早春由于AMDE或"北极霾"的影响并未呈现明显的变化趋势。也有学者通过测定积雪中GEM含量通过反演模型推测了近60年（1950~2000）北极大气GEM的变化趋势（Dommergue et al., 2016），结果表明自20世纪50年代开始呈逐渐增加趋势，于60年代末至70年代初达到峰值后呈降低趋势，至1995~2000年降到最低，与全球汞的排放趋势表现出较好的相关性。

7.2.2 极地地区陆地和水环境汞及其影响因素

汞通过大气干湿沉降进入地表生态系统进而参与地表环境的生物地球化学循环。冰雪是两极地区生态系统中最重要的环境介质之一，会直接影响汞的生物地球化学循环过程。大气单质汞通过干湿沉降过程进入雪冰中，由于雪冰中的氧化作用以及新降雪的覆盖等因素，汞会暂时储存在雪冰中。一部分汞通过扩散与通风以气态单质汞的形式返回到大气中，另一部分随着雪冰的融化进入到径流中。例如，对于北极生态系统，在冬季和春季大气中的颗粒、溶质和污染物等通过沉降作用输入冰雪中，并通过多种过程进行再分配，比如冬季融化事件和雪变质过程（Kuhn, 2001）。如表7-1所示，极地积雪中TGM浓度在几百到几十万 ng/m^3 之间，是极地地区汞的重要的库。积雪与雪冰中的无机汞，在太阳光的照射下可实现不同价态间的相互转化（Durnford and Dastoor, 2011）：其中氧化剂可促进GEM氧化为氧化态汞（Ferrari et al., 2008; Poulain et al., 2007b），减少再次返回到大气中单质汞的量；太阳光照射也可促进单质汞的还原，使释放到大气中的汞含量增加。因此，积雪既可以作为汞的"汇"，又可以作为汞的"源"，这主要受温度、太阳光辐射以及积雪中化学物质组成控制（Tseng et al.，2004）。对北极海洋系统的研究表明，融雪径流输入可能不是海水总汞和甲基汞的重要源（St Louis et al.，2007；St Louis et al.，2005），然而对于淡水系统，融雪季径流中汞的浓度有明显的升高，表明融雪过程是汞的一个重要来源（Douglas et al.，2017）。而随着全球变暖导致的海冰减少，更多的卤素物质将从海洋释放到大气中，促进AMDE 的发生，更多的汞也将沉降到积雪上，增加融雪径流输入的贡献（Douglas et al.，2012）。

融雪进入极地湖泊的汞一部分又会通过光化学或微生物还原过程再循环到大气中（Costa and Liss，1999；Moller et al.，2011），例如，在北极阿拉斯加的一些湖泊中，DGM 释放通量与湿沉降通量基本相当（Tseng et al.，2004）。初级生产过程能显著增加水体中汞的去除及沉积物中汞的累积，这是影响极地湖泊大气沉降汞存留的关键因素之一（Outridge et al.，2005；Outridge et al.，2007）。在很多研究中也发现沉积物中总汞和甲基汞浓度与有机质含量具有较好相关性，表明沉积物中有机质是控制极地湖泊沉积物汞水平和形态的重要环境因素（John et al.，2008；Outridge et al.，2007）。河流也是极地海洋系统汞重要的源，例如估算表明河流每年可以向北冰洋输入大约 13 t 的汞（Outridge et al.，2008）。

海洋是极地大气沉降汞的重要的汇，同时通过再挥发过程也是大气汞的重要的二次源。相比于其他海洋环境，海冰是极地海洋环境特殊的自然现象。极地地区海冰和上覆积雪中汞浓度一般很高，并且会阻断海气交换过程。一旦海冰融化，海冰中富集的汞会被释放出来（Lindberg et al.，2001）。关于北极总汞和甲基汞在不同介质中分配的估算结果（Soerensen et al.，2016）（图 7-2）显示，超过一半的总汞（3900 t）储存在陆架区和深海的沉积物活跃层中，另有约 35%在深海水体中，上层海洋总汞只占总量的约 7%；与总汞不同，沉积物中储存的甲基汞只占总量的约 2.2%，绝大部分甲基汞（>95%）储存在深海（450 t）和次表层（104 t）水体中。近年来越来越多的研究发现在全球大洋次表层存在甲基汞的一个高值区（Bowman et al.，2015；Cossa et al.，2011），类似的次表层甲基汞高值区在北极海域也显著存在（Wang et al.，2018），次表层甲基汞的生成和保留可能是控制极地汞生物富集及风险的关键因素。

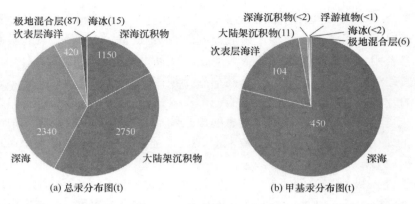

图 7-2　北极总汞和甲基汞在不同介质中分配情况 [根据 Soerensen 等（2016）图重新绘制]

7.2.3　极地地区生物及人体中汞及其影响因素

虽然极地地区汞本地污染源较少，但在极地尤其是北极鱼和哺乳类动物体内普遍检测到较高的汞浓度（表 7-1）（Braune et al.，2005；Riget et al.，2011；Su et al.，2011），给当地居民健康产生了很大威胁。例如，因纽特人妇女头发汞含量比美国妇女高一个数量级（Muckle et al.，2001），成年格陵兰人器官中汞的浓度甚至高于日本、韩国和欧洲等国家和地区的（Johansen et al.，2007）。汞随食物链的富集放大作用是极地地区较高营养级生物体内汞浓度较高的主要原因。通过测定不同生物肝组织中汞的含量（Bard，1999）发现，鱼中汞的含量大约为（0.005±0.569）μg/g，海鸟中增加为（0.046±2.67）μg/g，而海豹、鲸鱼和北极熊中汞的含量可高达 21.6 μg/g。此外，AMDE 导致的极地大气汞沉降的增加和气候变化导致的海洋生物地球化学过程以及食物链营养级结构的变化也可能导致极地生物体内汞含量的显著变化（Braune et al.，2015；Stern et al.，2012）。最近的一项研究（Wang et al.，2018）发现，加拿大北极区域海域次表层（100～300 m）有甲基汞的高值区，且含量呈自西向东递减的趋势，可以很好地解释生物体内汞的空间分布规律。由于该层是浮游动物和低营养级生物的栖息深度，生物体从次表层吸收甲基汞并进一步累积可能是控制北极区域生物体汞水平的关键。

工业革命之前极地地区汞有明显增加的趋势，例如北极和亚北极湖泊沉积物中汞的记录显示工业革命前的沉积通量为每年 0.7～54.35 μg/m^2（Landers et al.，1998），而工业革命之后北极地区大气汞沉降通量显著上升（Hermanson，1998）。在对生物体中汞的含量分布的研究也发现了类似的趋势，500 年前的海豹毛中汞含量大约为 0.6 μg/g，目前已增加到 2.6 μg/g。与之类似，居住在格陵兰地区的因纽特人头发中的含量水平从工业革命以前的 3.1 µg/g 增加到了现在的约 9.8 µg/g（Laughlin，1991）。为了使生物汞含量恢复到自然状态，需要对全球人为汞排放进行深度和持续地削减。然而，由于质量惯性和大气输入的主导作用较小，北冰洋生物汞浓度的下降将比减少排放的速度慢，而且比其他海洋和淡水的速度要慢。虽然极地大气汞的输入量自20世纪80年代开始呈逐渐降低的趋势（Lindberg et al.，2002），但北极地区生物圈中汞的含量水平依然呈现增加趋势。例如，20 世纪 90 年代白鲸肝中的汞含量水平是 80 年代的两倍（Wagemann et al.，1996），80 年代末 90 年代初环形海豹肝中的汞含量水平比 20 年前高出 3 倍。自 1975 年以来，五种海鸟蛋中汞的监测（Braune et al.，2016）结果表明20世纪 70 年代中期至 90 年代汞的浓度呈不断增加的趋势，而 20 世纪 90 年代至近年来基本维持不变。另有研究发现自 2004～2011 年以来南波弗特海北极熊体内汞含量呈明显降低的趋势，然而分析发现汞的降低并不是由于环境汞含量的变化，而是摄食和身体状况的变化（McKinney et al.，2017）。

7.3 极地地区汞的关键生物地球化学过程

7.3.1 汞大气化学过程

大气沉降是极地生态系统中汞的重要来源，大气中汞的运输和沉降受汞的形态分布、气象条件以及当地环境的物理化学性质所影响。不同形态的汞表现出不同的迁移性质（Mastromonaco et al.，2017；Zhang et al.，2009）。单质汞在大气中的存留时间为 0.5～2 a，可进行长距离运输；二价汞的存留时间较短，一般为 3 天左右，只能在污染源附近进行短距离迁移。大气中不同形态汞的相互转化对汞的全球生物地球化学循环起着重要的作用。GEM 水溶性低，反应不活跃，很难通过干湿沉降从大气中去除，也可被氧化转化为 RGM 或者吸附在颗粒态汞的表面，导致 GEM 浓度的降低。大气中的氧化物主要包括臭氧、羟基、卤素、过氧化氢以及硝酸根等（Sudo et al.，2002b）。此外 GEM 和 RGM 也可以吸附到大气颗粒物表面，大气颗粒物的存在也会影响大气汞的化学反应，RGM 吸附在颗粒物表面后会促进其还原反应，同时颗粒物对 GEM 的吸附过程也会影响其氧化过程的发生。

由于氧化剂（O_3，•OH，H_2O_2 和卤素）水平、光照等环境条件的差异，极地大气汞化学过程与低纬度和中纬度有明显不同。例如，根据全球大气化学模型估算结果，极地地区大气中 O_3、•OH、H_2O_2 的浓度有明显降低，大约为中低纬度地区的 1/10～1/4（Sudo et al.，2002a），而南北极 BrO 的浓度要明显高于中纬度地区（Yang et al.，2005）。大气中 GEM 的主要氧化途径是通过气相中与 O_3、•OH 或 H_2O_2

反应（Bauer et al.，2003；Biswajit and Ariya，2004；Pal and Ariya，2004；Tokos et al.，1998），而液相中 Hg(Ⅱ)的还原和 Hg(0)的氧化也会显著影响 GEM 的水平（Lin et al.，2007；Lin et al.，2006）。近年来，动力学和和热力学实验以及理论模型研究都表明 GEM 和卤素的气相反应可能在极地大气汞的去除中起着重要作用（Calvert and Lindberg，2003；Donohoue et al.，2006；Goodsite et al.，2004）。卤素物质的光化学反应是 AMDE 的起始反应，紫外光照射下，溴/氯单质生成的溴/氯自由基或与臭氧反应生成溴/氯氧化物。生成的溴/氯自由基或溴/氯氧化物可氧化气态单质汞生成活性气态汞（Lindberg et al.，2002）（图 7-1），并使大气中的汞以活性气态汞（RGM）和颗粒态汞[Hg(P)]的形式沉积在雪冰表面（Outridge et al.，2005；Steffen et al.，2014）。

7.3.2 水环境汞关键生物地球化学过程

1. 分子转化过程

如图 7-1 所示，氧化/还原和甲基化/去甲基化过程是极地海洋等水环境中汞的关键分子转化过程，显著影响到甲基汞的水平和水气界面汞的交换通量。环境中汞的氧化/还原过程包括光化学途径（Amyot et al.，1997；Krabbenhoft et al.，1998）或生物途径（Mason et al.，1995）。极地表层水体中汞的还原主要是光化学过程，而深层水体中主要是微生物过程（Mason et al.，1995；Poulain et al.，2004a）。光照强度（尤其是 UVA 和 UVB）及光可还原 Hg(Ⅱ)络合物的量是控制汞还原的关键因素（Amyot et al.，2010）。溶解有机质由于会影响到光的透过，也会显著影响水体汞的光还原过程。对北冰洋的研究表明，由于冬天海冰覆盖及雪、冰、溶解有机质、悬浮颗粒物等限制了光在水体中的透过（Granskog and Macdonald，2007），微生物还原过程可能在北极康沃利斯岛附近海域起着更重要的作用（Poulain et al.，2007a）。然而，对于何种微生物起关键作用目前还缺少清楚的认识。同时 Hg(0)也能被氧化为 Hg(Ⅱ)，其氧化反应和还原反应同时进行，是一个平衡的过程。有研究表明氯离子可以促进 Hg(0)的光化学氧化过程（Lalonde et al.，2001；Whalin et al.，2007）。

由于毒性大、易富集，甲基汞是最受关注、风险最大的汞形态，而汞的甲基化/去甲基化过程是控制水环境甲基汞水平的关键过程。极地地区冰雪消融后，陆地会出现淡水湿地系统。对加拿大北部 Ellesmere 岛池塘湿地的甲基汞循环研究发现，湿地内甲基汞的产生量远高于大气沉降等外部输入（Lehnherr et al.，2012a）。具体而言，底泥中甲基汞浓度主要由底泥汞甲基化潜势和生物可利用汞含量决定，而塘水中甲基汞浓度则取决于底泥甲基汞的输出、厌氧微生物活性及水层中的光致去甲基化作用强度（Lehnherr et al.，2012b）。比较北极苔原永久冻土区不同类型湿地的甲基汞浓度，发现冰楔槽池塘的甲基汞浓度高于多边形池塘，而湖泊的甲基汞浓度最低；甲基汞的浓度与有机物、营养元素（总磷、总氮）以及微生物活性有强的相关性（MacMillan et al.，2015）。极地海洋中的甲基汞来自海水、底泥和海冰中的甲基化以及陆地（海岛）的输入（Kirk and Louis，2009）。而海冰下方的海水中，甲基汞浓度平均为 0.06~0.1 ng/L（即 0.3~0.5 pmol/L），占总汞的比例较高（30%~45%），是北极海洋生态系统汞生物放大的重要来源（St. Louis et al.，2007）。北极海水中的甲基汞浓度也随海水深度而变化。例如，加拿大北部的极地海域中，表层水的甲基汞浓度通常很低（0.02 ng/L±0.01 ng/L），而中层和底层水的甲基汞浓度较高（0.07 ng/L±0.04 ng/L），且占总汞的比例较高（28%±16%；最高达 66%），可以推断北极海水中汞甲基化较为活跃（Kirk et al.，2008）。水层中发生的甲基化过程所产生的甲基汞占北极海水中甲基汞浓度的很大一部分（平均 47%±62%）（Lehnherr et al.，2011）。在北冰洋的 Beaufort 海中，次表层水中出现甲基汞浓度的最大值，该水层中同时具有较高的营养物质浓度和较低的溶解氧浓度，与其他海域观察到的规律相同，次表层甲基汞可能主要来自于原位的汞甲基化过程（Wang et al.，2012）。北极沿海底泥中也会发生硫酸盐还原和汞甲基化，并且这些过程受温度影响很大；但底泥中汞甲基化的强度总体较弱，不是北极海水中甲基汞的主要来源（St. Pierre et al.，2014）。除微生物甲基化过程外，有研究表明南极陆架海域叶绿素与二甲基汞浓度垂向分布上呈较好的正相关关系，并推测浮游植物有可能参与二甲基汞的生成（Pongratz and Heumann，1999）。硫酸盐还原菌（Han et al.，2010）、铁还原菌和产甲烷菌（Fleming et al.，2006；Hamelin et al.，2011；Kerin et al.，2006；Yu et al.，2012；Yu et al.，2013）被认为是自然环境中引起汞甲基化的关键微生物。近期的研究表明目前报道的能甲基化汞的微生物均含有 $hgcAB$ 基因，也扩展了环境中可甲基化汞微生物的范畴（Gilmour et al.，2013）。然而对于极地环境关键的汞甲基化微生物还缺少清楚的认识（Loseto et al.，2010）。

除甲基化过程外，去甲基化过程也是控制水环境甲基汞水平的关键过程。与氧化/还原过程类似，该过程也可以通过光化学或生物过程发生。有研究表明，80%~91%北极淡水湖泊输入的甲基汞会通过光去甲基化过程去除，是甲基汞的关键去除途径（Hammerschmidt et al.，2006）。海冰覆盖对于北极海水环境甲基汞的光降解有明显的抑制作用，通过海鸟同位素的研究结果表明气候变暖导致的海冰减少会增加表层海水甲基汞的降解（Point et al.，2011）。微生物去甲基化过程根据产物的不同包括氧化去甲基化和还原去甲基化两个途径（Schaefer et al.，2004），两个途径以哪个为主则取决于环境条件。有学者根据极地环境条件特点推测还原途径可能是极地甲基汞微生物降解的主要途径（Barkay and Poulain，2007），然而尚缺少相关证据。

2. 关键界面交换过程

除分子形态转化过程外，水环境外部的输入输出及内部的生物吸收、颗粒物吸附解吸附等界面交换过程也是影响极地水环境汞循环的关键过程。对于北冰洋汞收支的研究结果表明（Outridge et al.，2008），太平洋洋流、大西洋洋流、河流输入及海岸侵蚀每年分别向北极海域输入4 t、44 t、13 t和47 t的汞，这些大气输入之外汞的输入通量总和要高于大气沉降（98 t/a）（图7-3）。收支平衡上，每年的汞总输入量与总输出量基本持平。对于北极混合层甲基汞，大气沉降、河流输入和海岸侵蚀等外部输入过程输入的甲基汞要高于原位甲基汞的生成量，向次表层海水的颗粒物沉降和扩散过程也与原位降解过程相当（图7-3）。对于次表层海水，甲基汞主要来自于原位甲基化过程的生成，外源输入/输出的贡献较小，向北大西洋的输出是其主要的去除途径（Soerensen et al.，2016）。颗粒物的吸附解吸过程也是控制汞循环的关键过程，与浮游植物或细菌来源的颗粒物的结合可以加快汞向深海的沉降（Mason and Fitzgerald，1993；Sunderland et al.，2009）。全球海洋汞循环估算结果表明，大约一半的汞通过沉降过程传输到深海或沉积物（Sunderl and Mason，2007），北极区域的估算结果也基本一致（Outridge et al.，2008）。此外，沉积物也是湖泊、河口和陆架海域水体汞尤其是甲基汞的重要的源（Fitzgerald et al.，2007；Hammerschmidt et al.，2006）。

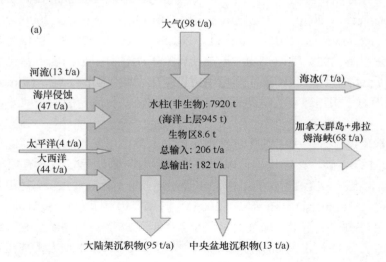

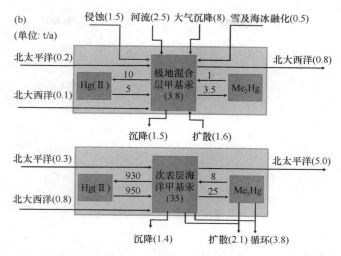

图 7-3　北冰洋总汞（a）和甲基汞（b）收支［根据 Outridge 等（2008）和 Soerensen 等（2016）图重新绘制］

7.3.3　雪、海冰中汞生物地球化学过程

雪和海冰是极地地区重要的环境介质，在汞等污染物循环中也起着重要的作用。极地地区雪和海冰也是污染物潜在的"过渡"汇，是污染物从大气向海洋等水生生态系统传输的重要纽带。此外，冰雪介质中也可能发生氧化/还原和甲基化/去甲基化等汞的形态转化，影响到极地地区汞的生物地球化学循环。越来越多的研究者认为，沉积的汞的大部分（但不一定全部）都被光还原并重新排放到大气。目前报道的雪中汞的还原速率大约在 0.006～1.6 pmol/（L·h）（Mann et al., 2014），不同研究得出的速率之间具有较大差异，可能主要是由于实验条件不同，这也限制了对于该过程通量的估算。雪的化学组成不仅影响雪中汞光还原的反应速率，也会影响光可还原形态汞的占比，是控制雪中汞光化学还原的关键因素（Braune et al., 2015）。相对于光化学还原过程，雪冰中汞的光氧化过程研究更少。有研究者（Lalonde et al., 2003）测定了光照条件下融化后的极地雪水中添加 Hg(0) 后其浓度的变化，发现随时间变化 Hg(0) 浓度有明显降低，表明有汞的氧化过程发生，而目前对于该过程的速率常数认识尚比较欠缺。

Beaufort 海的海冰中甲基汞浓度范围很大，最低浓度低于检测限（0.1 pmol/L），最高达 2.64 pmol/L；海冰中下部甲基汞占总汞的比例可以达到 40%，暗示海冰中可能发生原位汞甲基化（Beattie et al., 2014）。此外，在适宜的条件下，积雪中也可以发生汞的甲基化作用。例如，研究发现北极积雪中甲基汞的浓度与甲磺酸盐浓度呈显著正相关，并推断，北极积雪中可以发生好氧汞甲基化过程，而该过程的进行需要海洋浮游植物二甲巯基丙酸循环的代谢产物（Larose et al., 2010）。加拿大北部埃尔斯米尔（Ellesmere）岛的积雪中观测到较高浓度（0.28 ng/L 或 1.4 pmol/L）的甲基汞，发现甲基汞浓度与积雪中氯离子的浓度呈显著相关性，由此推断，积雪中的甲基汞来自极地海洋（St. Louis et al., 2005）。在极夜期间，北极苔原积雪中的甲基汞主要来自海洋气溶胶，而在融雪期间，海洋气溶胶的输入不能解释苔原积雪中较高浓度的甲基汞（高于 0.2 ng/L），推测积雪中可能存在汞甲基化过程（Constant et al., 2007）。

7.4 展　　望

汞是一种引起全球广泛关注的有毒污染物。作为地球极端环境的一个典型代表，极地地区的汞循环研究具有重要意义。此外，虽然极地地区受人类活动影响较少，但目前鱼和哺乳动物体内甲基汞浓度较高，已引起显著生态和环境风险。对于极地生物体内甲基汞浓度较高的原因，目前已有包括 AMDE 导致的极地大气汞沉降的增加、气候变化导致的海洋生物地球化学过程和食物链营养级结构的变化、海洋次表层甲基汞生成和富集等一些解释。然而目前尚缺乏较为直接证据，结合过程同位素示踪和同位素分馏等技术有可能揭示导致这一现象的化学和生物机制。积雪和海冰在极地汞循环中起着重要作用。全球气候变化将对极地积雪和海冰产生显著影响，了解全球气候变化背景下极地雪、冰介质汞的生物地球化学循环是预测未来极地汞循环及风险的关键。此外，目前对于极地地区汞生物地球化学循环的研究主要集中在北极地区，亟待在南极地区开展相关研究工作。

参 考 文 献

Amyot M, Lean D, Mierle G, 2010. Photochemical formation of volatile mercury in High Arctic lakes. Environmental Toxicology & Chemistry, 16: 2054-2063.

Amyot M, Mierle G, Lean D, McQueen D J, 1997. Effect of solar radiation on the formation of dissolved gaseous mercury in temperate lakes. Geochimica Et Cosmochimica Acta, 61(5): 975-987.

Angot H, Dastoor A, De Simone F, Gardfeldt K, Gencarelli C N, Hedgecock I M, Langer S, Magand O, Mastromonaco M N, Nordstrom C, Pfaffhuber K A, Pirrone N, Ryjkov A, Selin N E, Skov H, Song S J, Sprovieri F, Steffen A, Toyota K, Travnikov O, Yang X, Dommergue A, 2016. Chemical cycling and deposition of atmospheric mercury in polar regions: Review of recent measurements and comparison with models. Atmospheric Chemistry and Physics, 16(16): 10735-10763.

Bard S M, 1999. Global transport of anthropogenic contaminants and the consequences for the Arctic marine ecosystem. Marine Pollution Bulletin, 38(5): 356-379.

Bargagli R, Agnorelli C, Borghini F, Monaci F, 2005. Enhanced deposition and bioaccumulation of mercury in Antarctic terrestrial ecosystems facing a coastal polynya. Environmental Science & Technology, 39(21): 8150-8155.

Bargagli R, Battisti E, Focardi S, Formichi P, 1993. Preliminary data on environmental distribution of mercury in Northern Victoria Land, Antarctica. Antarctic Science, 5(1): 3-8.

Bargagli R, Monaci F, Sanchez-Hernandez J C, Cateni D, 1998. Biomagnification of mercury in an Antarctic marine coastal food web. Marine Ecology Progress Series, 169: 65-76.

Barkay T, Poulain A J, 2007. Mercury (micro) biogeochemistry in polar environments. FEMS Microbiology Ecology, 59(2): 232-241.

Bauer D, D'Ottone L, Campuzano-Jost P, Hynes A J, 2003. Gas phase elemental mercury: A comparison of LIF detection techniques and study of the kinetics of reaction with the hydroxyl radical. Journal of Photochemistry & Photobiology A Chemistry, 157(2): 247-256.

Beattie S A, Armstrong D, Chaulk A, Comte J, Gosselin M, Wang, F Y, 2014. Total and methylated mercury in Arctic multiyear sea ice. Environmental Science & Technology, 48(10): 5575-5582.

Berg T, Bartnicki J, Munthe J, Lattila H, Hrehoruk J, Mazur A, 2001. Atmospheric mercury species in the European Arctic: Measurements and modelling. Atmospheric Environment, 35(14): 2569-2582.

Berg T, Sekkesaeter S, Steinnes E, Valdal A K, Wibetoe G, 2003. Springtime depletion of mercury in the European Arctic as observed at Svalbard. Science of the Total Environment, 304(1-3): 43-51.

Biswajit P, Ariya P A, 2004. Gas-phase HO-initiated reactions of elemental mercury: Kinetics, product studies, and atmospheric implications. Environmental Science & Technology, 38(21): 5555.

Bond A L, Hobson K A, Branfireun B A, 2015. Rapidly increasing methyl mercury in endangered ivory gull (*Pagophila eburnea*) feathers over a 130 year record. Proceedings of the Royal Society B-Biological Sciences, 282(1805): 1-8.

Bowman K L, Hammerschmidt C R, Lamborg C H, Swarr G, 2015. Mercury in the North Atlantic Ocean: The US GEOTRACES zonal and meridional sections. Deep-Sea Research Part Ii-Topical Studies in Oceanography, 116: 251-261.

Braune B, Chetelat J, Amyot M, Brown T, Clayden M, Evans M, Fisk A, Gaden A, Girard C, Hare A, Kirk J, Lehnherr I, Letcher R, Loseto L, Macdonald R, Mann E, McMeans B, Muir D, O'Driscoll N, Poulain A, Reimer K, Stern G, 2015. Mercury in the marine environment of the Canadian Arctic: Review of recent findings. Science of the Total Environment, 509: 67-90.

Braune B M, Gaston A J, Mallory M L, 2016. Temporal trends of mercury in eggs of five sympatrically breeding seabird species in the Canadian Arctic. Environmental Pollution, 214: 124-131.

Braune B M, Outridge P M, Fisk A T, Muir D C G, Helm P A, Hobbs K, Hoekstra P F, Kuzyk Z A, Kwan M, Letcher R J, Lockhart W L, Norstrom R J, Stern G A, Stirling I, 2005. Persistent organic pollutants and mercury in marine biota of the Canadian Arctic: An overview of spatial and temporal trends. Science of the Total Environment, 351: 4-56.

Brooks S, Arimoto R, Lindberg S, Southworth G, 2008a. Antarctic polar plateau snow surface conversion of deposited oxidized mercury to gaseous elemental mercury with fractional long-term burial. Atmospheric Environment, 42(12): 2877-2884.

Brooks S, Lindberg S, Southworth G, Arimoto R, 2008b. Springtime atmospheric mercury speciation in the McMurdo, Antarctica coastal region. Atmospheric Environment, 42(12): 2885-2893.

Bubach D, Catan S P, Di Fonzo C, Dopchiz L, Arribere M, Ansaldo M, 2016. Elemental composition of *Usnea* sp. lichen from Potter Peninsula, 25 de Mayo (King George) Island, Antarctica. Environmental Pollution, 210: 238-245.

Calvert J G, Lindberg S E, 2004. The potential influence of iodine-containing compounds on the chemistry of the troposphere in the polar spring. II. Mercury depletion. Atmospheric Environment, 38(30): 5105-5116.

Calvert J G, Lindberg S E, 2003. A modeling study of the mechanism of the halogen-ozone-mercury homogeneous reactions in the troposphere during the polar spring. Atmospheric Environment, 37(32): 4467-4481.

Camacho A, Rochera C, Hennebelle R, Ferrari C, Quesada A, 2015. Total mercury and methyl-mercury contents and accumulation in polar microbial mats. Science of the Total Environment, 509: 145-153.

Chaulk A, Stern G A, Armstrong D, Barber D G, Wang F Y, 2011. Mercury distribution and transport across the ocean-sea-ice-atmosphere interface in the Arctic ocean. Environmental Science & Technology, 45(5): 1866-1872.

Chetelat J, Amyot M, Arp P, Blais J M, Depew D, Emmerton C A, Evans M, Gamberg M, Gantner N, Girard C, Graydon J, Kirk J, Lean D, Lehnherr I, Muir D, Nasr M, Poulain A J, Power M, Roach P, Stern G, Swanson H, Velden S V, 2015. Mercury in freshwater ecosystems of the Canadian Arctic: Recent advances on its cycling and fate. Science of the Total Environment, 509: 41-66.

Constant P, Poissant L, Villemur R, Yumvihoze E, Lean D, 2007. Fate of inorganic mercury and methyl mercury within the snow cover in the low arctic tundra on the shore of Hudson Bay (Quebec, Canada). Journal of Geophysical Research-Atmospheres, 112: D08309.

Cossa D, Heimburger L E, Lannuzel D, Rintoul S R, Butler E C V, Bowie A R, Averty B, Watson R J, Remenyi T, 2011. Mercury in the Southern Ocean. Geochimica Et Cosmochimica Acta, 75(14): 4037-4052.

Costa M, Liss P S, 1999. Photoreduction of mercury in sea water and its possible implications for Hg(0) air-sea fluxes. Marine Chemistry, 68(1-2): 87-95.

Dastoor A, Ryzhkov A, Dumford D, Lehnherr I, Steffen A, Morrison H, 2015. Atmospheric mercury in the Canadian Arctic. Part II: Insight from modeling. Science of the Total Environment, 509: 16-27.

Dastoor A P, Durnford D A, 2014. Arctic Ocean: Is it a sink or a source of atmospheric mercury? Environmental Science & Technology, 48(3): 1707-1717.

Dastoor A P, Larocque Y, 2004. Global circulation of atmospheric mercury: A modelling study. Atmospheric Environment, 38(1): 147-161.

Dommergue A, Ferrari C P, Gauchard P A., Boutron C F, Poissant L, Pilote M, Jitaru P, Adams F C, 2003. The fate of mercury species in a Sub-Arctic snowpack during snowmelt. Geophysical Research Letters, 30(12): 23.

Dommergue A, Larose C, Fain X, Clarisse O, Foucher D, Hintelmann H, Schneider D, Ferrari C P, 2010. Deposition of mercury species in the Ny-Alesund Area (79°N) and their transfer during snowmelt. Environmental Science & Technology, 44(3): 901-907.

Dommergue A, Martinerie P, Courteaud J, Witrant E, Etheridge D M, 2016. A new reconstruction of atmospheric gaseous elemental mercury trend over the last 60 years from Greenland firn records. Atmospheric Environment, 136: 156-164.

Donohoue D L, Dieter B, Brandi C, Hynes A J, 2006. Temperature and pressure dependent rate coefficients for the reaction of Hg with Br and the reaction of Br with Br: A pulsed laser photolysis-pulsed laser induced fluorescence study. Journal of Physical Chemistry A, 110(21): 6623-6632.

Dos Santos I R, Silva E V, Schaefer C, Sella S M, Silva C A, Gomes V, Passos M J D C R, Ngan P V, 2006. Baseline mercury and zinc concentrations in terrestrial and coastal organisms of Admiralty Bay, Antarctica. Environmental Pollution, 140(2): 304-311.

Douglas T A, Loseto L L, Macdonald R W, Outridge P, Dommergue A, Poulain A, Amyot M, Barkay, T, Berg T, Chetelat J, Constant P, Evans M, Ferrari C, Gantner N, Johnson M S, Kirk J, Kroer N, Larose C, Lean D, Nielsen T G, Poissant L, Rognerud S, Skov H, Sorensen S, Wang F Y, Wilson S, Zdanowicz C M, 2012. The fate of mercury in Arctic terrestrial and aquatic ecosystems: A review. Environmental Chemistry, 9(4): 321-355.

Douglas T A, Sturm M, Blum J D, Polashensld C, Stuefer S, Hiemstra C, Steffen A, Filhol S, Prevost R, 2017. A pulse of mercury and major ions in snowmelt runoff from a small Arctic Alaska watershed. Environmental Science & Technology, 51(19): 11145-11155.

Durnford D, Dastoor A, 2011. The behavior of mercury in the cryosphere: A review of what we know from observations. Journal of Geophysical Research-Atmospheres, 116: D06305.

Durnford D, Dastoor A, Figueras-Nieto D, Ryjkov A, 2010. Long range transport of mercury to the Arctic and across Canada. Atmospheric Chemistry and Physics, 10(13): 6063-6086.

Ebinghaus R, Kock H H, Temme C, Einax J W, Lowe A G, Richter A, Burrows J P, Schroeder W H, 2002. Antarctic springtime depletion of atmospheric mercury. Environmental Science & Technology, 36(6): 1238-1244.

Eisele F, Davis D D, Helmig D, Oltmans S J, Neff W, Huey G, Tanner D, Chen G, Crawford J, Arimoto R, Buhr M, Mauldin L, Hutterli M, Dibb J, Blake D, Brooks S B, Johnson B, Robert, J M, Wang Y H, Tan D, Flocke F, 2008. Antarctic Tropospheric Chemistry Investigation (ANTCI) 2003 overview. Atmospheric Environment, 42(12): 2749-2761.

Favaro Dit C G, Damato S R, Mazzilli B P, Braga E S, Bosquilha G E, 2004. Total mercury in bottom sediment samples from Admiralty Bay, King George Island, Antarctic region. Materials Geoenvironment, 51(2): 972.

Ferrari C P, Padova C, Fain X, Gauchard P A, Dommergue A, Aspmo K, Berg T, Cairns W, Barbante C, Cescon P, Kaleschke L, Richter A, Wittrock F, Boutron C, 2008. Atmospheric mercury depletion event stuy in Ny-Alesund (Svalbard) in spring 2005. Deposition and transformation of Hg in surface snow during springtime. Science of the Total Environment, 397(1-3), 167-177.

Fitzgerald W F, Lamborg C H, Hammerschmidt C R, 2007. Marine biogeochemical cycling of mercury. Cheminform, 38(20): 641-662.

Fleming E J, Mack E E, Green P G, Nelson D C, 2006. Mercury methylation from unexpected sources: Molybdate-inhibited freshwater sediments and an iron-reducing bacterium. Applied and Environmental Microbiology, 72(1): 457-464.

Fox A L, Hughes E A, Trocine R P, Trefry J H, Schonberg S V, McTigue N D, Lasorsa B K, Konar B, Cooper L W, 2014. Mercury in the northeastern Chukchi Sea: Distribution patterns in seawater and sediments and biomagnification in the benthic food web. Deep-Sea Research Part Ii-Topical Studies in Oceanography, 102: 56-67.

Gamberg M, Chetelat J, Poulain A J, Zdanowicz C T, Zheng J C, 2015. Mercury in the Canadian Arctic terrestrial environment: An update. Science of the Total Environment, 509: 28-40.

Garbarino J R, Snyder-Conn E, Leiker T J, Hoffman G L, 2002. Contaminants in Arctic snow collected over northwest Alaskan sea ice. Water Air and Soil Pollution, 139(1-4): 183-214.

Gilmour C C, Podar M, Bullock A L, Graham A M, Brown S D, Somenahally A C, Johs A, Hurt R A, Bailey K L, Elias D A, 2013. Mercury methylation by novel microorganisms from new environments. Environmental Science & Technology, 47(20): 11810-11820.

Goodsite M E, Plane J M C, Skov H, 2004. A theoretical study of the oxidation of Hg^0 to $HgBr_2$ in the troposphere. Environmental Science & Technology, 38(6): 1772-1776.

Goutte A, Cherel Y, Churlaud C, Ponthus J P, Masse G, Bustamante P, 2015. Trace elements in Antarctic fish species and the influence of foraging habitats and dietary habits on mercury levels. Science of the Total Environment, 538: 743-749.

Granskog M A, Macdonald R W, 2007. Distribution, characteristics and potential impacts of chromophoric dissolved organic matter (CDOM) in Hudson Strait and Hudson Bay, Canada. Continental Shelf Research, 27(15): 2032-2050.

Grigal D F, 2003. Mercury sequestration in forests and peatlands: A review. Journal of Environmental Quality, 32(2): 393-405.

Halbach K, Mikkelsen O, Berg T, Steinnes E, 2017. The presence of mercury and other trace metals in surface soils in the Norwegian Arctic. Chemosphere, 188: 567-574.

Hamelin S, Amyot M, Barkay T, Wang Y, Planas D, 2011. Methanogens: Principal methylators of mercury in lake periphyton. Environmental Science & Technology, 45(18): 7693-7700.

Hammerschmidt C R, Fitzgerald W F, Lamborg C H, Balcom P H, Mao T, 2006. Biogeochemical cycling of methylmercury in lakes and tundra watersheds of Arctic Alaska. Environmental Science & Technology, 40(4): 1204.

Han S, Narasingarao P, Obraztsova A, Gieskes J, Hartmann A C, Tebo B M, Allen E E, Deheyn D D, 2010. Mercury speciation in marine sediments under sulfate-limited conditions. Environmental Science & Technology, 44(10): 3752-3757.

Hansen J C T T, Muhs A G, 1989. Trace metals in humans and animal hair from the 15th century graves in Qilakitsoq, compared with recent samples. The mummies from Qilakitsoq Meddr Groenl, 12. In: Hansen J P, Gulløv H, editors. Man Soc, 7-161.

Henrik S, Christensen J H, Goodsite M E, Heidam N Z, Bjarne J, Peter W H, Gerald G, 2004. Fate of elemental mercury in the Arctic during atmospheric mercury depletion episodes and the load of atmospheric mercury to the Arctic. Environmental Science & Technology, 38(8): 2373-2382.

Hermanson M H, 1998. Anthropogenic mercury deposition to arctic lake sediments. Water Air and Soil Pollution, 101(1-4): 309-321.

Iversen T, Joranger E, 1985. Arctic air pollution and large scale atmospheric flows. Atmospheric Environment, 19(12): 2099-2108.

Johansen P, Mulvad G, Pedersen H S, Hansen J C, Riget F, 2007. Human accumulation of mercury in Greenland. Science of the Total Environment, 377(2-3): 173-178.

John C, Marc A, Louise C, Alexandre P, 2008. Metamorphosis in chironomids, more than mercury supply, controls methylmercury transfer to fish in High Arctic lakes. Environmental Science & Technology, 42(24): 9110.

Kellerhals M, Beauchamp S, Belzer W, Blanchard P, Froude F, Harvey B, McDonald K, Pilote M, Poissant L, Puckett K, Schroeder B, Steffen A, Tordon R, 2003. Temporal and spatial variability of total gaseous mercury in Canada: results from the Canadian Atmospheric Mercury Measurement Network (CAMNet). Atmospheric Environment, 37(7): 1003-1011.

Kerin E J, Gilmour C, Roden E, Suzuki M, Coates J, Mason R, 2006. Mercury methylation by dissimilatory iron-reducing bacteria. Applied and Environmental Microbiology, 72(12): 7919-7921.

Kirk J L, Louis V L S, 2009. Multiyear total and methyl mercury exports from two major Sub-Arctic rivers draining into Hudson Bay, Canada. Environmental Science & Technology, 43(7): 2254-2261.

Kirk J L, Louis V L S, Hintelmann H, Lehnherr I, Else B, Poissant L, 2008. Methylated mercury species in marine waters of the Canadian High and Sub-Arctic. Environmental Science & Technology, 42(22): 8367-8373.

Krabbenhoft D P, Hurley J P, Olson M L, Cleckner L B, 1998. Diel variability of mercury phase and species distributions in the Florida Everglades. Biogeochemistry, 40(2-3): 311-325.

Kuhn M, 2001. The nutrient cycle through snow and ice: a review. Aquatic Sciences, 63(2): 150-167.

Lalonde J D, Amyot M, Kraepiel A M, Morel F M, 2001. Photooxidation of Hg(0) in artificial and natural waters. Environmental Science & Technology, 35(7): 1367-1372.

Lalonde J D, Amyot M, Doyon M R, Auclair J C, 2003. Photo-induced Hg(II) reduction in snow from the remote and temperature Experimental Lakes Area(Ontario, Canada). Journal of Geophysical Research, 108(D6): 4200.

Landers D H, Gubala C, Verta M, Lucotte M, Johansson K, Vlasova T, Lockhart W L, 1998. Using lake sediment mercury flux ratios to evaluate the regional and continental dimensions of mercury deposition in arctic and boreal ecosystems. Atmospheric Environment, 32(5): 919-928.

Larose C, Dommergue A, De Angelis M, Cossa D, Averty B, Maruszak N, Soumis N, Schneider D, Ferrari C, 2010. Springtime changes in snow chemistry lead to new insights into mercury methylation in the Arctic. Geochimica Et Cosmochimica Acta, 74(22): 6263-6275.

Laughlin W S, 1991. The Greenland Mummies. *In*: Hansen P H, Meldgaard J, Nordqvist J. American Journal of Physical Anthropology, 86(4): 562-563.

Lehnherr I, St Louis V L, Emmerton C A, Barker J D, Kirk J L, 2012a. Methylmercury cycling in High Arctic wetland ponds: Sources and sinks. Environmental Science & Technology, 46(19): 10514-10522.

Lehnherr I, St Louis V L, Hintelmann H, Kirk J L, 2011. Methylation of inorganic mercury in polar marine waters. Nature Geoscience, 4(5): 298-302.

Lehnherr I, St Louis V L, Kirk J L, 2012b. Methylmercury cycling in High Arctic wetland ponds: Controls on sedimentary production. Environmental Science & Technology, 46(19): 10523-10531.

Li C S, Cornett J, Willie S, Lam J, 2009. Mercury in Arctic air: The long-term trend. Science of the Total Environment, 407(8): 2756-2759.

Lin C J, Pongprueksa P, Jr O R B, Lindberg S E, Pehkonen S O, Jang C, Braverman T, Ho T C, 2007. Scientific uncertainties in atmospheric mercury models II: Sensitivity analysis in the CONUS domain. Atmospheric Environment, 41(31): 6544-6560.

Lin C J, Pongprueksa P, Lindberg S E, Pehkonen S O, Byun D, Jang C, 2006. Scientific uncertainties in atmospheric mercury models I: Model science evaluation. Atmospheric Environment, 40(16): 2911-2928.

Lindberg S E, Brooks S, Lin C J, Scott K, Meyers T, Chambers L, Landis M, Stevens R, 2001. Formation of reactive gaseous mercury in the Arctic: Evidence of oxidation of Hg(0) to gas-phase Hg-II compounds after Arctic sunrise. Water Air & Soil Pollution Focus, 1(5-6): 295-302.

Lindberg S E, Brooks S, Lin C J, Scott K J, Landis M S, Stevens R K, Goodsite M, Richter A, 2002. Dynamic oxidation of gaseous mercury in the Arctic troposphere at polar sunrise. Environmental Science & Technology, 36(6): 1245-1256.

Lindberg S E, Brooks S, Lin C J, Scott K, Meyers T, Chambers L, Landis M, Stevens R, 2001. Formation of reactive gaseous mercury in the Arctic: Evidence of oxidation of Hg0 to gas-phase Hg-II compounds after arctic sunrise. Water, Air, and Soil Pollution, 1: 295-302.

Liu X D, Sun L G, Xie Z Q, Yin X B, Wang Y H, 2005. A 1300-year record of penguin populations at Ardley Island in the Antarctic, as deduced from the geochemical data in the ornithogenic lake sediments. Arctic Antarctic and Alpine Research, 37(4): 490-498.

Loseto L L, Siciliano S D, Lean D R S, 2010. Methylmercury production in High Arctic wetlands. Environmental Toxicology & Chemistry, 23(1): 17-23.

Lu J Y, Schroeder W H, Barrie L A, Steffen A, Welch H E, Martin K, Lockhart L, Hunt R V, Boila G, Richter A, 2001. Magnification of atmospheric mercury deposition to polar regions in springtime: The link to tropospheric ozone depletion chemistry. Geophysical Research Letters, 28(17): 3219-3222.

Moller A K, Tamar B, Waleed A S, Sorensen S R J, Henrik S, Niels K, 2011. Diversity and characterization of mercury-resistant bacteria in snow, freshwater and sea-ice brine from the High Arctic. Fems Microbiology Ecology, 75(3): 390-401.

MacMillan G A, Girard C, Chetelat J, Laurion I, Amyot M, 2015. High methylmercury in Arctic and Subarctic ponds is related to nutrient levels in the warming Eastern Canadian Arctic. Environmental Science & Technology, 49(13): 7743-7753.

Mann E, Ziegler S, Mallory M, O'Driscoll N, 2014. Mercury photochemistry in snow and implications for Arctic ecosystems. Environmental Reviews, 22(4): 331-345.

Mason R P, Fitzgerald W F, 1993. The distribution and biogeochemical cycling of mercury in the equatorial Pacific Ocean. Deep Sea Research Part I: Oceanographic Research Papers, 40(9): 1897-1924.

Mason R P, Morel F M M, Hemond H F, 1995. The role of microorganisms in elemental mercury formation in natural waters. Water Air & Soil Pollution, 80(1-4): 775-787.

Mastromonaco M G N, Gardfeldt K, Langer S, 2017. Mercury flux over West Antarctic seas during winter, spring and summer. Marine Chemistry, 193: 44-54.

Matsumoto G, Chikazawa K, Murayama H, Torii T, Fukushima H, Hanya T, 1983. Distribution and correlation of total organic carbon and mercury in Antarctic dry valleys soils, sediments and organisms. Geochemical Journal, 17: 247-255.

McKinney M A, Atwood T C, Pedro S, Peacock E, 2017. Ecological change drives a decline in mercury concentrations in southern Beaufort Sea Polar bears. Environmental Science & Technology, 51(14): 7814-7822.

McNamara J P, Kane D L, Hinzman L D, 1998. An analysis of streamflow hydrology in the Kuparuk River basin, Arctic Alaska: A nested watershed approach. Journal of Hydrology, 206(1-2): 39-57.

Muckle G, Ayotte P, Dewailly E, Jacobson S W, Jacobson J L, 2001. Prenatal exposure of the Northern Quebec Inuit infants to environmental contaminants. Environmental Health Perspectives, 109(12): 1291-1299.

Negri A, Burns K, Boyle S, Brinkman D, Webster N, 2006. Contamination in sediments, bivalves and sponges of McMurdo Sound, Antarctica. Environmental Pollution, 143(3): 456-467.

Nguyen H T, Kim K H, Shon Z H, Hong S, 2009. A review of atmospheric mercury in the Polar environment. Critical Reviews in Environmental Science and Technology, 39(7): 552-584.

Niemisto L, Perttila M, 1995. Trace-elements in the Weddell Sea-water and sediments in the continental-shelf area. Chemosphere, 31(7): 3643-3650.

O'Driscoll N J, Poissant L, Canario J, Lean D R S, 2008. Dissolved gaseous mercury concentrations and mercury volatilization in a frozen freshwater fluvial lake. Environmental Science & Technology, 42(14): 5125-5130.

O'Driscoll N J, Rencz A, Lean D R S, 2005. The biogeochemistry and fate of mercury in the environment. Biogeochemical Cycles of Elements, 43: 221-238.

Okelsrud A, Lydersen E, Fjeld E, 2016. Biomagnification of mercury and selenium in two lakes in southern Norway. Science of the Total Environment, 566: 596-607.

Outridge P M, Stern G A, Hamilton P B, Perciva J B, Mcneely R, Lodchart W L, 2005. Trace metal profiles in the varved sediment of an Arctic lake. Geochimica Et Cosmochimica Acta, 69(20): 4881.

Outridge P M, Macdonald R W, Wang F, Stern G A, Dastoor A P, 2008. A mass balance inventory of mercury in the Arctic Ocean. Environmental Chemistry, 5(2): 89-111.

Outridge P M, Sanei L H, Stern G A, Hamilton P B, Goodarzi F, 2007. Evidence for control of mercury accumulation rates in Canadian High Arctic lake sediments by variations of aquatic primary productivity. Environmental Science & Technology, 41(15): 5259-5265.

Pal B, Ariya P A, 2004. Studies of ozone initiated reactions of gaseous mercury: Kinetics, product studies, and atmospheric implications. Physical Chemistry Chemical Physics, 6(3): 572-579.

Point D, Sonke J E, Day R D, Roseneau D G, Hobson K A, Pol S S V, Moors A J, Pugh R S, Donard O F X, Becker P R, 2011. Methylmercury photodegradation influenced by sea-ice cover in Arctic marine ecosystems. Nature Geoscience, 4(3): 188-194.

Poissant L, Pilote M, 2003. Time series analysis of atmospheric mercury in Kuujjuarapik/Whapmagoostui (Québec). Journal De Physique IV, 107(107): 1079-1082.

Polito M J, Brasso R L, Trivelpiece W Z, Karnovsky N, Patterson W P, Emslie S D, 2016. Differing foraging strategies influence mercury (Hg) exposure in an Antarctic penguin community. Environmental Pollution, 218: 196-206.

Pongratz R, Heumann K G, 1999. Production of methylated mercury, lead, and cadmium by marine bacteria as a significant natural source for atmospheric heavy metals in polar regions. Chemosphere, 39(1): 89-102.

Poulain A J, Amyot M, Findlay D, Telor S, Barkay T, Hintelmann H, 2004a. Biological and photochemical production of dissolved gaseous mercury in a boreal lake. Limnology & Oceanography, 49(6): 2265-2275.

Poulain A J, Edenise G, Marc A, Campbell P G C, Farhad R, Ariya P A, 2007a. Biological and chemical redox transformations of mercury in fresh and salt waters of the High Arctic during spring and summer. Environmental Science & Technology, 41(6): 1883-1888.

Poulain A J, Lalonde J D, Amyot M, Shead J A, Raofie F, Ariya P A, 2004b. Redox transformations of mercury in an Arctic snowpack at springtime. Atmospheric Environment, 38(39): 6763-6774.

Poulain A J, Roy V, Amyot M, 2007b. Influence of temperate mixed and deciduous tree covers on Hg concentrations and photoredox transformations in snow. Geochimica Et Cosmochimica Acta, 71(10): 2448-2462.

Riget F, Braune B, Bignert A, Wilson S, Aars J, Born E, Dam M, Dietz R, Evans M, Evans T, Gamberg M, Gantner N, Green N, Gunnlaugsdottir H, Kannan K, Letcher R, Muir D, Roach P,

Sonne C, Stern G, Wiig O, 2011. Temporal trends of Hg in Arctic biota, an update. Science of the Total Environment, 409(18): 3520-3526.

Riva S D, Abelmoschi M L, Magi E, Soggia F, 2004. The utilization of the antarctic environmental specimen bank (BCAA) in monitoring Cd and Hg in an antarctic coastal area in Terra Nova Bay (Ross Sea-Northern Victoria Land). Chemosphere, 56(1): 59-69.

Schaefer J K, Jane Y, Reinfelder J R, Tamara C, Ellickson K M, Shoshana T O, Tamar B, 2004. Role of the bacterial organomercury lyase (MerB) in controlling methylmercury accumulation in mercury-contaminated natural waters. Environmental Science & Technology, 38(16): 4304-4311.

Schroeder W H, Anlauf K G, Barrie L A, Lu J Y, Steffen A, Schneeberger D R, Berg T, 1998. Arctic springtime depletion of mercury. Nature, 394(6691): 331-332.

Schroeder W H, Munthe J, 1998. Atmospheric mercury: An overview. Atmospheric Environment, 32(5): 809-822.

Semkin R G, Mierle G, Neureuther R J, 2005. Hydrochemistry and mercury cycling in a High Arctic watershed. Science of the Total Environment, 342(1-3): 199-221.

Skov H, Christensen J H, Goodsite M E, Heidam N Z, Jensen B, Wahlin P, Geernaert G, 2004. Fate of elemental mercury in the arctic during atmospheric mercury depletion episodes and the load of atmospheric mercury to the arctic. Environmental Science & Technology, 38(8): 2373-2382.

Slemr F, Brunke E G, Ebinghaus R, Temme C, Munthe J, Wangberg I, Schroeder W, Steffen A, Berg T, 2003. Worldwide trend of atmospheric mercury since 1977. Geophysical Research Letters, 30(10): 4.

SnyderConn E, Garbarino J R, Hoffman G L, Oelkers A, 1997. Soluble trace elements and total mercury in Arctic Alaskan snow. Arctic, 50(3): 201-215.

Soerensen A L, Jacob D J, Schartup A T, Fisher J A, Lehnherr I, St Louis V L, Heimburger L E, Sonke J E, Krabbenhoft D P, Sunderland E M, 2016. A mass budget for mercury and methylmercury in the Arctic ocean. Global Biogeochemical Cycles, 30(4): 560-575.

Sprovieri F, Pirrone N, 2000. A preliminary assessment of mercury levels in the Antarctic and Arctic troposphere. Journal of Aerosol Science, 31(Suppl 1): 757-758.

Sprovieri, F, Pirrone N, Hedgecock I M, Landis M S, Stevens R K, 2002. Intensive atmospheric mercury measurements at Terra Nova Bay in Antarctica during november and december 2000. Journal of Geophysical Research-Atmospheres, 107(D23): 4722.

St Louis V L, Hintelmann H, Graydon J A, Kirk J L, Barker J, Dimock B, Sharp M J, Lehnherr I, 2007. Methylated mercury species in Canadian High Arctic marine surface waters and snowpacks. Environmental Science & Technology, 41(18): 6433-6441.

St Louis V L, Sharp M J, Steffen A, May A, Barker J, Kirk J L, Kelly D J A, Arnott S E, Keatley B, Smol J P, 2005. Some sources and sinks of monomethyl and inorganic mercury on Ellesmere island in the Canadian High Arctic. Environmental Science & Technology, 39(8): 2686-2701.

St Louis V L, Hintelmann H, Graydon J A, Kirk J L, Barker J, Dimock B, Sharp M J, Lehnherr I, 2007. Methylated mercury species in Canadian High Arctic marine surface waters and snowpacks. Environmental Science & Technology, 41(18): 6433-6441.

St Louis V L, Sharp M J, Steffen A, May A, Barker J, Kirk J L, Kelly D J A, Arnott S E, Keatley B, Smol J P, 2005. Some sources and sinks of monomethyl and inorganic mercury on Ellesmere island in the Canadian High Arctic. Environmental Science & Technology, 39(8): 2686-2701.

St Pierre K A, Chetelat J, Yumvihoze E, Poulain A J, 2014. Temperature and the sulfur cycle control monomethylmercury cycling in High Arctic coastal marine sediments from Allen Bay, Nunavut,

Canada. Environmental Science & Technology, 48(5): 2680-2687.

Steffen A, Bottenheim J, Cole A, Ebinghaus R, Lawson G, Leaitch W R, 2014. Atmospheric mercury speciation and mercury in snow over time at Alert, Canada. Atmospheric Chemistry and Physics, 14(5): 2219-2231.

Steffen A, Douglas T, Amyot M, Ariya P, Aspmo K, Berg T, Bottenheim J, Brooks S, Cobbett F, Dastoor A, Dommergue A, Ebinghaus R, Ferrari C, Gardfeldt K, Goodsite M E, Lean D, Poulain A J, Scherz C, Skov H, Sommar J, Temme C, 2008. A synthesis of atmospheric mercury depletion event chemistry in the atmosphere and snow. Atmospheric Chemistry and Physics, 8(6): 1445-1482.

Steffen A, Schroeder W, Macdonald R, Poissant L, Konoplev A, 2005. Mercury in the Arctic atmosphere: An analysis of eight years of measurements of GEM at Alert (Canada) and a comparison with observations at Amderma (Russia) and Kuujjuarapik (Canada). Science of the Total Environment, 342(1-3): 185-198.

Stern G A, Macdonald R W, Outridge P M, Wilson S, Chetelat J, Cole A, Hintelmann H, Loseto L L, Steffen A, Wang F Y, Zdanowicz C, 2012. How does climate change influence arctic mercury? Science of the Total Environment, 414: 22-42.

Su Y S, Hung H L, Stern G, Sverko E, Lao R, Barresi E, Rosenberg B, Fellin P, Li H, Xiao H, 2011. Bias from two analytical laboratories involved in a long-term air monitoring program measuring organic pollutants in the Arctic: A quality assurance/quality control assessment. Journal of Environmental Monitoring, 13(11): 3111-3118.

Sudo K, Takahashi M, Akimoto H, 2002a. CHASER: A global chemical model of the troposphere 2. Model results and evaluation. Journal of Geophysical Research, 107(D21), ACH-1-ACH: 9-39.

Sudo K, Takahashi M, Akimoto H, 2002b. CHASER: A global chemical model of the troposphere-2 . Model results and evaluation. Journal of Geophysical Research-Atmospheres, 107(D21): DOI: 10.1029/2001JD001113.

Sunderl E M, Mason R P, 2007. Human impacts on open ocean mercury concentrations. Global Biogeochemical Cycles, 21(4): DOI: 10.1029/2006gb002876.

Sunderland E M, Krabbenhoft D P, Moreau J W, Strode S A, Landing W M, 2009. Mercury sources, distribution, and bioavailability in the North Pacific Ocean: Insights from data and models. Global Biogeochemical Cycles, 23(2): 1-14.

Tartu S, Bustamante P, Goutte A, Cherel Y, Weimerskirch H, Bustnes J O, Chastel O, 2014. Age-related mercury contamination and relationship with luteinizing hormone in a long-lived Antarctic bird. Plos One, 9(7): DOI: 10.1371/journal.pone.0103642.

Temme C, Blanchard P, Steffen A, Banic C, Beauchamp S, Poissant L, Tordon R, Wiens B, 2007. Trend, seasonal and multivariate analysis study of total gaseous mercury data from the Canadian Atmospheric Mercury Measurement network (CAMNet). Atmospheric Environment, 41(26): 5423-5441.

Tokos J J S, Hall B O, Calhoun J A, Prestbo E M, 1998. Homogeneous gas-phase reaction of Hg(0) with H_2O_2, O_3, CH_3I, and $(CH_3)_2S$: Implications for atmospheric Hg cycling. Atmospheric Environment, 32(5): 823-827.

Tseng C M, Lamborg C, Fitzgerald W F, Engstrom D R, 2004. Cycling of dissolved elemental mercury in Arctic Alaskan lakes. Geochimica Et Cosmochimica Acta, 68(6): 1173-1184.

Tvinnereim H M, Eide R, Riise T, 2000. Heavy metals in human primary teeth: Some factors influencing the metal concentrations. Science of the Total Environment, 255(1-3): 21-27.

Wagemann R, Innes S, Richard P R, 1996. Overview and regional and temporal differences of heavy metals in Arctic whales and ringed seals in the Canadian Arctic. Science of the Total Environment, 186(1-2): 41-66.

Wang F Y, Macdonald R W, Armstrong D A, Stern G A, 2012. Total and methylated mercury in the Beaufort Sea: The role of local and recent organic remineralization. Environmental Science & Technology, 46(21): 11821-11828.

Wang K, Munson K M, Beaupre-Laperriere A, Mucci A, Macdonald R, Wang F Y, 2018. Subsurface seawater methylmercury maximum explains biotic mercury concentrations in the Canadian Arctic. Scientific Reports, 8: 5.

Whalin L, Kim E, Mason R, 2007. Factors influencing the oxidation, reduction, methylation and demethylation of mercury species in coastal waters. Marine Chemistry, 107(3): 278-294.

Wintle N J P, Sleadd I M, Gundersen D T, Kohl K, Buckley B A, 2015. Total mercury in six Antarctic notothenioid fishes. Bulletin of Environmental Contamination and Toxicology, 95(5): 557-560.

Wojtun B, Kolon K, Samecka-Cymerman A, Jasion M, Kempers A J, 2013. A survey of metal concentrations in higher plants, mosses, and lichens collected on King George Island in 1988. Polar Biology, 36(6): 913-918.

Yang X, Cox R A, Warwick N J, Pyle J A, Carver G D, O'Connor F M, Savage N H, 2005. Tropospheric bromine chemistry and its impacts on ozone: A model study. Journal of Geophysical Research Atmospheres, 110: D23.

Yu J, Xie Z Q, Kang H, Li Z, Sun C, Bian LG, Zhang P F, 2014. High variability of atmospheric mercury in the summertime boundary layer through the central Arctic Ocean. Scientific Reports, 4: 7.

Yu R Q, Flanders J R, Mack E E, Turner R, Mirza M B, Barkay T, 2012. Contribution of coexisting sulfate and iron reducing bacteria to methylmercury production in freshwater river sediments. Environmental Science & Technology, 46(5): 2684-2691.

Yu R Q, Reinfelder J R, Hines M E, Barkay T, 2013. Mercury methylation by the methanogen *Methanospirillum hungatei*. Applied and Environmental Microbiology, 79(20): 6325-6330.

Zdanowicz C, Krummel E M, Lean D, Poulain A J, Yumvihoze E, Chen J B, Hintelmann H, 2013. Accumulation, storage and release of atmospheric mercury in a glaciated Arctic catchment, Baffin Island, Canada. Geochimica Et Cosmochimica Acta, 107: 316-335.

Zhang L M, Wright L P, Blanchard P, 2009. A review of current knowledge concerning dry deposition of atmospheric mercury. Atmospheric Environment, 43(37): 5853-5864.

Zverina O, Coufalik P, Brat K, Cervenka R, Kuta J, Mikes O, Komarek J, 2017. Leaching of mercury from seal carcasses into Antarctic soils. Environmental Science and Pollution Research, 24(2): 1424-1431.

Zverina O, Laska K, Cervenka R, Kuta J, Coufalik P, Komarek J, 2014. Analysis of mercury and other heavy metals accumulated in lichen *Usnea antarctica* from James Ross Island, Antarctica. Environmental Monitoring and Assessment, 186(12): 9089-9100.

附录 缩略语(英汉对照)

AE	assimilation efficiency,同化效率
AMDE	atmospheric mercury depletion event,大气汞"亏损"事件
CWT	concentration-weighted trajectory,浓度加权轨迹法
DGM	dissolved gaseous mercury,溶解气态汞
DMHg	dimethylmercury,二甲基汞
DOC	dissolved organic carbon,溶解性有机碳
DOM	dissolved organic matter,溶解有机质
EC	elemental carbon,元素碳
EPS	extracellular polymeric substance,胞外分泌物
even-MIF	mass-independent fractionation of even isotopes,偶数汞同位素的非质量分馏
FeRB	iron-reducing bacteria,铁还原菌
GEM	gaseous elemental mercury,气态单质汞
GFD	gridded frequency distribution,轨迹网格化概率分布
GOM	gaseous oxidized mercury,活性气态汞
GSH	glutathione,谷胱甘肽
HYSPLIT	hybrid single particle Lagrangian integrated trajectory,混合单粒子拉格朗日轨迹
LD_{50}	median lethal dose,半数致死量
MDF	mass-dependent fractionation,质量分馏
MDN	Mercury Deposition Network,汞沉降观测网
MIE	magnetic isotope effect,磁同位素效应
MMHg	monomethylmercury,单甲基汞
NOM	natural organic matter,天然有机物
NVE	nuclear volume effect,核体积效应
OC	organic carbon,有机碳
odd-MIF	mass-independent fractionation of odd isotopes,奇数同位素的非质量分馏

PAR	photosynthetically active radiation,	光合有效辐射
PBM	particulate bound mercury,	颗粒态汞
PCA	principal component analysis,	主成分分析法
PMF	positive matrix factorization,	正交矩阵因子分析
PSCF	potential source contribution function,	潜在污染源贡献函数
RGM	reactive gaseous mercury,	活性气态汞
ROS	reactive oxygen species,	活性氧自由基
SRB	sulfate-reducing bacteria,	硫酸盐还原菌
TGM	total gaseous mercury,	气态总汞
UVA	ultraviolet A,	紫外线 A
UVB	ultraviolet B,	紫外线 B
UVC	ultraviolet C,	紫外线 C

索 引

B
冰冻 32

C
臭氧 20
磁铁矿 45
磁同位素效应 143

D
碘物种 23
大气化学过程 285

E
二元羧酸 38
二价汞 19

F
放大效应 94
分馏系数 140
分子转化过程 286
非质量分馏 140

G
《关于汞的水俣公约》 11
过氧化氢 20
过氧化氢酶 34
汞产量 9
汞还原酶 55
汞同位素模型 176

H
还原 19
还原态有机质 30
环境中汞的来源 0
环境分布 85
亨利定律常数 19
活性卤物种 28
核体积效应 143
混合卤素物种 23
黄铁矿 45

J
甲基化 33, 99
甲基汞 82
界面交换过程 288

K
可还原汞 42
扩展 X 射线吸收精细结构 34
颗粒物 39
颗粒结合态零价汞 59

L
绿锈 45
菱铁矿 45
氯物种 22
硫化汞 42
零价汞 19

Q
气相均相汞氧化 20
去甲基化 110
羟基自由基 20
巯基 34
巯基配体 30

R
人体暴露与健康风险 83
溶解气态汞 28
溶解有机质 28

溶解度 19
溶解氧 38, 42

S

生物积累 94
释汞通量 47

T

土壤 47
同位素示踪 30
同位素 139
铁矿物 38
碳酸根自由基 29

W

微生物 55
微生物可利用性 106
微藻 57

X

悬浮颗粒物 44

溴物种 22
雪、海冰中汞生物地球化学 289

Y

一价汞 31
有机汞裂解酶 57
氧化 19
氧化亚铁硫杆菌 56

Z

直接光解 37
紫外线 A 40
紫外线 B 28
质量分馏 140

其他

NO_3^- 20
pH 43
X 射线吸收近边结构 33

彩 图

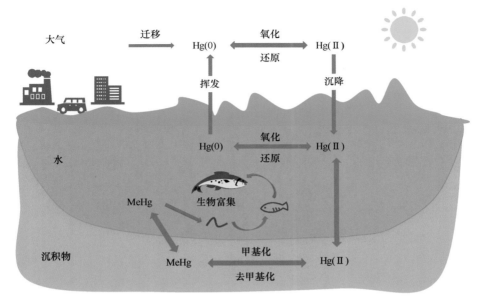

图 1-1 汞的形态转化

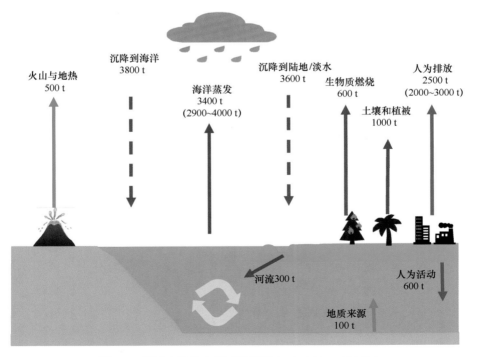

图 1-3 环境中汞的来源 [数据引自（UNEP，2019）]

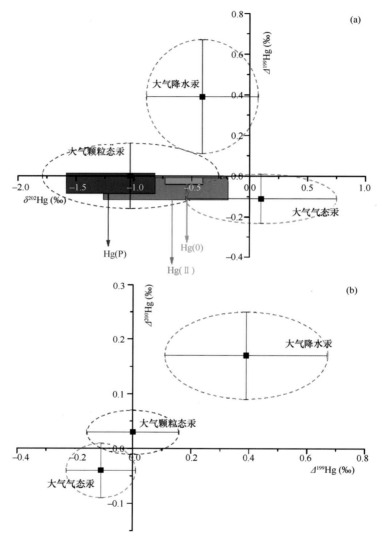

图 4-6 已剔除特殊采样位点和异常值的世界不同地区大气气态汞、大气降水汞、大气颗粒态汞的质量分馏值（δ^{202}Hg）与奇数同位素非质量分馏（Δ^{199}Hg）图解（a）以及奇数同位素非质量分馏值（Δ^{199}Hg）与偶数同位素非质量分馏（Δ^{200}Hg）图解（b）。图中的椭圆代表不同形态汞的均值±1SD 的范围；矩形代表人为源大气形态汞排放的同位素变化区间（Sun et al., 2016b）

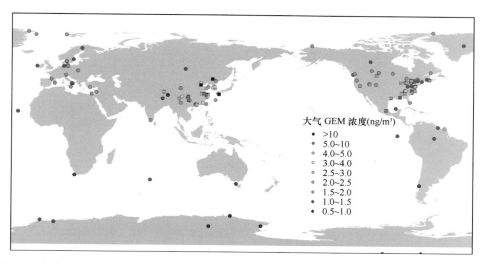

图 5-1　全球地表大气气态单质汞（GEM）浓度的分布特征（圆形表示背景区大气气态单质汞浓度，方形表示城市大气气态单质汞浓度）（Amouroux et al.，1999；Wangberg et al.，2001；Lindberg et al.，2002a；Liu et al.，2002；Berg et al.，2003；Kellerhals et al.，2003；Malcolm et al.，2003；Munthe et al.，2003；Temme et al.，2003；Weiss-Penzias et al.，2003；Fang et al.，2004；Han et al.，2004；Gabriel et al.，2005；Lynam and Keeler，2005；Poissant et al.，2005；Zielonka et al.，2005；Swartzendruber et al.，2006；Laurier and Mason，2007；Nguyen et al.，2007；Valente et al.，2007；Wang et al.，2007；Chand et al.，2008；Choi et al.，2008；Fu et al.，2008a；Fain et al.，2009；Yang et al.，2009；Fu et al.，2010；Liu et al.，2010；Sheu et al.，2010；Brooks et al.，2011；Ci et al.，2011；Friedli et al.，2011；Fu et al.，2011；Nguyen et al.，2011；Steen et al.，2011；Cheng et al.，2012；Fu et al.，2012b；Lan et al.，2012；Li，2012；Zhu et al.，2012；Chen et al.，2013；Zhang et al.，2013；Wang et al.，2014a；Slemr et al.，2015；Xu et al.，2015；Yu et al.，2015；Zhang et al.，2015a；Fu et al.，2016c；Liu et al.，2016；Sprovieri et al.，2016；Zhang et al.，2016a；李舒等，2016）

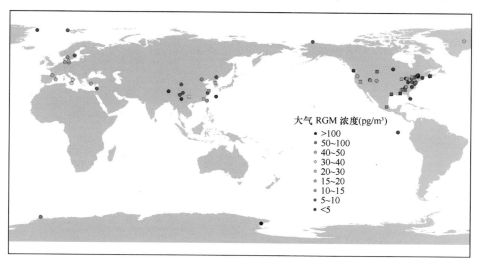

图 5-2　全球大气活性气态汞（RGM）浓度分布特征（圆形表示背景区活性气态汞浓度，方形表示城市活性气态汞浓度）（Sprovieri et al.，2002；Berg et al.，2003；Munthe et al.，2003；Temme et al.，2003；Weiss-Penzias et al.，2003；Han et al.，2004；Gabriel et al.，2005；Poissant et al.，2005；Swartzendruber et al.，2006；Laurier and Mason，2007；Chand et al.，2008；Fu et al.，2008b；Choi et al.，2009；Fain et al.，2009；Liu et al.，2010；Sheu et al.，2010；Brooks et al.，2011；Fu et al.，2011；Steen et al.，2011；Cheng et al.，2012；Fu et al.，2012a；Lan et al.，2012；Zhang et al.，2013；Wang et al.，2014a；Fu et al.，2015b；Xu et al.，2015；Yu et al.，2015；Zhang et al.，2015a；de Foy et al.，2016；Fu et al.，2016c；Liu et al.，2016；Zhang et al.，2016a；李舒等，2016）

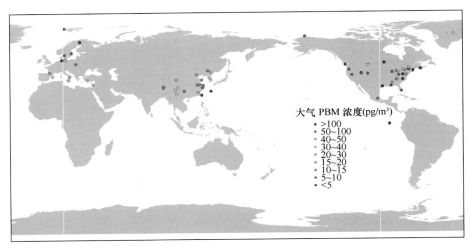

图 5-3　全球大气颗粒态汞（PBM）浓度的分布特征（圆形表示背景区颗粒态汞浓度，方形表示城市颗粒态汞浓度）（Fang et al., 2001; Lindberg et al., 2002a; Malcolm et al., 2003; Munthe et al., 2003; Weiss-Penzias et al., 2003; Gabriel et al., 2005; Poissant et al., 2005; Swartzendruber et al., 2006; Valente et al., 2007; Chand et al., 2008; Choi et al., 2008; Fu et al., 2008b; Fain et al., 2009; Liu et al., 2010; Sheu et al., 2010; Brooks et al., 2011; Fu et al., 2011; Steen et al., 2011; Cheng et al., 2012; Fu et al., 2012a; Lan et al., 2012; Xu et al., 2013; Zhang et al., 2013; Wang et al., 2014a; Zhu et al., 2014; Fu et al., 2015b; Xu et al., 2015; Yu et al., 2015; Zhang et al., 2015a; Zhang et al., 2015c; de Foy et al., 2016; Fu et al., 2016c; Huang et al., 2016; Zhang et al., 2016a; 李舒等, 2016）

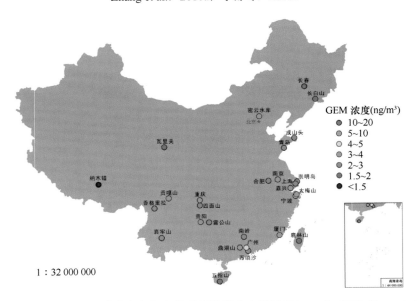

图 5-4　我国城市和背景区地表大气气态单质汞浓度分布特征（Liu et al., 2002; Fang et al., 2004; Wang et al., 2007; Fu et al., 2008a; Yang et al., 2009; Fu et al., 2010; Sheu et al., 2010; Ci et al., 2011; Friedli et al., 2011; Fu et al., 2011; Nguyen et al., 2011; Fu et al., 2012b; Li, 2012; Zhu et al., 2012; Chen et al., 2013; Zhang et al., 2013; Xu et al., 2015; Yu et al., 2015; Zhang et al., 2015a; Liu et al., 2016; Zhang et al., 2016a; 李舒等, 2016）

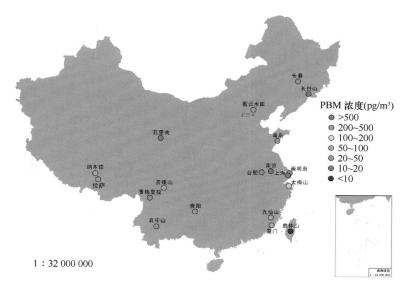

图 5-5 我国大气颗粒态汞（PBM）浓度分布特征（Fang et al.，2001；Fu et al.，2008b；Sheu et al.，2010；Fu et al.，2011；Fu et al.，2012a；Xu et al.，2013；Zhang et al.，2013；Zhu et al.，2014；Fu et al.，2015b；Xu et al.，2015；Yu et al.，2015；Zhang et al.，2015a；Zhang et al.，2015c；de Foy et al.，2016；Huang et al.，2016；Zhang et al.，2016a；李舒等，2016）

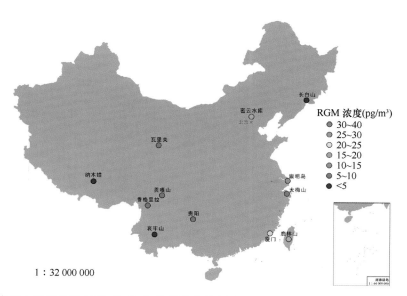

图 5-6 中国大气活性气态汞（RGM）浓度分布特征（Sheu et al.，2010；Fu et al.，2011；Fu et al.，2012a；Zhang et al.，2013；Fu et al.，2015b；Xu et al.，2015；Yu et al.，2015；Zhang et al.，2015a；de Foy et al.，2016；Zhang et al.，2016a；李舒等，2016）

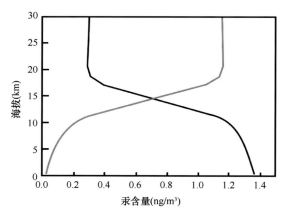

图 6-1 模型计算显示的大气气态单质汞和氧化态汞（包括活性气态汞和颗粒态汞）的垂直分布模式（黑线表示气态单质汞，红线表示氧化态汞）

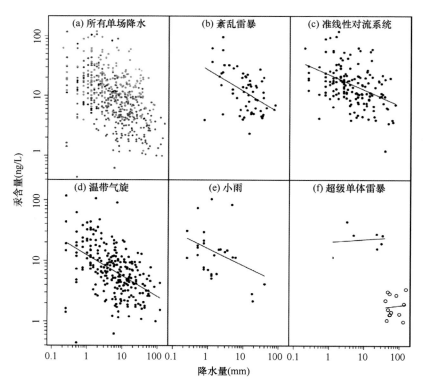

图 6-4 不同降水类型中汞含量与降水量关系图［引用并修改自（Kaulfus et al.，2017）］
（a）单场降水事件散点图：黄色代表超级单体雷暴，红色代表紊乱雷暴，绿色代表准线形对流系统，蓝色代表温带气旋，黑色代表小雨，灰色代表登陆热带气旋；（b~f）不同降水类型汞浓度

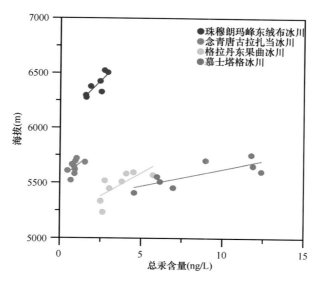

图 6-5 青藏高原四条典型冰川表层雪中汞浓度随海拔变化趋势
[引用并修改自（Huang et al., 2012b）]

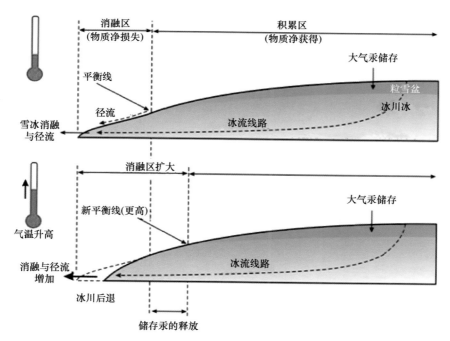

图 6-9 气候变暖背景下冰川消融释汞过程简图 [引用并修改自（Stern et al., 2012）]

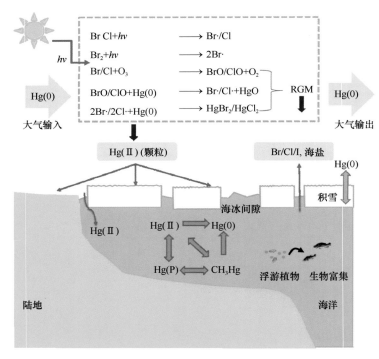

图 7-1 极地汞循环图 [根据 Steffen 等 (2008) 图重新绘制]

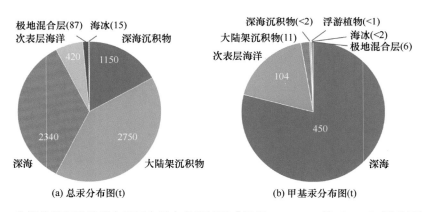

图 7-2 北极总汞和甲基汞在不同介质中分配情况 [根据 Soerensen 等 (2016) 图重新绘制]